J. R. Pasek

APPLIED OPTIMAL ESTIMATION

APPLIED OPTIMAL ESTIMATION

written by Technical Staff
THE ANALYTIC SCIENCES CORPORATION

edited by Arthur Gelb

principal authors
Arthur Gelb
Joseph F. Kasper, Jr.
Raymond A. Nash, Jr.
Charles F. Price
Arthur A. Sutherland, Jr.

THE M.I.T. PRESS
Massachusetts Institute of Technology
Cambridge, Massachusetts, and London, England

This book was designed by TASC.

Second printing, February 1975

Library of Congress Catalog Card Number: 74-1604
ISBN 0 262 20027-9 (hardcover)
ISBN 0 262 70008-5 (paperback)

FOREWORD

Estimation is the process of extracting information from data – data which can be used to infer the desired information and may contain errors. Modern estimation methods use known relationships to compute the desired information from the measurements, taking account of measurement errors, the effects of disturbances and control actions on the system, and prior knowledge of the information. Diverse measurements can be blended to form "best" estimates, and information which is unavailable for measurement can be approximated in an optimal fashion. The intent of this book is to enable readers to achieve a level of competence that will permit their participation in the design and evaluation of *practical* estimators. Therefore, the text is oriented to the *applied* rather than theoretical aspects of optimal estimation. It is our intent throughout to provide a simple and interesting picture of the central issues underlying modern estimation theory and practice. Heuristic, rather than theoretically elegant, arguments are used extensively, with emphasis on physical insights and key questions of practical importance.

The text is organized into three principal parts. Part I introduces the subject matter and provides brief treatments of the underlying mathematics. Chapter 1 presents a brief overview, including a historical perspective; Chapters 2 and 3 treat the mathematics underlying random process theory and state-space characterization of linear dynamic systems, both of which are essential prerequisites to understanding optimal estimation theory. Part II provides derivations, interpretations and examples pertinent to the theory of optimal estimation. Thus, Chapters 4 and 5 address optimal linear filtering and smoothing, respectively, while Chapter 6 addresses the subject of nonlinear filtering and smoothing. Part III treats those practical issues which often mean the difference between success or failure of the implemented optimal estimator. The practical and often pivotal issues of suboptimal filtering, sensitivity analysis and implementation considerations are discussed at some length in Chapters 7 and 8. Additional topics of practical value are presented in Chapter 9; these include refinements and other viewpoints of estimation theory, and the close connection of the mathematics which underly both optimal linear estimation theory and optimal linear control theory.

Many illustrative examples have been interspersed throughout the text to assist in effective presentation of the theoretical material. Additionally, problems with "built-in" answers have been included at the end of each chapter, to further enable self-study of the subject matter.

This book is the outgrowth of a course taught by The Analytic Sciences Corporation (TASC) at a number of U.S. Government facilities. The course notes were, in turn, based on the considerable practical experience of TASC in applying modern estimation theory to large-scale systems of diverse nature. Thus, virtually all Members of the Technical Staff of TASC have, at one time or another, contributed to the material contained herein. It is a pleasure to specifically acknowledge those members of TASC who, in addition to the principal authors, have *directly* contributed to the writing of this book. Bard S. Crawford and William F. O'Halloran, Jr., provided complete sections; Robert G. Bellaire, Julian L. Center, Jr., Joseph A. D'Appolito, Norman H. Josephy (now at the University of Wisconsin), Bahar J. Uttam and Ronald S. Warren contributed through providing text additions, technical comments and insights of a very diverse nature. Frances A. Smith typed most of the original manuscript, William R. Sullivan and Vicky M. Koczerga created all the artwork. Willie J. Smith gave the manuscript a thorough reading for clarity, and organized production of the final text copy. Professors Renwick E. Curry, John J. Deyst, Jr., and Wallace E. Vander Velde of M.I.T. contributed through technical discussions and by providing some problems for inclusion at the end of several chapters. Professor Charles E. Hutchinson, University of Massachusetts, contributed through his participation in early TASC work which set the stage for this book. Special acknowledgement is due Harry B. Silverman of TASC for his encouragement of the project from its inception.

THE ANALYTIC SCIENCES CORPORATION — **Arthur Gelb**
Reading, Massachusetts — **14 January 1974**

CONTENTS

APPLIED OPTIMAL ESTIMATION

1. INTRODUCTION

HISTORICAL PERSPECTIVE

The development of data processing methods for dealing with random variables can be traced to Gauss (circa 1800), who invented the technique of deterministic least-squares and employed it in a relatively simple orbit measurement problem (Ref. 1). The next significant contribution to the broad subject of estimation theory occurred more than 100 years later when Fisher (circa 1910), working with probability density functions, introduced the approach of maximum likelihood estimation (Ref. 2). Utilizing random process theory, Wiener (circa 1940) set forth a procedure for the frequency domain design of statistically optimal filters (Refs. 3, 4). The technique addressed the continuous-time problem in terms of correlation functions and the continuous filter impulse response. It was limited to statistically stationary processes and provided optimal estimates only in the steady-state regime. In the same time period, Kolmogorov treated the discrete-time problem (Ref. 5). During the next 20 years, Wiener's work was extended – in a way which often required cumbersome calculations – to include nonstationary and multiport systems (Refs. 6-8). Kalman and others (circa 1960) advanced optimal recursive filter techniques based on state-space, time domain formulations (Refs. 9-13). This approach, now known as the Kalman filter, is ideally suited for digital computer implementation. Indeed, it is the very foundation for data mixing in modern multisensor systems.

It is interesting to see the many similarities between Gauss' work and the more "modern" approaches. As is pointed out in Ref. 14, Gauss notes the need for redundant data to eliminate the influence of measurement errors; he raises the issue of dynamic modeling of the system under study; he refers to the inaccuracy of observations and thereby sets the stage for probabilistic considerations; he refers to the "suitable combination" of observations which will provide the most accurate estimates and thus touches upon the questions of estimator structure and performance criterion definition; he refers to the number of observations that are absolutely required for determination of the unknown quantities and thus addresses the subject currently referred to as "observability" of the system. Other similarities can also be cited. In fact, it can be argued that the Kalman filter is, in essence, a *recursive* solution* to Gauss' original least-squares problem.

It is also interesting to note two underlying differences between the classical and modern techniques; namely, the use of random process vs. deterministic signal descriptions and the use of high-speed digital computers to generate numerical solutions vs. the requirement for closed-form "pencil and paper" solutions. The former consideration enables the modern mathematics to more closely characterize physical situations being treated; the latter tremendously broadens the range of problems which may be studied.

OPTIMAL ESTIMATION

An optimal estimator is a computational algorithm that processes measurements to deduce a minimum error† estimate of the state of a system by utilizing: knowledge of system and measurement dynamics, assumed statistics of system noises and measurement errors, and initial condition information. Among the presumed advantages of this type of data processor are that it minimizes the estimation error in a well defined statistical sense and that it utilizes *all* measurement data plus prior knowledge about the system. The corresponding potential disadvantages are its *sensitivity* to erroneous *a priori* models and statistics, and the inherent *computational burden.* The important concepts embodied in these statements are explored in the sequel.

The three types of estimation problems of interest are depicted in Fig. 1.0-1. When the time at which an estimate is desired coincides with the last measurement point, the problem is referred to as *filtering*; when the time of interest falls within the span of available measurement data, the problem is termed *smoothing*; and when the time of interest occurs after the last available measurement, the problem is called *prediction*.

Probably the most common optimal filtering technique is that developed by Kalman (Ref. 10) for estimating the state of a *linear* system. This technique, depicted in Fig. 1.0-2, provides a convenient example with which to illustrate the capabilities and limitations of optimal estimators. For instance, given a linear system model and any measurements of its behavior, plus statistical models which characterize system and measurement errors, plus initial condition

*A solution that enables *sequential*, rather than *batch*, processing of the measurement data.
†In accordance with some stated criterion of optimality.

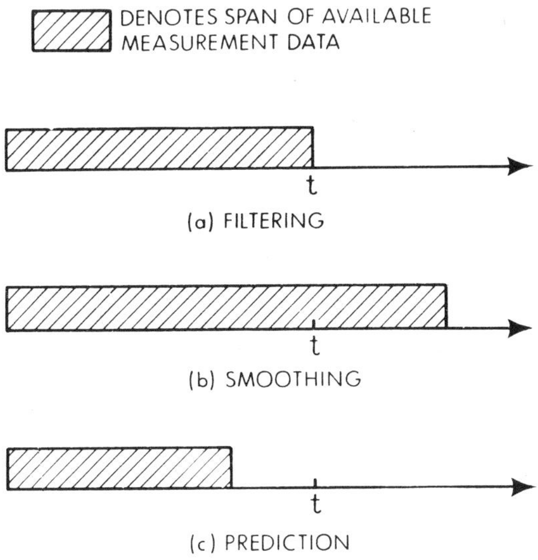

Figure 1.0-1 Three Types of Estimation Problems (estimate desired at time t)

information, the Kalman filter describes how to process the measurement data. However, the Kalman filter *per se* does not solve the problem of establishing an optimal measurement schedule, or of design in the presence of parameter uncertainties, or of how to deal with computational errors. Other design criteria, in addition to those used to derive the filtering algorithm, must be imposed to resolve these questions.

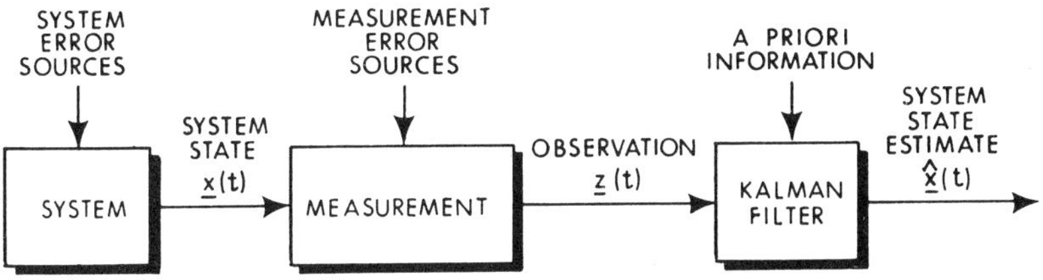

Figure 1.0-2 Block Diagram Depicting System, Measurement and Estimator

APPLICATIONS OF OPTIMAL ESTIMATION THEORY

The theory of optimal estimation has application to a tremendously broad range of problem areas. To cite a few, we have: tracer studies in nuclear medicine, statistical image enhancement, estimation of traffic densities, chemical process control, estimation of river flows, power system load forecasting, classification of vectorcardiograms, satellite orbit estimation and nuclear reactor parameter identification.

The estimation problem may be posed in terms of a single sensor making measurements on a single process or, more generally, in terms of multiple sensors and multiple processes. The latter case is referred to as a *multisensor system*. Suppose there are ℓ sensors, which provide measurements on m physical processes. Some of the sensors may measure the same quantity, in which case simple *redundant* measurements are provided; others may measure quantities related only *indirectly* to the processes of interest. The estimation problem, in the context of this multisensor system, is to process the sensor outputs such that "best" estimates of the processes of interest are obtained. A computer-implemented data processing algorithm operates on the sensor data to provide the desired estimates. These estimates may be used to drive displays and also to serve as control signals for the physical systems under observation, as illustrated in Fig. 1.0-3. In modern multisensor systems, the data processing algorithm is very often derived using optimal filter theory.

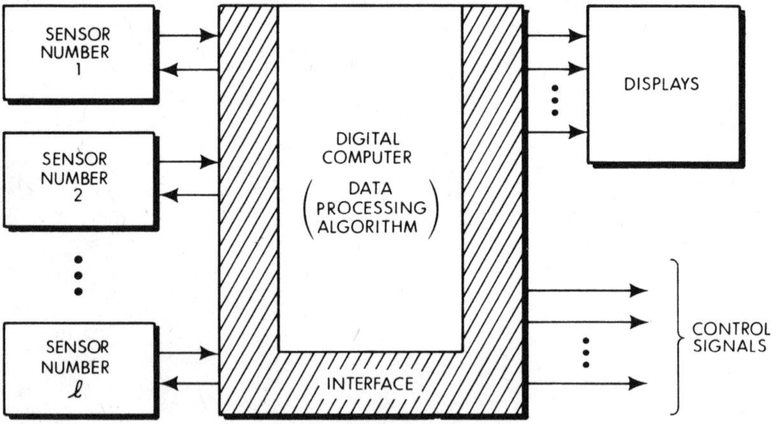

Figure 1.0-3 Modern Multisensor System

Some outstanding examples of multisensor systems occur in the field of navigation. External measurements were originally used to update navigation variables in a deterministic manner; for example, system position indication was changed to agree with the results of an external position measurement. This approach ignored two important facts. First, external measurements themselves contain random errors that may be significant, when compared to the navigation system errors. Secondly, navigation system errors are primarily caused by random, time-varying navigation sensor errors. The optimal use of external measurements, together with those provided by the navigation system, can provide a resulting navigation accuracy which is better than that obtainable from either external measurements or the navigation system alone.

Application of modern estimation techniques to multisensor navigation systems began in the mid-1960's, shortly after optimal recursive filter theory was developed and published. Because the errors in a typical navigation system propagate in essentially a linear manner and linear combinations of these errors can be detected from external measurements, the Kalman filter is ideally suited for their estimation. It also provides useful estimates of all system error sources with significant correlation times. In addition, the Kalman filter provides improved design and operational flexibility. As a time-varying filter, it can accommodate nonstationary error sources when their statistical behavior is known. Configuration changes in the navigation system are relatively easy to effect by programming changes. The Kalman filter provides for optimal use of any number, combination, and sequence of external measurements. It is a technique for systematically employing all available external measurements, regardless of their errors, to improve the accuracy of navigation systems. This application of the theory, among others, is often made in the following chapters.

Perhaps the essential non-hardware issues in any practical application of optimal estimation are those of *modeling* and *realistic performance projections.* These issues are, of course, related and their proper treatment is a prerequisite to successful system operation in a real environment. Based upon this perspective, it becomes clear that a reasonable use of estimation theory in any operational system is: *first* – design and computer evaluation of the "optimal" system behavior; *second* – design of a suitable "suboptimal" system with cost constraints, sensitivity characteristics, computational requirements, measurement schedules, etc., in mind; and *third* – construction and test of a prototype system, making final adjustments or changes as warranted.

Example 1.0-1

Consider a system comprised of two sensors, each making a single measurement, $z_i(i = 1,2)$, of a constant *but unknown* quantity, x, in the presence of random, independent, unbiased measurement errors, $v_i(i = 1,2)$. Design a data processing algorithm that combines the two measurements to produce an optimal estimate of x.

The measurements are described by

$$z_1 = x + v_1 \quad \text{and} \quad z_2 = x + v_2 \tag{1.0-1}$$

In the absence of any other information, we might seek an estimate of x which is a linear function of the measurements in the form (superscript "^" denotes estimate)

$$\hat{x} = k_1 z_1 + k_2 z_2 \tag{1.0-2}$$

where k_1 and k_2 remain to be specified. Defining the estimation error, $\tilde{x}$, as

$$\tilde{x} = \hat{x} - x$$

we seek to minimize the mean square value of $\tilde{x}$ as the criterion of optimality. Furthermore, we require that our choice of k_1 and k_2 be independent of the value of x; this condition* will hold if the estimate is *unbiased* – i.e., if

$$E[\tilde{x}] = E[k_1(x + v_1) + k_2(x + v_2) - x] = 0 \tag{1.0-3}$$

where E denotes the ensemble expectation† or average. Performing the indicated expectation, with $E[v_1] = E[v_2] = 0$ and $E[x] = x$ since x is "nonrandom", we obtain

$$k_2 = 1 - k_1 \tag{1.0-4}$$

Combining Eqs. (1.0-1) through (1.0-4), the mean square error is computed in the form

$$E[\tilde{x}^2] = k_1^2 \sigma_1^2 + (1 - k_1)^2 \sigma_2^2 \tag{1.0-5}$$

where σ_1^2 and σ_2^2 denote the variances of v_1 and v_2, respectively. Differentiating this quantity with respect to k_1 and setting the result to zero yields

$$2k_1\sigma_1^2 - 2(1 - k_1)\sigma_2^2 = 0$$

or

$$k_1 = \frac{\sigma_2^2}{\sigma_1^2 + \sigma_2^2}$$

The corresponding minimum mean square estimation error is

$$E[\tilde{x}^2] = \left(\frac{1}{\sigma_1^2} + \frac{1}{\sigma_2^2}\right)^{-1} \tag{1.0-6}$$

It can be seen that the mean square estimation error is smaller than either of the mean square measurement errors. The algorithm [Eq. (1.0-2)]

$$\hat{x} = \left(\frac{\sigma_2^2}{\sigma_1^2 + \sigma_2^2}\right) z_1 + \left(\frac{\sigma_1^2}{\sigma_1^2 + \sigma_2^2}\right) z_2 \tag{1.0-7}$$

makes sense in the various limits of interest – i.e., if $\sigma_1^2 = \sigma_2^2$, the measurements are averaged; if one measurement is perfect (σ_1 or σ_2 equal to zero) the other is rejected; etc.

In Example 1.0-1, the physical quantity of interest was a scalar constant and only two linear measurements of that constant were available. In general, we shall be interested in estimates of vector time-varying quantities where many measurements are provided.

---

*This condition is imposed because x is unknown; hence, the gains k_1 and k_2 *must* be independent of x.

†The expectation operator is discussed in Section 2.2.

Physical systems and measurements can be categorized according to whether they are *static* or *dynamic, continuous* or *discrete*, and *linear* or *nonlinear.* Furthermore, we shall be interested in real-time data processing algorithms *(filtering* and *prediction)* and post-experiment algorithms *(smoothing)*. Due to its notational simplicity and theoretical power, subsequent treatment of these physical systems and data processing algorithms is couched in terms of state space representation, using the mathematics of vectors and matrices and the notions of random process theory. These topics are discussed in Chapters 2 and 3.

REFERENCES

1. Gauss, K.F., *Theoria Motus*, 1809; also, *Theory of the Motion of the Heavenly Bodies About the Sun in Conic Sections*, Dover, New York, 1963.
2. Fisher, R.A., "On an Absolute Criterion for Fitting Frequency Curves," *Messenger of Math.*, Vol. 41, 1912, p. 155.
3. Wiener, N., "The Extrapolation, Interpolation and Smoothing of Stationary Time Series," OSRD 370, Report to the Services 19, Research Project DIC-6037, MIT, February 1942.
4. Wiener, N., *The Extrapolation, Interpolation and Smoothing of Stationary Time Series*, John Wiley & Sons, Inc., New York, 1949.
5. Kolmogorov, A.N., "Interpolation and Extrapolation von Stationaren Zufalligen Folgen," *Bull. Acad. Sci. USSR*, Ser. Math. 5, 1941, pp. 3-14.
6. Zadeh, L.A., and Ragazzini, J.R., "An Extension of Wiener's Theory of Prediction," *J. Appl. Phys.* Vol. 21, July 1950, pp. 645-655.
7. Laning, J.H., Jr. and Battin, R.H., *Random Processes in Automatic Control*, McGraw-Hill Book Co., Inc., New York, 1956, Chapter 8.
8. Shinbrot, M., "Optimization of Time-Varying Linear Systems with Nonstationary Inputs," *Trans. ASME*, Vol. 80, 1958, pp. 457-462.
9. Swerling, P., "First Order Error Propagation in a Stagewise Smoothing Procedure for Satellite Observations," *J. Astronautical Sciences*, Vol. 6, 1959, pp. 46-52.
10. Kalman, R.E., "A New Approach to Linear Filtering and Prediction Problems," *J. Basic Eng.*, March 1960, pp. 35-46.
11. Kalman, R.E., and Bucy, R.S., "New Results in Linear Filtering and Prediction Theory," *J. Basic Eng.*, March 1961, pp. 95-108.
12. Blum, M., "A Stagewise Parameter Estimation Procedure for Correlated Data," *Numer. Math.*, Vol. 3, 1961, pp. 202-208.
13. Battin, R.H., "A Statistical Optimizing Navigation Procedure for Space Flight," *ARS Journal*, Vol. 32, 1962, pp. 1681-1696.
14. Sorenson, H.W., "Least-Squares Estimation: From Gauss to Kalman," *IEEE Spectrum*, July 1970, pp. 63-68.

PROBLEMS

Problem 1-1

Repeat Example 1.0-1 for the case where measurement errors are correlated; that is, $E[v_1v_2] = \rho\sigma_1\sigma_2$ where ρ is a correlation coefficient ($|\rho| \leqslant 1$). Show that

$$k = \frac{\sigma_2^2 - \rho\sigma_1\sigma_2}{\sigma_1^2 + \sigma_2^2 - 2\rho\sigma_1\sigma_2}$$

and

$$E[\tilde{x}^2] = \frac{\sigma_1^2\sigma_2^2(1-\rho^2)}{\sigma_1^2 + \sigma_2^2 - 2\rho\sigma_1\sigma_2}$$

for the optimal estimate. Physically interpret the meaning of $E[\tilde{x}^2] = 0$ for $\rho = \pm 1$.

Problem 1-2

Compute $E[\tilde{x}^2]$ from Eqs. (1.0-1) and (1.0-2) with the gains k_1 and k_2 unrestricted. Show that values of the gains which minimize $E[\tilde{x}^2]$ are functions of x, using the fact that $E[x] = x$ and $E[x^2] = x^2$.

Problem 1-3

Consider a case similar to Example 1.0-1, but in which three independent measurements are available instead of two. Argue why an estimator should be sought in the form

$$\hat{x} = k_1 z_1 + k_2 z_2 + (1 - k_1 - k_2)\, z_3$$

Develop optimum values for k_1 and k_2 and use these to show that

$$E[\tilde{x}^2] = \left(\frac{1}{\sigma_1^2} + \frac{1}{\sigma_2^2} + \frac{1}{\sigma_3^2}\right)^{-1}$$

for the optimal estimate.

Problem 1-4

The concentration of a substance in solution decreases exponentially during an experiment. Noisy measurements of the concentration are made at times t_1 and t_2, such that $(i = 1,2)$

$$z_i = x_0\, e^{-t_i} + v_i$$

An estimate of the initial concentration, x_0, is desired. Demonstrate that an unbiased estimator is given by

$$\hat{x}_0 = (k e^{t_1})\, z_1 + [(1-k)\, e^{t_2}]\, z_2$$

where k, not yet specified, is a constant. Show that the value of k, which minimizes the mean square estimation error, is

$$k = \frac{\sigma_2^2}{\sigma_1^2 e^{-2(t_2 - t_1)} + \sigma_2^2}$$

and that the corresponding mean square estimation error is

$$E[(\hat{x}_0 - x_0)^2] = \left(\frac{1}{\sigma_1^2} e^{-2t_1} + \frac{1}{\sigma_2^2} e^{-2t_2}\right)^{-1}$$

where σ_1^2 and σ_2^2 are measurement error variances.

2. REVIEW OF UNDERLYING MATHEMATICAL TECHNIQUES

In this chapter, mathematical techniques utilized in the development and application of modern estimation theory are reviewed. These techniques include vector and matrix operations and their application to least-squares estimation. Also included is a presentation on probability, random variables and random processes. Of these, the latter is the most important for our work and, consequently, is given the greatest emphasis. The specific goal is to provide the mathematical tools necessary for development of state vector and covariance matrix methods and their subsequent application.

Note that the material in this chapter is neither a complete discussion of the topics, nor mathematically rigorous. However, more detailed material is provided in the references cited.

2.1 VECTORS, MATRICES, AND LEAST SQUARES

This section contains pertinent definitions and operations for the application of vector and matrix methods in modern estimation. The reader requiring a more complete discussion is referred to Refs. 1-3.

VECTOR OPERATIONS

An array of elements, arranged in a column, is designated by a lowercase letter with an underbar and is referred to as a *vector* (more precisely, a column vector). The number of elements in the vector is its dimension. Thus, an n-dimensional vector $\underline{x}$ is:

$$\underline{x} = \begin{bmatrix} x_1 \\ x_2 \\ \cdot \\ \cdot \\ \cdot \\ x_n \end{bmatrix} \tag{2.1-1}$$

Vector Addition – Addition of two vectors is defined by

$$\underline{x} + \underline{y} \triangleq \begin{bmatrix} x_1 + y_1 \\ x_2 + y_2 \\ \cdot \\ \cdot \\ \cdot \\ x_n + y_n \end{bmatrix} \tag{2.1-2}$$

Vector subtraction is defined in a similar manner. In both vector addition and subtraction, $\underline{x}$ and $\underline{y}$ must have the same dimension.

Scalar Multiplication – A vector may be multiplied by a scalar, k, yielding

$$k\underline{x} = \begin{bmatrix} kx_1 \\ kx_2 \\ \cdot \\ \cdot \\ \cdot \\ kx_n \end{bmatrix} \tag{2.1-3}$$

Zero Vector – The vector $\underline{0}$ is defined as the vector in which every element is the number 0.

Vector Transpose – The transpose of a vector is defined such that, if $\underline{x}$ is the column vector

$$\underline{x} = \begin{bmatrix} x_1 \\ x_2 \\ \cdot \\ \cdot \\ \cdot \\ x_n \end{bmatrix} \tag{2.1-4}$$

its transpose is the row vector

$$\underline{x}^T = [x_1 x_2 \ldots x_n] \tag{2.1-5}$$

Inner Product – The quantity $\underline{x}^T\underline{y}$ is referred to as the inner product or dot product and yields the *scalar*

$$\underline{x}^T\underline{y} = x_1y_1 + x_2y_2 + \ldots + x_ny_n \tag{2.1-6}$$

If $\underline{x}^T\underline{y} = 0$, $\underline{x}$ and $\underline{y}$ are said to be *orthogonal*. In addition, $\underline{x}^T\underline{x}$, the squared length of the vector $\underline{x}$, is

$$\underline{x}^T\underline{x} = {x_1}^2 + {x_2}^2 + \ldots + {x_n}^2 \tag{2.1-7}$$

The length of the vector $\underline{x}$ is denoted by

$$|\underline{x}| = \sqrt{\underline{x}^T\underline{x}} \tag{2.1-8}$$

Outer Product – The quantity $\underline{x}\underline{y}^T$ is referred to as the outer product and yields the *matrix*

$$\underline{x}\underline{y}^T = \begin{bmatrix} x_1y_1 & x_1y_2 & \ldots & x_1y_n \\ x_2y_1 & x_2y_2 & \ldots & x_2y_n \\ \cdot & \cdot & & \cdot \\ \cdot & \cdot & & \cdot \\ \cdot & \cdot & & \cdot \\ x_ny_1 & x_ny_2 & \ldots & x_ny_n \end{bmatrix}$$

Similarly, we can form the matrix $\underline{x}\underline{x}^T$ as

$$\underline{x}\underline{x}^T = \begin{bmatrix} {x_1}^2 & x_1x_2 & \ldots & x_1x_n \\ x_2x_1 & {x_2}^2 & \ldots & x_2x_n \\ \cdot & \cdot & & \cdot \\ \cdot & \cdot & & \cdot \\ \cdot & \cdot & & \cdot \\ x_nx_1 & x_nx_2 & \ldots & {x_n}^2 \end{bmatrix} \tag{2.1-9}$$

where $\underline{x}\underline{x}^T$ is called the *scatter matrix* of the vector $\underline{x}$.

Vector Derivative – By using previously defined operations, the derivative of a vector may be defined. For continuous vectors $\underline{x}(t)$ and $\underline{x}(t + \Delta t)$ we get

$$\underline{x}(t + \Delta t) - \underline{x}(t) = \begin{bmatrix} x_1(t + \Delta t) \\ x_2(t + \Delta t) \\ \cdot \\ \cdot \\ \cdot \\ x_n(t + \Delta t) \end{bmatrix} - \begin{bmatrix} x_1(t) \\ x_2(t) \\ \cdot \\ \cdot \\ \cdot \\ x_n(t) \end{bmatrix} = \begin{bmatrix} x_1(t + \Delta t) - x_1(t) \\ x_2(t + \Delta t) - x_2(t) \\ \cdot \\ \cdot \\ \cdot \\ x_n(t + \Delta t) - x_n(t) \end{bmatrix} \tag{2.1-10}$$

Multiplying both sides by the scalar $1/\Delta t$ and taking the limit as $\Delta t \rightarrow 0$ yields

$$\dot{\underline{x}}(t) = \begin{bmatrix} \dot{x}_1(t) \\ \dot{x}_2(t) \\ \cdot \\ \cdot \\ \cdot \\ \dot{x}_n(t) \end{bmatrix} \tag{2.1-11}$$

which is the desired result. The integral of a vector is similarly described – i.e., in terms of the integrals of its elements.

MATRIX OPERATIONS

A *matrix* is an $m \times n$ rectangular array of elements in m rows and n columns, and will be designated by a capital letter. The matrix A, consisting of m rows and n columns, is denoted as:

$$A = [a_{ij}] \tag{2.1-12}$$

where a_{ij} is the element in the ith row and j^{th} column, for $i = 1,2, \ldots, m$ and $j = 1,2, \ldots, n$. For example, if $m = 2$ and $n = 3$, A is a 2×3 matrix

$$A = \begin{bmatrix} a_{11} & a_{12} & a_{13} \\ a_{21} & a_{22} & a_{23} \end{bmatrix}$$

Note that a *column vector* may be thought of as a matrix of n rows and 1 column; a *row vector* may be thought of as a matrix of 1 row and n columns; and a *scalar* may be thought of as a matrix of 1 row and 1 column. A matrix of n rows and n columns is *square*; a square matrix in which all elements off the main diagonal are zero is termed a *diagonal matrix*. The main diagonal starts in the upper left corner and ends in the lower right; its elements are $a_{11}, a_{22}, \ldots, a_{nn}$.

Matrix Addition – Matrix addition is defined *only* when the two matrices to be added are of identical dimensions, i.e., they have the same number of rows and the same number of columns. Specifically.

$$A + B \triangleq [a_{ij} + b_{ij}] \tag{2.1-13}$$

and for $m = 3$ and $n = 2$:

$$A + B \triangleq \begin{bmatrix} a_{11} + b_{11} & a_{12} + b_{12} \\ a_{21} + b_{21} & a_{22} + b_{22} \\ a_{31} + b_{31} & a_{32} + b_{32} \end{bmatrix}$$

Scalar Multiplication – The matrix A may be multiplied by a scalar k. Such scalar multiplication is denoted by kA where

$$kA = [ka_{ij}] \tag{2.1-14}$$

Thus, for matrix subtraction, $A - B = A + (-1)B$, that is, one simply subtracts corresponding elements.

Matrix Multiplication – The product of two matrices, AB, read A times B, in that order, is defined by the matrix

$$AB = C = [c_{ij}] \tag{2.1-15}$$

The product AB is defined only when A and B are *conformable*, that is, the number of columns of A is equal to the number of rows of B. Where A is m X p and B is p X n, the product matrix $[c_{ij}]$ has m rows and n columns and c_{ij} is given by

$$c_{ij} = \sum_{k=1}^{p} a_{ik}b_{kj} = a_{i1}b_{1j} + a_{i2}b_{2j} + \ldots + a_{ip}b_{pj} \tag{2.1-16}$$

For example, with A and B as previously defined, AB is given by

$$AB = [c_{ij}] = \begin{bmatrix} a_{11}b_{11} + a_{12}b_{21} + a_{13}b_{31} & a_{11}b_{12} + a_{12}b_{22} + a_{13}b_{32} \\ a_{21}b_{11} + a_{22}b_{21} + a_{23}b_{31} & a_{21}b_{12} + a_{22}b_{22} + a_{23}b_{32} \end{bmatrix}$$

for m = 2, p = 3, n = 2. It is noted that two matrices are equal if, and only if, all of their corresponding elements are equal. Thus A = B implies $a_{ij} = b_{ij}$ for all i (1,2, . . . ,m) and all j (1,2, . . . ,n). For square matrices A and B of equal dimension, the products AB and BA are both defined, but in general AB ≠ BA – i.e., matrix multiplication is noncommutative.

Vector-Matrix Product – If a vector $\underline{x}$ and matrix A are conformable, the product

$$\underline{y} = A\underline{x} \tag{2.1-17}$$

is defined such that

$$y_i = \sum_{j=1}^{n} a_{ij}x_j \tag{2.1-18}$$

Matrix Derivative and Integral – Using the operations of addition and multiplication by a scalar, the derivative and integral of a matrix may be formulated. Analogous to the vector operations, we obtain

$$\dot{A}(t) = [\dot{a}_{ij}(t)] \tag{2.1-19}$$

$$\int A(t)\, dt = \left[\int a_{ij}(t)\, dt\right] \tag{2.1-20}$$

Zero Matrix – The matrix [0], herein simply denoted 0, is defined as the matrix in which every element is the number 0.

Identity Matrix – The identity matrix I is a square matrix with 1 located in each position down the main diagonal of the matrix and 0's elsewhere – i.e.,

$$I = \begin{bmatrix} 1 & 0 & \dots & 0 \\ 0 & 1 & \dots & 0 \\ \cdot & & & \cdot \\ \cdot & & & \cdot \\ \cdot & & & \cdot \\ 0 & 0 & \dots & 1 \end{bmatrix}$$

$$= [\delta_{ij}] \tag{2.1-21}$$

where δ_{ij} is the Kronecker delta defined by

$$\delta_{ij} = \begin{cases} 1 \text{ if } i = j \\ 0 \text{ if } i \neq j \end{cases} \tag{2.1-22}$$

The result of multiplication by the identity matrix is AI = IA = A, that is, no change.

Matrix Determinant – The square matrix A has a determinant denoted by |A|. The determinant is a scalar defined by

$$|A| = \sum_{i=1}^{n} \sum_{\substack{j=1 \\ j\neq i}}^{n} \dots \sum_{\substack{\ell=1 \\ \ell\neq i,j\dots,k}}^{n} a_{1i}a_{2j}\dots a_{n\ell} \tag{2.1-23}$$

In each term the second subscripts i,j, . . . , ℓ are permutations of the numbers 1,2, . . . ,n. Terms whose subscripts are even permutations are given plus signs, and those with odd permutations are given minus signs. The determinant of the product AB is

$$|AB| = |A| \cdot |B| \tag{2.1-24}$$

The determinant of the identity matrix is unity. One common use of the determinant is in solving a set of linear algebraic equations by means of Cramer's Rule (Ref. 1). However, modern computer algorithms utilize other techniques, which do not rely on calculating the matrix determinant for solving linear equations. For present purposes, the matrix determinant is used as part of a criterion for matrix invertibility – a subject which is discussed in subsequent paragraphs.

The Inverse of a Matrix – In considering the inverse of a matrix we must restrict our discussion to square matrices. If A is a square matrix, its inverse is denoted by A^{-1} such that

$$A^{-1} A = AA^{-1} = I \tag{2.1-25}$$

That is, the multiplication of a matrix by its inverse is commutative. For square matrices A and B of the same dimension, it is easily shown that

$$(AB)^{-1} = B^{-1} A^{-1} \tag{2.1-26}$$

It is noted that all square matrices do not have inverses – only those that are *nonsingular* have an inverse. For purposes of clarification, a nonsingular matrix may be defined as follows:

No row (column) is a linear combination of other rows (columns)

or

The determinant of the matrix is not zero – i.e., $|A| \neq 0$.

If $|A| = 0$, two or more of the rows (columns) of A are linearly dependent. Use of the inverse matrix A^{-1} can be illustrated by its role in solving the matrix equation

$$A\underline{x} = \underline{y} \tag{2.1-27}$$

where A is an n X n matrix, and $\underline{x}$ and $\underline{y}$ are n-row column vectors (n X 1). Premultiplication by A^{-1} (assuming that it exists) yields

$$\underline{x} = A^{-1}\underline{y} \tag{2.1-28}$$

which is the solution to Eq. (2.1-27). Computing inverses for large matrices is a time-consuming operation; however, it is suitable for computer solution. Results for 2 X 2 and 3 X 3 matrices are given below. If

$$A = \begin{bmatrix} a & b \\ c & d \end{bmatrix} \tag{2.1-29}$$

then

$$A^{-1} = \frac{1}{|A|}\begin{bmatrix} d & -b \\ -c & a \end{bmatrix} \tag{2.1-30}$$

where

$$|A| = ad - bc \tag{2.1-31}$$

If

$$A = \begin{bmatrix} a & b & c \\ d & e & f \\ g & h & i \end{bmatrix} \tag{2.1-32}$$

then

$$A^{-1} = \frac{1}{|A|}\begin{bmatrix} ei - fh & ch - bi & bf - ec \\ gf - di & ai - gc & dc - af \\ dh - ge & gb - ah & ae - bd \end{bmatrix} \tag{2.1-33}$$

where

$$|A| = aei + bfg + cdh - ceg - bdi - afh \tag{2.1-34}$$

It can be shown that, in general,

$$A^{-1} = \frac{1}{|A|}\,\mathrm{adj}A$$

where $|A|$ is the determinant of the square matrix A. and adjA (the *adjoint* of A) is the matrix formed by replacing each element with its cofactor and transposing the result [the *cofactor* of a_{ij} is the determinant of the submatrix formed by deleting from A the ith row and j^{th} column, multiplied by $(-1)^{i+j}$]. Clearly, hand computation of the inverse of a large matrix is tedious, since a number of successively smaller submatrices are obtained and the determinant for each of these must also be computed.

The Transpose of a Matrix – The transpose of a matrix is obtained by interchanging its rows and columns. For example, if

$$A = \begin{bmatrix} a_{11} & a_{12} & a_{13} \\ a_{21} & a_{22} & a_{23} \end{bmatrix}$$

then

$$A^T = \begin{bmatrix} a_{11} & a_{21} \\ a_{12} & a_{22} \\ a_{13} & a_{23} \end{bmatrix}$$

or, in general

$$A = [a_{ij}], \quad A^T = [a_{ji}] \tag{2.1-35}$$

Thus, an m × n matrix has an n × m transpose. For square matrices, if $A = A^T$, then A is said to be *symmetric*. If $A^T = A^{-1}$, then A is said to be *orthogonal*. The determinant of an orthogonal matrix is unity. If $A^T = -A$, A is said to be *skew-symmetric*. A property of a skew-symmetric matrix is that all elements on the main diagonal are zero.

For matrices A and B, of appropriate dimension, it can be shown that

$$(AB)^T = B^T A^T \tag{2.1-36}$$

If A is invertible then

$$(A^{-1})^T = (A^T)^{-1} \tag{2.1-37}$$

Trace – The trace of square matrix A is the scalar sum of its diagonal elements. Thus

$$\text{trace}[A] = \sum_{i=1}^{n} a_{ii} \tag{2.1-38}$$

For square matrices A and B

$$\text{trace}\,[AB] = \text{trace}\,[BA] \tag{2.1-39}$$

Rank – The rank of matrix A is the dimension of the largest square matrix contained in A (formed by deleting rows and columns) which has a nonzero determinant. It follows from the discussion of invertibility that a nonsingular $n \times n$ matrix has rank n.

Matrix Pseudoinverse – The matrix inverse is defined for square matrices only; it is used in the solution of sets of linear equations – of the form $A\underline{x} = \underline{y}$ – in which the number of equations is equal to the number of unknowns. For nonsquare matrices used to describe systems of equations where the number of equations does not equal the number of unknowns, the equivalent operator is the *pseudoinverse* (Ref. 4).

If a matrix A has more rows than columns, the pseudoinverse is defined as

$$A^{\#} = (A^T A)^{-1} A^T \tag{2.1-40}$$

for nonsingular $(A^T A)$. In the solution of linear equations this is the so-called *overdetermined* case – where there are more equations than unknowns. The resulting solution, $\underline{x} = A^{\#}\underline{y}$ is best in a least-squares sense. If A has more columns than rows, the pseudoinverse is defined as

$$A^{\#} = A^T(AA^T)^{-1} \tag{2.1-41}$$

This corresponds to the *underdetermined* case – there are fewer equations than unknowns. Typically, such a situation leads to an infinite number of least-squares solutions (consider the least-squares fit of a straight line to a single point). The solution resulting from the pseudoinverse is also best in a least-squares sense, and the vector $\underline{x} = A^{\#}\underline{y}$ is the solution of minimum length.

Several useful relations concerning the pseudoinverse are given below:

$$AA^{\#}A = A \tag{2.1-42}$$

$$A^{\#}AA^{\#} = A^{\#} \tag{2.1-43}$$

$$(A^{\#}A)^T = A^{\#}A \tag{2.1-44}$$

$$(AA^{\#})^T = AA^{\#} \tag{2.1-45}$$

Functions of Square Matrices – With λ as a scalar variable, the equation

$$f(\lambda) = |\lambda I - A| = 0 \tag{2.1-46}$$

is called the *characteristic equation* of the square matrix A. Values of λ that satisfy this equation are the *eigenvalues* of the square matrix A. By expanding the determinant, it can be seen that $f(\lambda)$ is a polynomial in λ. The Cayley-Hamilton Theorem states that, for the same polynomial expression,

$$f(A) = 0 \tag{2.1-47}$$

That is, every square matrix satisfies its own characteristic equation.

It is possible to define special polynomial functions of a square matrix A, two of which are:

$$e^A = I + A + \frac{1}{2!}A^2 + \frac{1}{3!}A^3 + \ldots \tag{2.1-48}$$

$$\sin A = A - \frac{1}{3!}A^3 + \frac{1}{5!}A^5 - \ldots \tag{2.1-49}$$

where

$$A^2 = AA \tag{2.1-50}$$

$$A^3 = AAA \quad \text{etc.} \tag{2.1-51}$$

The matrix exponential e^A occurs in the study of constant coefficient matrix differential equations, and is utilized in Chapter 3. Coddington and Levinson (Ref. 5) provide a number of useful relations for the matrix exponential. Among them are:

$$e^{A+B} = e^A e^B, \quad \text{if } AB=BA \tag{2.1-52}$$

$$e^{TFT^{-1}} = Te^F T^{-1}, \text{ for } |T| \neq 0 \tag{2.1-53}$$

$$|e^F| = e^{\text{trace}[F]} \tag{2.1-54}$$

VECTOR-MATRIX OPERATIONS

Vectors and matrices can be combined in mathematical expressions in various ways. Since a vector of dimension n can be thought of as an n X 1 matrix, the rules developed above for matrix operations can be readily applied. Several of the more common operations are briefly considered in the following discussion.

Quadratic Forms – An n X n symmetric matrix A and the n-dimensional vector $\underline{x}$ can be used to define the scalar quantity

$$J = \underline{x}^T A \underline{x} \tag{2.1-55}$$

or

$$J = a_{11}x_1^2 + a_{22}x_2^2 + \ldots + a_{nn}x_n^2$$
$$+ 2(a_{12}x_1x_2 + a_{13}x_1x_3 + \ldots + a_{n-1,n}x_{n-1}x_n)$$

This scalar expression is referred to as a *quadratic form*. An orthogonal matrix Q can always be found such that (Ref. 1)

$$A' = Q^T A Q \tag{2.1-56}$$

is diagonal with $a'_{ii} = \lambda_i$, where the λ_i are the eigenvalues of A. It can be shown that the quadratic form reduces to

$$J = \lambda_1 x_1'^2 + \lambda_2 x_2'^2 + \ldots + \lambda_n x_n'^2 \tag{2.1-57}$$

where

$$\underline{x}' = Q^T \underline{x} \tag{2.1-58}$$

Definite Forms – The quadratic form is further used to define properties of the matrix A:

If $\underline{x}^T A \underline{x} > 0$ for all real $\underline{x}$, A is said to be *positive definite*.

If $\underline{x}^T A \underline{x} \geqslant 0$ for all real $\underline{x}$, A is said to be *positive semidefinite*.

If $\underline{x}^T A \underline{x} \leqslant 0$ for all real $\underline{x}$, A is said to be *negative semidefinite*.

If $\underline{x}^T A \underline{x} < 0$ for all real $\underline{x}$, A is said to be *negative definite*.

Norm – The matrix-associated quantity, analogous to the length of a vector, is called the *norm* and is defined by

$$\|A\| = \max_{\text{all } \underline{x}} \frac{|A\underline{x}|}{|\underline{x}|} \tag{2.1-59}$$

With vector length as defined in Eq. (2.1-8), the norm is readily computed as

$$\|A\| = \sqrt{\lambda_1} \tag{2.1-60}$$

where λ_1 is the maximum eigenvalue of the matrix product $A^T A$ (Ref. 6).

Gradient Operations – Differentiation of vector-matrix expressions, with respect to scalar quantities such as time, has been previously discussed. Rules for differentiation, with respect to vector and matrix quantities, are given below. The *gradient* or derivative of a scalar function z, with respect to a vector $\underline{x}$, is the vector

$$\frac{\partial z}{\partial \underline{x}} = \underline{a} \tag{2.1-61}$$

with

$$a_i = \frac{\partial z}{\partial x_i} \tag{2.1-62}$$

A case of special importance is the vector gradient of the inner product. We have

$$\frac{\partial}{\partial \underline{x}}(\underline{y}^T \underline{x}) = \underline{y} \tag{2.1-63}$$

and

$$\frac{\partial}{\partial \underline{x}}(\underline{x}^T \underline{y}) = \underline{y} \tag{2.1-64}$$

The second partial of a scalar z, with respect to a vector $\underline{x}$, is a matrix denoted by

$$\frac{\partial^2 z}{\partial \underline{x}^2} = A \tag{2.1-65}$$

with

$$a_{ij} = \frac{\partial^2 z}{\partial x_i \partial x_j} \tag{2.1-66}$$

The determinant of A is called the *Hessian* of z.

In general, the vector gradient of a vector is the matrix defined as

$$\frac{\partial \underline{z}^T}{\partial \underline{x}} = A \tag{2.1-67}$$

with

$$a_{ij} = \frac{\partial z_j}{\partial x_i} \tag{2.1-68}$$

If $\underline{z}$ and $\underline{x}$ are of equal dimension, the determinant of A can be found and is called the *Jacobian* of $\underline{z}$. The matrix gradient of a scalar z is defined by

$$\frac{\partial z}{\partial A} = B \tag{2.1-69}$$

with

$$b_{ij} = \frac{\partial z}{\partial a_{ij}} \tag{2.1-70}$$

Two scalar functions of special note are the matrix trace and determinant. A tabulation of gradient functions for the trace and determinant is provided in Ref. 7. Some of particular interest for square matrices A, B, and C are given below:

$$\frac{\partial}{\partial A} \text{trace}[A] = I \tag{2.1-71}$$

$$\frac{\partial}{\partial A} \text{trace}[BAC] = B^T C^T \tag{2.1-72}$$

$$\frac{\partial}{\partial A} \text{trace}[ABA^T] = 2AB \tag{2.1-73}$$

$$\frac{\partial}{\partial A} \text{trace}[e^A] = e^A \tag{2.1-74}$$

$$\frac{\partial}{\partial A} |BAC| = |BAC| (A^{-1})^T \tag{2.1-75}$$

LEAST-SQUARES TECHNIQUES

Vector and matrix methods are particularly convenient in the application of least-squares estimation techniques. A specific example of least-squares estimation occurs in curve-fitting problems, where it is desired to obtain a functional form of some chosen order that best fits a given set of measurements. The criterion for goodness of fit is to minimize the sum of squares of differences between measurements and the "estimated" functional form or curve.

The linear least-squares problem involves using a set of measurements, $\underline{z}$, which are linearly related to the unknown quantities $\underline{x}$ by the expression

$$\underline{z} = H\underline{x} + \underline{v} \tag{2.1-76}$$

where $\underline{v}$ is a vector of measurement "noise." The goal is to find an *estimate* of the unknown, denoted by $\hat{\underline{x}}$. In particular, given the vector difference

$$\underline{z} - H\hat{\underline{x}}$$

we wish to find the $\hat{\underline{x}}$ that minimizes the sum of the squares of the elements of $\underline{z} - H\hat{\underline{x}}$. Recall that the vector inner product generates the sum of squares of a vector. Thus, we wish to minimize the scalar cost function J, where

$$J = (\underline{z}-H\hat{\underline{x}})^T (\underline{z}-H\hat{\underline{x}}) \tag{2.1-77}$$

Minimization of a scalar, with respect to a vector, is obtained when

$$\frac{\partial J}{\partial \hat{\underline{x}}} = \underline{0} \tag{2.1-78}$$

and the Hessian of J is positive semidefinite

$$\left| \frac{\partial^2 J}{\partial \hat{\underline{x}}^2} \right| \geqslant 0 \tag{2.1-79}$$

Differentiating J and setting the result to zero yields

$$H^T H\hat{\underline{x}} = H^T \underline{z} \tag{2.1-80}$$

It is readily shown that the second derivative of J, with respect to $\hat{\underline{x}}$, is positive semidefinite; and thus Eq. (2.1-80) does, indeed, define a minimum. When $H^T H$ possesses an inverse (i.e., when it has a nonzero determinant), the least-squares estimate is

$$\hat{\underline{x}} = (H^T H)^{-1} H^T \underline{z} \tag{2.1-81}$$

Note the correspondence between this result and that given in the discussion of the pseudoinverse. This derivation verifies the least-squares property of the pseudoinverse.

Implied in the preceding discussion is that all available measurements are utilized together at one time – i.e., in a so-called *batch processing* scheme. Subsequent discussion of optimal filtering is based on the concept of *recursive processing* in which measurements are utilized sequentially, as they become available.

2.2 PROBABILITY AND RANDOM PROCESSES*

This section contains a brief summation of material on probability, random variables and random processes with particular emphasis on application to modern estimation. The reader desiring a more extensive treatment of this material is referred to any one of the following excellent books on the subject: Davenport and Root (Ref. 9), Laning and Battin (Ref. 10) and Papoulis (Ref. 11).

*This material closely follows a similar presentation in Appendix H of Ref. 8.

PROBABILITY

Consider an event E, which is a possible outcome of a random experiment. We denote the *probability* of this event by Pr(E) and intuitively think of it as the limit, as the number of trials becomes large, of the ratio of the number of times E occurred to the number of times the experiment was attempted. If all possible outcomes of the experiment are denoted by E_i, i=1,2, . . . ,n, then

$$0 \leqslant Pr(E_i) \leqslant 1 \tag{2.2-1}$$

and

$$\sum_{i=1}^{n} Pr(E_i) = 1 \tag{2.2-2}$$

prescribe the limits of the probability.

The joint event that A and B and C, etc., occurred is denoted by ABC . . . , and the probability of this joint event, by Pr(ABC . . .). If the events A,B,C, etc., are mutually *independent* – which means that the occurrence of any one of them bears no relation to the occurrence of any other – the probability of the joint event is the product of the probabilities of the simple events. That is,

$$Pr(ABC\ldots) = Pr(A)Pr(B)Pr(C)\ldots \tag{2.2-3}$$

if the events A,B,C, etc., are mutually independent. Actually, the mathematical definition of independence is the reverse of this statement, but the result of consequence is that independence of events and the multiplicative property of probabilities go together.

The event that either A or B or C, etc., occurs is denoted by A+B+C and the probability of this event, by Pr(A+B+C). If these events are mutually *exclusive* – which means that the occurrence of one precludes the occurrence of any other – the probability of the total event is the sum of the probabilities of the simple events. That is,

$$Pr(A+B+C+\ldots) = Pr(A) + Pr(B) + Pr(C) + \ldots \tag{2.2-4}$$

if events A,B,C, etc., are mutually exclusive. If two events A and B are not mutually exclusive, then

$$Pr(A+B) = Pr(A) + Pr(B) - Pr(AB) \tag{2.2-5}$$

Clearly, if two events are mutually exclusive their joint probability Pr(AB) is zero.

For events which are not independent, the concept of *conditional probability* can provide added information. The probability of event A occurring, given that event B has occurred, is denoted by $\Pr(A \mid B)$. This probability is defined by

$$\Pr(A|B) = \frac{\Pr(AB)}{\Pr(B)} \tag{2.2-6}$$

It is apparent that, if events A and B are independent, the conditional probability reduces to the simple probability Pr(A). Since A and B are interchangeable in Eq. (2.2-6), it follows that

$$\Pr(A|B)\Pr(B) = \Pr(B|A)\Pr(A) \tag{2.2-7}$$

from which we reformulate Eq. (2.2-6) as

$$\Pr(A|B) = \frac{\Pr(B|A)\Pr(A)}{\Pr(B)} \tag{2.2-8}$$

Let us consider possible outcomes A_i, i=1,2, ... ,n, given that B has occurred,

$$\Pr(A_i|B) = \frac{\Pr(B|A_i)\Pr(A_i)}{\Pr(B)} \tag{2.2-9}$$

But

$$\Pr(B) = \Pr(B|A_1)\Pr(A_1) + \Pr(B|A_2)\Pr(A_2) + \cdots + \Pr(B|A_n)\Pr(A_n) \tag{2.2-10}$$

so that

$$\Pr(A_i|B) = \frac{\Pr(B|A_i)\Pr(A_i)}{\sum_{i=1}^{n} \Pr(B|A_i)\Pr(A_i)} \tag{2.2-11}$$

Equation (2.2-11) is a statement of *Bayes' theorem*. We shall have the occasion to utilize this result in subsequent chapters.

Example 2.2-1

The probability concepts defined above are best illustrated by using a family of simple examples. In particular, the commonly used experiment involving the roll of a die is utilized.

Probability

- Experiment – roll a die
- Event – value of the die

- Possible values – 1, 2, 3, 4, 5, 6
- Pr(value = j; j = 1, 2, 3, 4, 5, 6) = 1/6.

Joint (Independent) Event

- Experiment – roll two dice, A and B
- Events – value of A and value of B
- Joint event – values of A and B
- Possible joint event values – (1,1), (1,2), . . . , (6,6)
- Pr(joint event) $\triangleq$ Pr(AB) = 1/36 = Pr(A) Pr(B).

Mutually Exclusive Events

- Experiment – roll a die
- Event – value is either 1 or 2
- Pr(1 + 2) = Pr(1) + Pr(2) = 1/6 + 1/6 = 1/3.

Non-Mutually Exclusive Events

- Experiment – roll two dice, A and B
- Event – value of either A or B is 1
- $\Pr(A = 1 \text{ or } B = 1) = \Pr(A = 1) + \Pr(B = 1) - \Pr(A = 1 \text{ and } B = 1)$
  $= 1/6 + 1/6 - 1/36 = 11/36.$

Conditional Probability

- Experiment – roll three dice A, B and C
- Event E_1 – obtain exactly two 1's
  E_2 – A = 1, B and C any value
- $$\begin{aligned}\Pr(E_1) &= \Pr(A = 1) \cdot \Pr(B = 1) \cdot \Pr(C \neq 1)\\ &\quad + \Pr(A \neq 1) \cdot \Pr(B = 1) \cdot \Pr(C = 1)\\ &\quad + \Pr(A = 1) \cdot \Pr(B \neq 1) \cdot \Pr(C = 1)\\ &= 3(1/6 \times 1/6 \times 5/6) = 5/72\end{aligned}$$

  $$\Pr(E_2) = 1/6 \times 1 \times 1 = 1/6$$

  $$\begin{aligned}\Pr(E_1E_2) &= \Pr(A = 1) \cdot \Pr(B = 1) \cdot \Pr(C \neq 1)\\ &\quad + \Pr(A = 1) \cdot \Pr(B \neq 1) \cdot \Pr(C = 1)\\ &= 2(1/6 \times 1/6 \times 5/6) = 5/108\end{aligned}$$

  $$\Pr(E_1 | E_2) = \frac{\Pr(E_1E_2)}{\Pr(E_2)} = \frac{\frac{5}{108}}{\frac{1}{6}} = \frac{5}{18}$$

- Thus, given that A = 1, the probability of E_1 occurring is four times greater.

RANDOM VARIABLES

A random variable X is, in simplest terms, a variable which takes on values at random; and may be thought of as a function of the outcomes of some random experiment. The manner of specifying the probability with which different values are taken by the random variable is by the *probability distribution function* F(x), which is defined by

$$F(x) = \Pr(X \leqslant x) \tag{2.2-12}$$

or by the *probability density function* f(x), which is defined by

$$f(x) = \frac{dF(x)}{dx} \tag{2.2-13}$$

The inverse of the defining relationship for the probability density function is

$$F(x) = \int_{-\infty}^{x} f(u)\, du \tag{2.2-14}$$

An evident characteristic of any probability distribution or density function is

$$F(\infty) = \int_{-\infty}^{\infty} f(u)\, du = 1 \tag{2.2-15}$$

From the definition, the interpretation of f(x) as the density of probability of the event that X takes a value in the vicinity of x is clear:

$$\begin{aligned} f(x) &= \lim_{dx \to 0} \frac{F(x + dx) - F(x)}{dx} \\ &= \lim_{dx \to 0} \frac{\Pr(x < X \leqslant x + dx)}{dx} \end{aligned} \tag{2.2-16}$$

This function is finite if the probability that X takes a value in the infinitesimal interval between x and x + dx (the interval closed on the right) is an infinitesimal of order dx. This is usually true of random variables which take values over a continuous range. If, however, X takes a set of discrete values x_i – with nonzero probabilities p_i – f(x) is infinite at these values of x. This is expressed as a series of Dirac delta functions weighted by the appropriate probabilities:

$$f(x) = \sum_{i} p_i\, \delta(x - x_i) \tag{2.2-17}$$

An example of such a random variable is the outcome of the roll of a die. A suitable definition of the *delta function*, δ(x), for the present purpose is a function which is zero everywhere except at x = 0, where it is infinite in such a way that the integral of the function across the singularity is unity. An important property of the delta function, which follows from this definition, is

$$\int_{-\infty}^{\infty} G(x)\,\delta(x - x_0)\,dx = G(x_0) \tag{2.2-18}$$

if G(x) is a finite-valued function which is continuous at $x = x_0$.

A random variable may take values over a continuous range and, in addition, take a discrete set of values with nonzero probability. The resulting probability density function includes both a finite function of x and an additive set of probability-weighted delta functions; such a distribution is called *mixed.*

The simultaneous consideration of more than one random variable is often necessary or useful. In the case of two, the probability of the occurrence of pairs of values in a given range is prescribed by the *joint probability distribution function*

$$F_2(x,y) = \Pr(X \leqslant x \text{ and } Y \leqslant y) \tag{2.2-19}$$

where X and Y are the two random variables under consideration. The corresponding *joint probability density function* is

$$f_2(x,y) = \frac{\partial^2 F_2(x,y)}{\partial x \partial y} \tag{2.2-20}$$

It is clear that the individual probability distribution and density functions for X and Y can be derived from the joint distribution and density functions. For the distribution of X,

$$F(x) = F_2(x,\infty) \tag{2.2-21}$$

$$f(x) = \int_{-\infty}^{\infty} f_2(x,y)dy \tag{2.2-22}$$

Corresponding relations give the distribution of Y. These concepts extend directly to the description of the joint behavior of more than two random variables.

If X and Y are independent, the event $(X \leqslant x)$ is independent of the event $(Y \leqslant y)$; thus the probability for the joint occurrence of these events is the product of the probabilities for the individual events. Equation (2.2-19) then gives

$$\begin{aligned} F_2(x,y) &= \Pr(X \leqslant x \text{ and } Y \leqslant y) \\ &= \Pr(X \leqslant x)\Pr(Y \leqslant y) \\ &= F_X(x)F_Y(y) \end{aligned} \tag{2.2-23}$$

From Eq. (2.2-20) the joint probability density function is, then,

$$f_2(x,y) = f_X(x)f_Y(y) \tag{2.2-24}$$

Subscripts X and Y are used to emphasize the fact that the distributions are different functions of different random variables.

Expectations and Statistics of Random Variables – The *expectation* of a random variable is defined as the sum of all values the random variable may take, each weighted by the probability with which the value is taken. For a random variable that takes values over a continuous range, the summation is done by integration. The probability, in the limit as $dx \rightarrow 0$, that X takes a value in the infinitesimal interval of width dx near x is given by Eq. (2.2-16) as f(x) dx. Thus, the expectation of X, which we denote by E[X] is

$$E[X] = \int_{-\infty}^{\infty} x\, f(x)dx \tag{2.2-25}$$

This is also called the *mean value* of X, the mean of the distribution of X, or the *first moment* of X. This is a precisely defined number toward which the average of a number of observations of X tends, in the probabilistic sense, as the number of observations increases. Equation (2.2-25) is the analytic definition of the expectation, or mean, of a random variable. This expression is valid for random variables having a continuous, discrete, or mixed distribution if the set of discrete values that the random variable takes is represented by impulses in f(x) according to Eq. (2.2-17).

It is frequently necessary to find the expectation of a function of a random variable. If Y is defined as some function of the random variable X – e.g., Y=g(X) – then Y is itself a random variable with a distribution derivable from the distribution of X. The expectation of Y is defined by Eq. (2.2-25) where the probability density function for Y would be used in the integral. Fortunately, this procedure can be abbreviated. The expectation of any function of X can be calculated directly from the distribution of X by the integral

$$E[g(X)] = \int_{-\infty}^{\infty} g(x)f(x)dx \tag{2.2-26}$$

An important statistical parameter descriptive of the distribution of X is its *mean squared value*. Using Eq. (2.2-26), the expectation of the square of X is written

$$E[X^2] = \int_{-\infty}^{\infty} x^2 f(x)dx \tag{2.2-27}$$

The quantity $E[X^2]$ is also called the *second moment* of X. The root-mean-squared (rms) value of X is the square root of $E[X^2]$. The *variance* of a random variable is the mean squared deviation of the random variable from its mean; it is denoted by σ^2 where

$$\sigma^2 = \int_{-\infty}^{\infty} (x - E[X])^2 f(x)dx = E[X^2] - E[X]^2 \tag{2.2-28}$$

The square root of the variance, or σ, is the *standard deviation* of the random variable. The rms value and standard deviation are equal only for a zero-mean random variable.

Other functions whose expectations are of interest are sums and products of random variables. It is easily shown that the expectation of the sum of random variables is equal to the sum of the expectations,

$$E[X_1 + X_2 + \ldots + X_n] = E[X_1] + E[X_2] + \ldots + E[X_n] \tag{2.2-29}$$

whether or not the variables are independent, and that the expectation of the product of random variables is equal to the product of the expectations,

$$E[X_1 X_2 \ldots X_n] = E[X_1] \cdot E[X_2] \cdot \ldots \cdot E[X_n] \tag{2.2-30}$$

if the variables are independent. It is also true that the variance of the sum of random variables is equal to the sum of the variances if the variables are independent – i.e., if

$$X = \sum_{i=1}^{n} X_i$$

then, for independent X_i,

$$\sigma_X{}^2 = \sum_{i=1}^{n} \sigma_{X_i}{}^2 \tag{2.2-31}$$

A very important concept is that of statistical correlation between random variables. A partial indication of the degree to which one variable is related to another is given by the *covariance*, which is the expectation of the product of the deviations of two random variables from their means,

$$E\left[(X-E[X])(Y-E[Y])\right] = \int_{-\infty}^{\infty} dx \int_{-\infty}^{\infty} dy (x-E[X])(y-E[Y]) f_2(x,y)$$

$$= E[XY] - E[X]\,E[Y] \tag{2.2-32}$$

In the expression, the term E[XY] is the second moment of X and Y. The covariance, normalized by the standard deviations of X and Y, is called the *correlation coefficient*, and is expressed as

$$\rho = \frac{E[XY] - E[X]E[Y]}{\sigma_X \sigma_Y} \tag{2.2-33}$$

The correlation coefficient is a measure of the degree of linear dependence between X and Y. If X and Y are independent, ρ is zero (the inverse is not true); if Y is a linear function of X, ρ is ± 1. If an attempt is made to approximate Y by some linear function of X, the minimum possible mean-squared error in the approximation is $\sigma_Y^2 (1 - \rho^2)$. This provides another interpretation of ρ as a measure of the degree of linear dependence between random variables.

One additional quantity associated with the distribution of a random variable is the *characteristic function*. It is defined by

$$\begin{aligned} g(t) &= E[\exp(jtX)] \\ &= \int_{-\infty}^{\infty} \exp(jtx)\, f(x)dx \end{aligned} \tag{2.2-34}$$

A property of the characteristic function that largely explains its value is that the characteristic function of a random variable which is the sum of independent random variables is the product of the characteristic functions of the individual variables. If the characteristic function of a random variable is known, the probability density function can be determined from

$$f(x) = \frac{1}{2\pi} \int_{-\infty}^{\infty} \exp(-jtx)\, g(t)\, dt \tag{2.2-35}$$

Notice that Eqs. (2.2-34) and (2.2-35) are in the form of a Fourier transform pair. Another useful relation is

$$E[X^n] = j^{-n} \left. \frac{d^n g(t)}{dt^n} \right|_{t=0} \tag{2.2-36}$$

Thus, the *moments* of x can be generated directly from the derivatives of the characteristic function.

The Uniform and Normal Probability Distributions – Two important specific forms of probability distribution are the uniform and normal distributions. The *uniform distribution* is characterized by a uniform (constant) probability density, over some finite interval. The magnitude of the density function in this interval is the reciprocal of the interval width as required to make the integral of the function unity. This function is shown in Fig. 2.2-1. The *normal probability density function*, shown in Fig. 2.2-2, has the analytic form

$$f(x) = \frac{1}{\sqrt{2\pi}\,\sigma} \exp\left[-\frac{(x-m)^2}{2\sigma^2}\right] \tag{2.2-37}$$

where the two parameters that define the distribution are m, the mean, and σ, the standard deviation. The integral of the function, or area under the curve, is unity. The area within the $\pm 1\sigma$ bounds centered about the mean is approximately 0.68. Within the $\pm 2\sigma$ bounds the area is 0.95. As an interpretation of the meaning of these values, the probability that a normally distributed random value is *outside* $\pm 2\sigma$ is approximately 0.05.

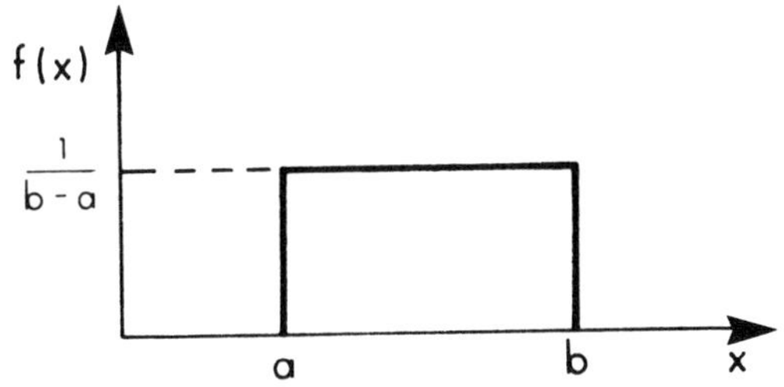

Figure 2.2-1 Uniform Probability Density Function

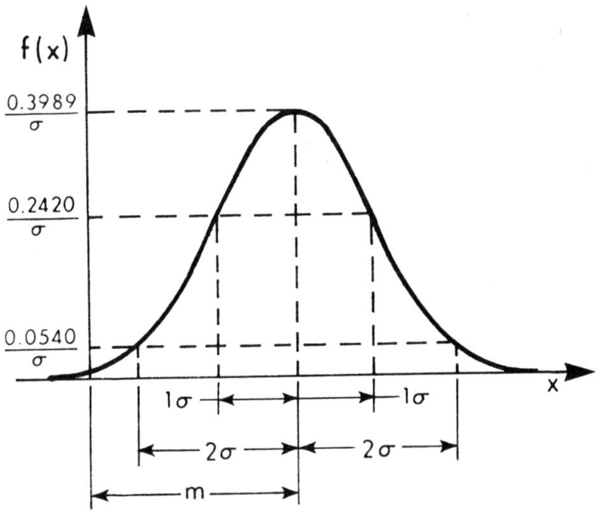

Figure 2.2-2 Normal Probability Density Function

By calculating the characteristic function for a normally distributed random variable, one can immediately show that the distribution of the sum of independent normally distributed variables is also normal. Actually, this remarkable property of preserving the distribution form is true of the sum of normally distributed random variables, whether they are independent or not. Even more remarkable is the fact that under certain circumstances the distribution of the sum of independent random variables, each having an *arbitrary* distribution, tends toward the normal distribution as the number of variables in the sum tends toward infinity. This statement, together with the conditions under which the result can be proved, is known as the *central limit theorem*. The conditions are rarely tested in practical situations, but the empirically observed fact is that a great many random variables display a distribution which closely approximates the normal. The reason for the common occurrence of normally distributed random variables is certainly stated in the central limit theorem and the fact that superposition is common in nature.

We are often interested in two random variables which possess *a bivariate normal distribution*. The form of the joint probability density function for zero-mean variables, written in terms of statistical parameters previously defined, is

$$f_2(x_1,x_2) = \frac{1}{2\pi\,\sigma_1\sigma_2\sqrt{1-\rho^2}}\;\exp -\left[\frac{\dfrac{x_1^{\,2}}{\sigma_1^{\,2}} - 2\rho\dfrac{x_1}{\sigma_1}\dfrac{x_2}{\sigma_2} + \dfrac{x_2^{\,2}}{\sigma_2^{\,2}}}{2(1-\rho^2)}\right] \tag{2.2-38}$$

For n random variables, the *multidimensional* or *multivariate normal distribution* is

$$f_n(x_1,x_2,\ldots,x_n) = \frac{1}{(2\pi)^{n/2}\,|P|^{1/2}}\exp\left[-\frac{1}{2}(\underline{x}-\underline{m})^T P^{-1}(\underline{x}-\underline{m})\right] \tag{2.2-39}$$

with

$$\underline{x}^T \triangleq (x_1\; x_2\; \ldots\; x_n) \tag{2.2-40}$$

The quantities

$$\underline{m} = E[\underline{x}]$$

and

$$P = E[(\underline{x}-\underline{m})(\underline{x}-\underline{m})^T]$$

are the *mean* and *covariance* of the vector $\underline{x}$, respectively. Vector and matrix functions of random variables adhere to the operational definitions and rules established in Section 2.1. Thus, the expected value of a vector (matrix) is the vector (matrix) containing expected values of the respective elements.

RANDOM PROCESSES

A random process may be thought of as a collection, or *ensemble*, of functions of time, any one of which might be observed on any trial of an experiment. The ensemble may include a finite number, a countable infinity, or a noncountable infinity of such functions. We shall denote the ensemble of functions by $\{x(t)\}$, and any observed member of the ensemble by x(t). The value of the observed member of the ensemble at a particular time, say t_1, as shown in Fig. 2.2-3, is a random variable. On repeated trials of the experiment, $x(t_1)$ takes different values at random. The probability that $x(t_1)$ takes values in a certain range is given by the probability distribution function, as it is for any random variable. In this case the dependence on the time of observation is shown explicitly in the notation, viz.:

$$F(x_1,t_1) = \Pr[x(t_1) \leqslant x_1] \tag{2.2-41}$$

The corresponding probability density function is

$$f(x_1,t_1) = \frac{dF(x_1,t_1)}{dx_1} \tag{2.2-42}$$

These functions are adequate to define, in a probabilistic sense, the range of amplitudes which the random process displays. To gain a sense of how quickly

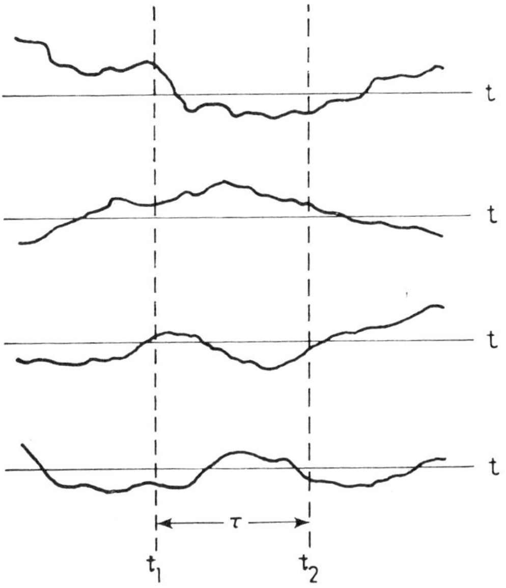

Figure 2.2-3 Members of the Ensemble $\{x(t)\}$

members of the ensemble are likely to vary, one has to observe the same member function at more than one time. The probability for the occurrence of a pair of values in certain ranges is given by the second-order joint probability distribution function

$$F_2(x_1,t_1;x_2,t_2) = \Pr[x(t_1) \leqslant x_1 \text{ and } x(t_2) \leqslant x_2] \tag{2.2-43}$$

and the corresponding joint probability density function

$$f_2(x_1,t_1;x_2,t_2) = \frac{\partial^2 F_2(x_1,t_1;x_2,t_2)}{\partial x_1 \partial x_2} \tag{2.2-44}$$

Higher-ordered joint distribution and density functions can be defined following this pattern, but only rarely does one attempt to deal with more than the second-order statistics of random processes.

If two random processes are under consideration, the simplest distribution and density functions that provide some indication of their joint statistical characteristics are the second-order functions

$$F_2(x,t_1;y,t_2) = \Pr[x(t_1) \leqslant x \text{ and } y(t_2) \leqslant y] \tag{2.2-45}$$

$$f_2(x,t_1;y,t_2) = \frac{\partial^2 F_2(x,t_1;y,t_2)}{\partial x \partial y} \tag{2.2-46}$$

Correlation Functions – Actually, the characterization of random processes, in practice, is usually limited to even less information than that given by the second-order distribution or density functions. Only the first moments of these distributions are commonly measured. These moments are called autocorrelation and cross-correlation functions. The *autocorrelation function* is defined as

$$\varphi_{xx}(t_1,t_2) = E[x(t_1)\,x(t_2)] = \int_{-\infty}^{\infty} dx_1 \int_{-\infty}^{\infty} dx_2\, x_1 x_2 f_2(x_1,t_1;x_2,t_2) \tag{2.2-47}$$

and the *cross-correlation function* as

$$\varphi_{xy}(t_1,t_2) = E[x(t_1)\,y(t_2)] = \int_{-\infty}^{\infty} dx \int_{-\infty}^{\infty} dy\; xy\, f_2(x,t_1;y,t_2) \tag{2.2-48}$$

In the case where $E[x(t_1)]$, $E[x(t_2)]$, and $E[y(t_2)]$ are all zero, these correlation functions are the covariances of the indicated random variables. If they are then normalized by the corresponding standard deviations, according to Eq. (2.2-33). they become correlation coefficients which measure on a scale from -1 to $+1$ the degree of linear dependence between the variables.

Stationarity – A *stationary random process* is one whose statistical properties are invariant in time. This implies that the first probability density function for the process, $f(x_1,t_1)$, is independent of the time of observation t_1. Then all the moments of this distribution, such as $E[x(t_1)]$ and $E[x^2(t_1)]$, are also independent of time – they are constants. The second probability density function is not in this case dependent on the absolute times of observation, t_1 and t_2, but still depends on the difference between them. So if t_2 is written as

$$t_2 = t_1 + \tau \tag{2.2-49}$$

$f_2(x_1,t_1;x_2,t_2)$ becomes $f_2(x_1,t_1;x_2,t_1 + \tau)$, which is independent of t_1, but still a function of τ. The correlation functions are then functions only of the single variable τ, viz.:

$$\varphi_{xx}(\tau) = E[x(t_1)\, x(t_1 + \tau)] \tag{2.2-50}$$

$$\varphi_{xy}(\tau) = E[x(t_1)\, y(t_1 + \tau)] \tag{2.2-51}$$

We note the following properties of these correlation functions:

$$\varphi_{xx}(0) = E[x^2] \tag{2.2-52}$$

$$\varphi_{xx}(-\tau) = \varphi_{xx}(\tau) \tag{2.2-53}$$

$$\varphi_{xy}(-\tau) = \varphi_{yx}(\tau) \tag{2.2-54}$$

$$\varphi_{xx}(0) \geqslant |\varphi_{xx}(\tau)| \tag{2.2-55}$$

Ergodicity – One further concept associated with stationary random processes is the *ergodic hypothesis*. This hypothesis claims that any statistic calculated by averaging over all members of an ergodic ensemble at a fixed time can also be calculated by averaging over all time on a single representative member of the ensemble. The key to this notion is the word "representative." If a particular member of the ensemble is to be statistically representative of all, it must display at various points in time the full range of amplitude, rate of change of amplitude, etc., which are to be found among all the members of the ensemble. A classic example of a stationary ensemble which is *not* ergodic is the ensemble of functions which are constant in time. The failing in this case is that no member of the ensemble is representative of all.

In practice, almost all empirical results for stationary processes are derived from tests on a single function, under the assumption that the ergodic hypothesis holds. In this case, the common statistics associated with a random process are written

$$E[x] = \lim_{T \to \infty} \frac{1}{2T} \int_{-T}^{T} x(t)\, dt \tag{2.2-56}$$

$$E[x^2] = \lim_{T \to \infty} \frac{1}{2T} \int_{-T}^{T} x^2(t)\, dt \tag{2.2-57}$$

$$\varphi_{xx}(\tau) = \lim_{T \to \infty} \frac{1}{2T} \int_{-T}^{T} x(t)\, x(t+\tau)\, dt \tag{2.2-58}$$

$$\varphi_{xy}(\tau) = \lim_{T \to \infty} \frac{1}{2T} \int_{-T}^{T} x(t)\, y(t+\tau)\, dt \tag{2.2-59}$$

Example 2.2-2

An example of a stationary ergodic random process is the ensemble of sinusoids, of given amplitude and frequency, with a uniform distribution of phase. The member functions of this ensemble are all of the form

$$x(t) = A \sin(\omega t + \theta)$$

where θ is a random variable, uniformly distributed over the interval $(0,2\pi)$ radians. Any average taken over the members of this ensemble at any fixed time would find all phase angles represented with equal probability density. But the same is true of an average over all time on any one member. For this process, then, all members of the ensemble qualify as "representative." Note that any distribution of the phase angle θ other than the uniform distribution over an integral number of cycles would define a nonstationary process. The relevant calculations are given below. For the *ensemble average* autocorrelation function $(f(\theta) = 1/2\pi$ for $0 \leqslant \theta \leqslant 2\pi)$ we get (Eq. (2.2-47))

$$\varphi_{xx}(\tau) = \int_0^{2\pi} A \sin(\omega t + \theta)\, A \sin(\omega t + \omega\tau + \theta)\, \frac{1}{2\pi}\, d\theta$$

$$= \frac{A^2}{4\pi} \int_0^{2\pi} [\cos \omega\tau - \cos(2\omega t + \omega\tau + 2\theta)]\, d\theta$$

$$= \frac{A^2}{2} \cos \omega\tau$$

while the time average autocorrelation function is (Eq. (2.2-58))

$$\varphi_{xx}(\tau) = \lim_{T \to \infty} \frac{1}{2T} \int_{-T}^{T} A \sin(\omega t + \theta)\, A \sin(\omega t + \omega\tau + \theta)\, dt$$

$$\varphi_{xx}(\tau) = \frac{A^2}{2} \lim_{T\to\infty} \frac{1}{2T} \int_{-T}^{T} [\cos\omega\tau - \cos(2\omega t + \omega\tau + 2\theta)]\, dt$$

$$= \frac{A^2}{2} \cos\omega\tau$$

The two results are equivalent and, thus, x(t) is an example of an ergodic process.

Gaussian Processes – A gaussian process is one characterized by the property that its joint probability distribution functions of all orders are multidimensional normal distributions. For a gaussian process, then, the distribution of x(t) for any time t is the normal distribution, for which the density function is expressed by

$$f(x,t) = \frac{1}{\sqrt{2\pi}\,\sigma} \exp\left[-\frac{(x-m)^2}{2\sigma^2}\right] \tag{2.2-60}$$

The joint distribution of $x(t_1)$ and $x(t_2)$ is the bivariate normal distribution; higher-order joint distributions are given by the multivariate normal distribution. If $\underline{x}(t)$ is an n-dimensional gaussian vector then the distribution of $\underline{x}(t)$ is the normal distribution expressed by

$$f(\underline{x},t) = \frac{1}{(2\pi)^{n/2}\,|P|^{1/2}} \exp\left[-\frac{1}{2}(\underline{x}-\underline{m})^T P^{-1}(\underline{x}-\underline{m})\right] \tag{2.2-61}$$

All the statistical properties of a gaussian random process are defined by the first and second moments of the distribution. Equivalently, the statistics of the process are all contained in the autocorrelation function of the process. Clearly, this property is a great boon to analytic operations. Additionally, the output of a linear system whose input is gaussian is also gaussian.

As a consequence of the central limit theorem, gaussian processes are those most frequently encountered in actual systems; as a consequence of their analytic convenience, they are also most frequently encountered in system analysis. It is, therefore, appropriate to introduce a shorthand notation to contain the information in Eq. (2.2-61). The notation, which is used extensively in the sequel, is

$$\underline{x} \sim N(\underline{m}, P) \tag{2.2-62}$$

which indicates that $\underline{x}$ is a gaussian (normal) random vector with mean $\underline{m}$ and covariance P. By way of example, for a one-dimensional random process x with mean m and standard deviation σ, we would write

$$x \sim N(m, \sigma^2) \tag{2.2-63}$$

Power Spectral Density Functions – The input-output relation for a linear system may be written (Ref. 11, also, Section 3.3 herein)

$$y(t) = \int_{-\infty}^{t} x(\tau)\, w(t,\tau)\, d\tau \qquad (2.2\text{-}64)$$

where x(t) is the input function, y(t) is the output, and w(t,τ) is the system *weighting function*, the response at time t to a unit impulse input at time τ. If the system is time-invariant, this *superposition integral* reduces to

$$y(t) = \int_{0}^{\infty} w(\tau)\, x(t-\tau)\, d\tau \qquad (2.2\text{-}65)$$

This expression is also referred to as the *convolution integral*. Using Eq. (2.2-65), the statistics of the output process can be written in terms of those of the input. If the input process is stationary, the output process is also stationary in the steady state. Manipulation of Eq. (2.2-65) in this instance leads to several expressions for the relationships between the moments of x(t) and y(t),

$$E[y] = E[x] \int_{0}^{\infty} w(t)\, dt \qquad (2.2\text{-}66)$$

$$E[y^2] = \int_{0}^{\infty} d\tau_1\, w(\tau_1) \int_{0}^{\infty} d\tau_2\, w(\tau_2)\, \varphi_{xx}(\tau_1 - \tau_2) \qquad (2.2\text{-}67)$$

$$\varphi_{yy}(\tau) = \int_{0}^{\infty} d\tau_1\, w(\tau_1) \int_{0}^{\infty} d\tau_2\, w(\tau_2)\, \varphi_{xx}(\tau + \tau_1 - \tau_2) \qquad (2.2\text{-}68)$$

$$\varphi_{xy}(\tau) = \int_{0}^{\infty} w(\tau_1)\, \varphi_{xx}(\tau - \tau_1)\, d\tau_1 \qquad (2.2\text{-}69)$$

Analytic operations on linear time-invariant systems are facilitated by the use of integral transforms which transform the convolution input-output relation of Eq. (2.2-65) into the algebraic operation of multiplication. Since members of stationary random ensembles must necessarily be visualized as existing for all negative and positive time, the two-sided Fourier transform is the appropriate transformation to employ. The Fourier transforms of the correlation functions defined above then appear quite naturally in analyses. The Fourier transform of the autocorrelation function

$$\Phi_{xx}(\omega) = \int_{-\infty}^{\infty} \varphi_{xx}(\tau) \exp(-j\omega\tau)\, d\tau \tag{2.2-70}$$

is called the *power spectral density function*, or power density spectrum of the random process $\{x(t)\}$. The term "power" is here used in a generalized sense, indicating the expected squared value of the members of the ensemble. $\Phi_{xx}(\omega)$ is indeed the spectral distribution of power density for $\{x(t)\}$ in that integration of $\Phi_{xx}(\omega)$ over frequencies in the band from ω_1 to ω_2 yields the mean-squared value of the process whose autocorrelation function consists only of those harmonic components of $\varphi_{xx}(\tau)$ that lie between ω_1 and ω_2. In particular, the mean-squared value of $\{x(t)\}$ itself is given by integration of the power density spectrum for the random process over the full range of ω. This last result is seen as a specialization of the inverse transform relation corresponding to Eq. (2.2-70), namely

$$\varphi_{xx}(\tau) = \frac{1}{2\pi} \int_{-\infty}^{\infty} \Phi_{xx}(\omega) \exp(j\omega\tau)\, d\omega \tag{2.2-71}$$

$$E[x^2] = \varphi_{xx}(0) = \frac{1}{2\pi} \int_{-\infty}^{\infty} \Phi_{xx}(\omega)\, d\omega \tag{2.2-72}$$

The Fourier transform of the cross-correlation function is called the *cross power spectral density function*

$$\Phi_{xy}(\omega) = \int_{-\infty}^{\infty} \varphi_{xy}(\tau) \exp(-j\omega\tau)\, d\tau \tag{2.2-73}$$

The desired input-output algebraic relationships corresponding to Eqs. (2.2-68) and (2.2-69) are

$$\Phi_{xy}(\omega) = | W(j\omega) | \Phi_{xx}(\omega) \tag{2.2-74}$$

and

$$\Phi_{yy}(\omega) = | W(j\omega) |^2 \Phi_{xx}(\omega) \tag{2.2-75}$$

where W is the system transfer function, defined as the Laplace transform of the system weighting function (with $s = j\omega$, see Ref. 13)

$$W(s) = \int_{0}^{\infty} w(\tau)\, e^{-s\tau}\, d\tau \tag{2.2-76}$$

White Noise – A particularly simple form for the power density spectrum is a constant, $\Phi_{xx}(\omega) = \Phi_0$. This implies that power is distributed uniformly over all frequency components in the full infinite range. By analogy with the corresponding situation in the case of white light, such a random process, usually a noise, is called a *white noise*. The autocorrelation function for white noise is a delta function

$$\varphi_{nn}(\tau) = \frac{1}{2\pi} \int_{-\infty}^{\infty} \Phi_0 \exp(j\omega\tau)\, d\omega = \Phi_0\, \delta(\tau) \tag{2.2-77}$$

We recall the definition of the Dirac delta $\delta(\tau)$ and note that the mean-squared value of white noise, $\varphi_{nn}(0) = \Phi_0\delta(0)$, is infinite. Thus the process is not physically realizable. White noise is an idealized concept which does, however, serve as a very useful *approximation* to situations in which a disturbing noise is wideband compared with the bandwidth of a system. A familiar physical process closely approximated by white noise is the *shot effect* which is a mathematical description of vacuum-tube circuit output fluctuation.

White noise is also quite useful in analytic operations; integration properties of the Dirac delta $\delta(\tau)$ can many times be used to advantage. Additionally, a number of random processes can be generated by passing white noise through a suitable filter. Illustrations of the autocorrelation functions and power spectral density functions for several common random processes are shown in Fig. 2.2-4. Applications of such random processes will be considered in subsequent chapters.

Gauss-Markov Processes – A special class of random processes which can be generated by passing white noise through simple filters is the family of gauss-markov processes. A continuous process x(t) is *first-order markov* if for every k and

$$t_1 < t_2 < \ldots < t_k$$

it is true that

$$F\,[x(t_k) \mid x(t_{k-1}), \ldots, x(t_1)] = F\,[x(t_k) \mid x(t_{k-1})] \tag{2.2-78}$$

That is, the probability distribution for the process $x(t_k)$ is dependent only on the value at one point immediately in the past, $x(t_{k-1})$. If the continuous process x(t) is first-order markov, it can be associated with the differential equation

$$\frac{dx}{dt} + \beta_1(t)\, x = w \tag{2.2-79}$$

where w is white noise (for a discrete first-order markov process, the associated relation is a first-order difference equation). If we add the restriction that the

PROCESS	AUTOCORRELATION FUNCTION	POWER SPECTRAL DENSITY
WHITE NOISE	$\phi_{xx}(\tau) = \phi_0 \delta(\tau)$	$\Phi_{xx} = \Phi_0$
MARKOV PROCESS	$\phi_{xx}(\tau) = \sigma^2 e^{-\beta|\tau|}$	$\Phi_{xx} = \dfrac{2\beta\sigma^2}{\omega^2 + \beta^2}$
SINUSOID	$\phi_{xx}(\tau) = \dfrac{A^2}{2} \cos \omega_1 \tau$	$\Phi_{xx} = \dfrac{\pi}{2} A^2 \left[\delta(\omega - \omega_1) + \delta(\omega + \omega_1) \right]$
RANDOM BIAS	$\phi_{xx}(\tau) = m^2$	$\Phi_{xx} = 2\pi m^2 \delta(\omega)$

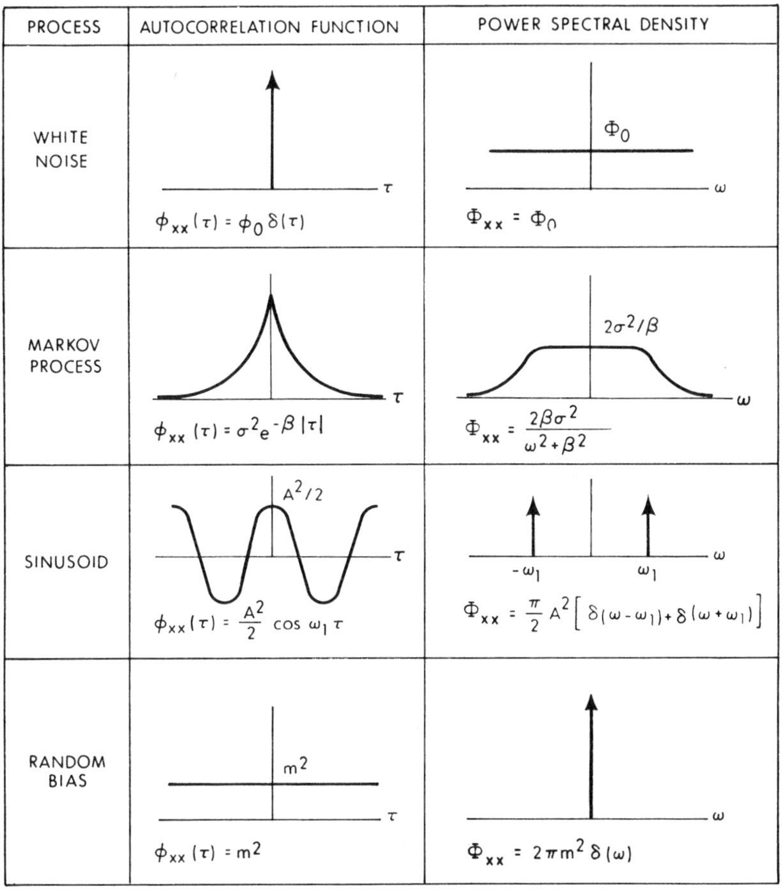

Figure 2.2-4   Descriptions of Common Random Processes

probability density functions of w and consequently x also are gaussian, the process x(t) is a *gauss-markov process*. The statistics of a stationary gauss-markov process are completely described by the autocorrelation function

$$\varphi_{xx}(\tau) = \sigma^2 e^{-\beta|\tau|} + m^2 \tag{2.2-80}$$

The so-called *correlation time* (1/e point) is $1/\beta_1$. The spectral density of the white noise, w, which generates the process described by Eq. (2.2-80) is given in terms of the variance of x as $2\beta_1 \sigma^2 \delta(\tau)$. The autocorrelations of many physical phenomena are well-described by Eq. (2.2-80).

A continuous process x(t) is *second-order markov*, if for every k and

$$t_1 < t_2 < \ldots < t_k$$

it is true that

$$F[x(t_k) \mid x(t_{k-1}), \ldots, x(t_1)] = F[x(t_k) \mid x(t_{k-1}), x(t_{k-2})] \qquad (2.2\text{-}81)$$

Equivalently, the probability distribution of $x(t_k)$ is dependent only on the conditions at two points immediately in the past. An associated differential equation is

$$\frac{d^2x}{dt^2} + 2\beta_2(t)\frac{dx}{dt} + \beta_2{}^2(t)\, x = w \qquad (2.2\text{-}82)$$

TABLE 2.2-1 CHARACTERISTICS OF STATIONARY MARKOV PROCESSES*

Order of Markov Process	Power Spectral Density, $\Phi_{xx}(\omega)$	Autocorrelation Function, $\varphi_{xx}(\tau)$	Correlation Time
1	$\dfrac{2\beta_1\sigma^2}{\omega^2 + \beta_1^2}$	$\sigma^2 e^{-\beta_1\|\tau\|}$	$\dfrac{1}{\beta_1}$
2	$\dfrac{4\beta_2^3\sigma^2}{\left(\omega^2 + \beta_2^2\right)^2}$	$\sigma^2 e^{-\beta_2\|\tau\|}\left\{1 + \beta_2\|\tau\|\right\}$	$\dfrac{2.146}{\beta_2}$
3	$\dfrac{16\beta_3^5\sigma^2}{3\left(\omega^2 + \beta_3^2\right)^3}$	$\sigma^2 e^{-\beta_3\|\tau\|}\left\{1 + \beta_3\|\tau\| + 3\beta_3^2\|\tau\|^2\right\}$	$\dfrac{2.903}{\beta_3}$
n	$\dfrac{(2\beta_n)^{2n-1}[\Gamma(n)]^2}{(2n-2)!\left(\omega^2 + \beta_n^2\right)^n}$	$\sigma^2 e^{-\beta_n\|\tau\|}\sum_{k=0}^{n-1}\dfrac{\Gamma(n)(2\beta_n\|\tau\|)^{n-k-1}}{(2n-2)!\,k!\,\Gamma(n-k)}$	Solved Arithmetically for each n
$n\to\infty$	$2\pi\sigma^2\delta(\omega)$	σ^2	∞

*$\Gamma(n)$ is the Gamma function.

If the density function of w is gaussian, the process x(t) is second-order gauss-markov. If x(t) has mean m and is stationary, its autocorrelation function has the form

$$\varphi_{xx}(\tau) = \sigma^2(1 + \beta_2 |\tau|)e^{-\beta_2 |\tau|} + m^2 \tag{2.2-83}$$

The correlation time of this process is approximately $2.146/\beta_2$; the spectral density of the white noise w is defined as $4\beta_2{}^3\sigma^2\delta(\tau)$. For the second-order gauss-markov process the derivative of $\varphi_{xx}(\tau)$ is zero at τ=0; this characteristic is appealing for many physical phenomena (see Example 3.8-1).

Definition of an *n^{th}-order gauss-markov process* proceeds directly from the above. The characteristics of such processes are given in Table 2.2-1 and Fig. 2.2-5. For the nth order gauss-markov process as $n \rightarrow \infty$, the result is a bias. Heuristically, white noise can be thought of as a "zeroth-order" gauss-markov process.

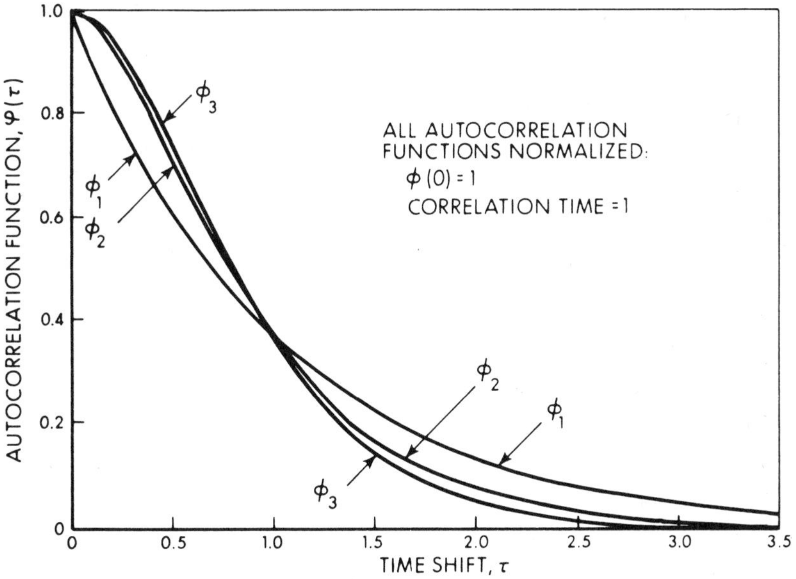

Figure 2.2-5 Autocorrelation Functions of Gauss-Markov Processes

REFERENCES

1. Hildebrand, F.B., *Methods of Applied Mathematics*, Prentice-Hall, Englewood Cliffs, 1961.
2. Hoffman, K. and Kunze, R., *Linear Algebra*, Prentice-Hall, Englewood Cliffs, 1961.
3. Luenberger, D., *Optimization by Vector Space Methods*, John Wiley & Sons, Inc., New York, 1969.
4. Penrose, R., "A Generalized Inverse for Matrices," *Proc. Cambridge Phil. Soc., 51*, 406-413.
5. Coddington, E.A. and Levinson, N., *Theory of Ordinary Differential Equations*, McGraw-Hill Book Co., Inc., New York, 1955.
6. Fadeev, D.K. and Fadeeva, V.N., *Computational Methods of Linear Algebra*, W.H. Freeman, San Francisco, 1963.
7. Athans, M., "Gradient Matrices and Matrix Calculations," Lincoln Laboratory, Lexington, Mass., November 1965 (AD624426).
8. Gelb, A. and VanderVelde, W.E., *Multiple-Input Describing Functions and Nonlinear System Design*, McGraw-Hill Book Co., Inc., New York, 1968.
9. Davenport, W.B., Jr. and Root, W.L., *An Introduction to Random Signals and Noise*, McGraw-Hill Book Co., Inc., New York, 1958.
10. Laning, J.H., Jr. and Battin, R.H., *Random Processes in Automatic Control*, McGraw-Hill Book Co., Inc., New York, 1956.
11. Papoulis, A., *Probability, Random Variables and Stochastic Processes*, McGraw-Hill Book Co., Inc., New York, 1965.
12. Singer, R.A., "Estimating Optimal Tracking Filter Performance for Manned Maneuvering Targets," *IEEE Trans. Aerosp. Elect. Syst., AES-6*, 473-483, July 1970.
13. Truxal, J.G., *Automatic Feedback Control System Synthesis*, McGraw-Hill Book Co., Inc., New York, 1955.
14. Schweppe, F.C., *Uncertain Dynamic Systems*, Prentice-Hall, Englewood Cliffs, N.J., 1973.

PROBLEMS

Problem 2-1

Show that $\dot{P}^{-1} = -P^{-1}\dot{P}P^{-1}$

Problem 2-2

For the matrix

$$A = \begin{bmatrix} 3 & -1 & -4 \\ 2 & 2 & 2 \\ -4 & 0 & -2 \end{bmatrix}$$

show that the eigenvalues are

$$\lambda_1 = 2, \quad \lambda_2 = -4, \quad \lambda_3 = 5.$$

Problem 2-3

Show that A is positive definite if, and only if, all of the eigenvalues of A are positive.

Problem 2-4

If R(t) is a time-varying orthogonal matrix, and

$$\frac{dR(t)}{dt} R^T(t) = S(t)$$

show that S(t) must be skew-symmetric.

Problem 2-5

Consider the matrix A

$$A = \begin{bmatrix} 1 & 2 \\ 3 & 4 \end{bmatrix}$$

(a) Use the Cayley-Hamilton theorem to show that

$$A^2 - 5A - 2I = 0$$

What are the eigenvalues (λ_1, λ_2) of A? (b) Use the result in (a) and the matrix exponential series expansion to show that

$$e^{At} = I + At + \frac{t^2}{2}(5A + 2I) + \frac{t^3}{6}(27A + 10I) + \frac{t^4}{24}(145A + 54I) + \ldots$$

(c) Collecting terms in the expression in (b) yields

$$e^{At} = I\left[1 + t^2 + \frac{5t^3}{3} + \frac{9t^4}{4} + \ldots\right] + A\left[t + \frac{5t^2}{2} + \frac{9t^3}{4} + \frac{145t^4}{24} + \ldots\right]$$

or

$$e^{At} = a_1(t)I + a_2(t)A$$

Closed form values for the two series $a_1(t)$ and $a_2(t)$ are not immediately apparent. However, the Cayley-Hamilton theorem can be used to obtain closed form expressions. Demonstrate that these solutions are of the form

$$a_1(t) = \frac{\lambda_2 e^{\lambda_1 t} - \lambda_1 e^{\lambda_2 t}}{\lambda_1 - \lambda_2} \quad \text{and} \quad a_2(t) = \frac{e^{\lambda_1 t} - e^{\lambda_2 t}}{\lambda_1 - \lambda_2}$$

Problem 2-6

If $\underline{r}$ is a three-dimensional position vector, the locus of points for which

$$\underline{r}^T E^{-1} \underline{r} \leqslant 1$$

where E is a positive definite symmetric matrix defines an ellipsoid. Show that the volume of this ellipsoid is

$$V = \frac{4\pi}{3}\sqrt{|E|}$$

Problem 2-7

The least-squares derivation in Section 2.1 is based on the assumption that all measurements $\underline{z}$ are of equal quality. If, in fact, it is known that it is reasonable to apply different weights to the various measurements comprising $\underline{z}$, the least-squares estimator should be appropriately modified. If the ith measurement z_i has a relative weight of w_i, it is reasonable to construct a matrix

$$W = \begin{bmatrix} {w_1}^2 & & & 0 \\ & {w_2}^2 & & \\ & & \ddots & \\ 0 & & & {w_m}^2 \end{bmatrix}$$

with which to define a weighted cost function

$$J = (\underline{z} - H\underline{\hat{x}})^T W (\underline{z} - H\underline{\hat{x}})$$

Show that the weighted least-squares estimate is

$$\underline{\hat{x}} = (H^T W H)^{-1} H^T W \underline{z}$$

Problem 2-8

For an arbitrary random process x(t), show that the cross-correlation between x(t) and its derivative $\dot{x}(t)$ is

$$\varphi_{x\dot{x}}(t_1,t_2) = \frac{\partial}{\partial t_2} \varphi_{xx}(t_1,t_2)$$

Problem 2-9

If X, Y, and Z are independent random variables each uniformly distributed in the interval (−1,1), show that the probability density function for the variable w = (X + Y + Z)/3 is

$$p_w(w) = \begin{cases} \frac{9}{8}(1 - 3w^2) & , \quad 0 \leqslant |w| < \frac{1}{3} \\ \frac{27}{16}(1 - w)^2 & , \quad \frac{1}{3} \leqslant |w| < 1 \\ 0 & , \quad |w| \geqslant 1 \end{cases}$$

Sketch $p_w(w)$ and comment on its shape.

Problem 2-10

The *Poisson* or *exponential random process* is frequently used to describe random arrival rates, or random service times in queueing problems. If μ is the average number of arrivals, the probability density function describing the probability of x successes is

$$f(x) = \frac{e^{-\mu}\mu^x}{x!}$$

Find the Poisson probability distribution function F(x). If the average arrival rate is 0.4/minute, show that the probability of exactly 4 arrivals in 10 minutes is 0.195 and the probability of no more than four arrivals in ten minutes is 0.629.

Problem 2-11

If x and y are independent normal random variables with zero mean and equal variance σ^2, show that the variable

$$z = \sqrt{x^2 + y^2}$$

has probability density described by

$$f(z) = \begin{cases} \frac{z}{\sigma^2} e^{-z^2/\sigma^2} & , \quad z \geq 0 \\ 0 & , \quad z < 0 \end{cases}$$

The random variable z is called a *Rayleigh process*. Show that its mean and variance are

$$\sigma\sqrt{\frac{\pi}{2}} \qquad \text{and} \qquad \left(2 - \frac{\pi}{2}\right)\sigma^2$$

respectively.

Problem 2-12

The probability density for the acceleration of a particular maneuvering vehicle is a combination of discrete and continuous functions. The probability of no acceleration is P_0. The probability of acceleration at a maximum rate $A_{max}(-A_{max})$ is P_{max}. For all other acceleration values, acceleration probability is described by a uniform density function. Sketch the combined probability density function. Show that the magnitude of the uniform density function is

$$b = \frac{1 - (P_0 + 2P_{max})}{2A_{max}}$$

and that the variance of the acceleration is (Ref. 12)

$$\sigma^2 = \frac{A_{max}^2}{3}\,[1 + 4P_{max} - P_0]$$

Problem 2-13

Find the mean, mean square and variance of a random variable uniformly distributed in the interval [a,b].

Problem 2-14

If x_1 and x_2 are zero-mean gaussian random variables with a joint probability density function given by Eq. (2.2-38), show that their sum $z = x_1 + x_2$ is gaussian. Note that this result holds even though x_1 and x_2 are dependent.

Problem 2-15

In Section 2.2 it is noted that white noise is physically unrealizable. However, it is often said that a first-order gauss-markov process

$$\dot{x}(t) = -\beta x(t) + w(t)$$

with

$$\varphi_{xx}(\tau) = \sigma_x^2 e^{-\beta |\tau|}$$

is a physically realizable process. Since a physically realizable process cannot have a derivative with infinite variance, show that $\dot{x}(t)$ is just as physically unrealizable as white noise.

Thus, it is suggested that none of the continuous stochastic processes treated in this book are physically realizable. However, as noted in Ref. 14, this fact merely serves to emphasize the point that mathematical models are approximate but highly useful representations of the real world.

3. LINEAR DYNAMIC SYSTEMS

Application of optimal estimation is predicated on the description of a physical system under consideration by means of differential equations. In this chapter, *state-space* notation is introduced to provide a convenient formulation for the required mathematical description, and techniques for solving the resultant vector-matrix differential equation are also presented. Although the initial discussion is concerned with *continuous* physical systems, results are extended to the *discrete* case in which information is available or desired only at specified time intervals. *Controllability* and *observability*, two properties based on system configuration and dynamics, are defined and illustrated. Equations for time propagation of system error state *covariance* are obtained for both discrete and continuous systems. A number of *models* for noise sources that affect the system error state covariance are discussed, with special attention given to their statistical properties and the manner in which they can be described in state-space notation. Finally, some considerations for empirical determination of error models are presented.

3.1 STATE-SPACE NOTATION

Early work in control and estimation theory involved system description and analysis in the *frequency domain*. In contrast to these efforts, most of the recent advances – work by Pontryagin, Bellman, Lyapunov, Kalman and others – have involved system descriptions in the *time domain*. The formulation used employs state-space notation which offers the advantage of mathematical and notational convenience. Moreover, this approach to system description is closer to physical reality than any of the frequency-oriented transform techniques. It is particularly useful in providing statistical descriptions of system behavior.

The dynamics of linear, lumped parameter systems can be represented by the first-order vector-matrix differential equation

$$\underline{\dot{x}}(t) = F(t)\underline{x}(t) + G(t)\underline{w}(t) + L(t)\underline{u}(t) \tag{3.1-1}$$

where $\underline{x}(t)$ is the system *state vector*, $\underline{w}(t)$ is a *random forcing function*, $\underline{u}(t)$ is a *deterministic (control) input*, and F(t), G(t), L(t) are matrices arising in the formulation. This is the continuous form ordinarily employed in modern estimation and control theory. Figure 3.1-1 illustrates the equation. The state vector for a dynamic system is composed of any set of quantities sufficient to completely describe the unforced motion of that system. Given the state vector at a particular point in time and a description of the system forcing and control functions from that point in time forward, the state at any other time can be computed. The state vector is not a unique set of variables; any other set $\underline{x}'(t)$ related to $\underline{x}(t)$ by a nonsingular transformation

$$\underline{x}'(t) = A(t)\underline{x}(t) \tag{3.1-2}$$

fulfills the above requirement.

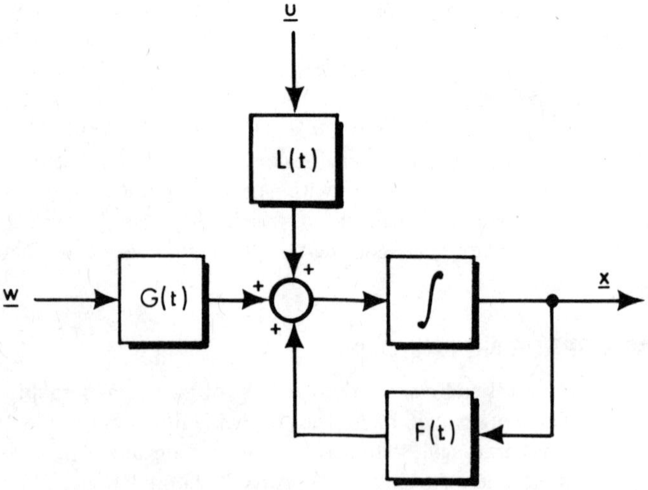

Figure 3.1-1 Block Diagram of Continuous Representation of Linear Dynamics Equation

Given an nth-order linear differential equation

$$[D^n + a_{n-1}(t)D^{n-1} + \ldots a_1(t)D + a_0(t)]\, y(t) = w(t) \qquad (3.1\text{-}3)$$

where $D = d/dt$, we can define a set of state variables $x_1(t), \ldots, x_n(t)$ by

$$\begin{aligned} x_1(t) &\triangleq y(t) \\ x_2(t) &\triangleq \dot{x}_1(t) \\ &\vdots \\ x_n(t) &\triangleq \dot{x}_{n-1}(t) \end{aligned} \qquad (3.1\text{-}4)$$

These relations can be written as a set of n first-order linear differential equations:

$$\begin{aligned} \dot{x}_1(t) &= x_2(t) \\ \dot{x}_2(t) &= x_3(t) \\ &\vdots \\ \dot{x}_n(t) &= -a_0(t)x_1(t) - a_1(t)x_2(t) - \ldots - a_{n-1}(t)x_n(t) + w(t) \end{aligned} \qquad (3.1\text{-}5)$$

The first n–1 of these equations follow from the state variable definitions; the last is obtained using the definitions and Eq. (3.1-3). Expressing the equations in vector-matrix form as in Eq. (3.1-1) yields

$$\begin{bmatrix} \dot{x}_1 \\ \dot{x}_2 \\ \vdots \\ \dot{x}_{n-1} \\ \dot{x}_n \end{bmatrix} = \begin{bmatrix} 0 & 1 & 0 & \ldots & 0 & 0 \\ 0 & 0 & 1 & \ldots & 0 & 0 \\ \vdots & \vdots & \vdots & & \vdots & \vdots \\ 0 & 0 & 0 & \ldots & 0 & 1 \\ -a_0 & -a_1 & -a_2 & \ldots & -a_{n-2} & -a_{n-1} \end{bmatrix} \begin{bmatrix} x_1 \\ x_2 \\ \vdots \\ x_{n-1} \\ x_n \end{bmatrix} + \begin{bmatrix} 0 \\ 0 \\ \vdots \\ 0 \\ w \end{bmatrix} \qquad (3.1\text{-}6)$$

This is called the *companion form* of Eq. (3.1-3). The system dynamics matrix F is square with dimension n, corresponding to the order of the original differential equation. Equation (3.1-6) is illustrated in block diagram form in Fig. 3.1-2; note that in this formulation the state variables are the outputs of integrators.

In many linear systems of interest the forcing and control functions are multivariable – i.e., $\underline{w}(t)$ and $\underline{u}(t)$ in Eq. (3.1-1) are composed of *several* nonzero

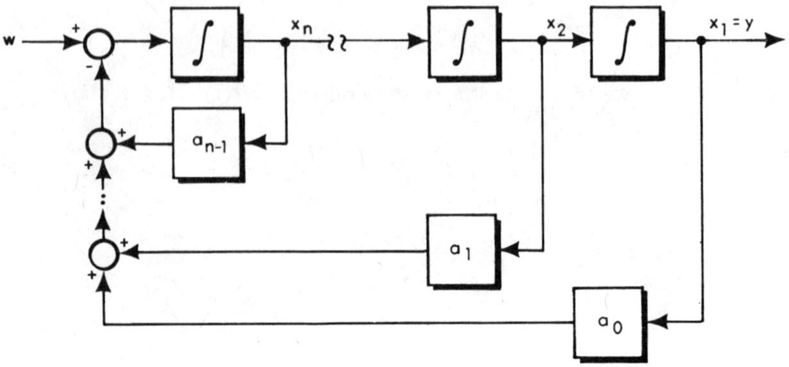

Figure 3.1-2 Block Diagram Representation of Eq. (3.1-6)

functions. Also, the individual elements of $\underline{w}(t)$ and $\underline{u}(t)$ may drive several state variables simultaneously, causing G(t) and L(t) to be matrices with significant elements at locations other than those on the main diagonal. Ordinarily, system dynamics are determined directly from the physics of the problem. A block diagram of the physical system may be sketched and the first-order vector-matrix differential equation determined by inspection. The outputs of each of the various integrators would constitute a convenient set of state variables. The system dynamics equations can be written in the form of Eq. (3.1-1) as

$$\begin{bmatrix} \dot{x}_1 \\ \dot{x}_2 \\ \vdots \\ \dot{x}_n \end{bmatrix} = \begin{bmatrix} f_{11} & f_{12} & \cdots & f_{1n} \\ f_{21} & f_{22} & \cdots & f_{2n} \\ \vdots & \vdots & & \vdots \\ f_{n1} & f_{n2} & \cdots & f_{nn} \end{bmatrix} \begin{bmatrix} x_1 \\ x_2 \\ \vdots \\ x_n \end{bmatrix} + \begin{bmatrix} g_{11} & g_{12} & \cdots & g_{1r} \\ g_{21} & g_{22} & \cdots & g_{2r} \\ \vdots & \vdots & & \vdots \\ g_{n1} & g_{n2} & \cdots & g_{nr} \end{bmatrix} \begin{bmatrix} w_1 \\ w_2 \\ \vdots \\ w_r \end{bmatrix}$$

$$+ \begin{bmatrix} \ell_{11} & \ell_{12} & \cdots & \ell_{1s} \\ \ell_{21} & \ell_{22} & \cdots & \ell_{2s} \\ \vdots & \vdots & & \vdots \\ \ell_{n1} & \ell_{n2} & \cdots & \ell_{ns} \end{bmatrix} \begin{bmatrix} u_1 \\ u_2 \\ \vdots \\ u_s \end{bmatrix} \tag{3.1-7}$$

The functions $\underline{w}$ and $\underline{u}$ need not be of dimension n; in the equation shown the dimension of $\underline{w}$ is r and that of $\underline{u}$ is s. However, it is required that products $G\underline{w}$ and $L\underline{u}$ be of dimension n. Reference 1 further demonstrates the steps required to convert a high-order differential equation into a set of state variables driven by a multivariable forcing function. Several examples of the application of state-space notation to physical systems are given below.

Example 3.1-1

Consider the mass m shown in Fig. 3.1-3; it is connected to the left wall by a spring with spring constant k and a damper with damping coefficient c. Frictionless wheels are assumed. Displacement x is measured (positive-left) between the indicators; the entire container is subject to acceleration w(t) which is positive to the right. This is a one-dimensional translation-motion-only system and, consequently, displacement x and velocity $\dot{x}$ are suitable state variables.

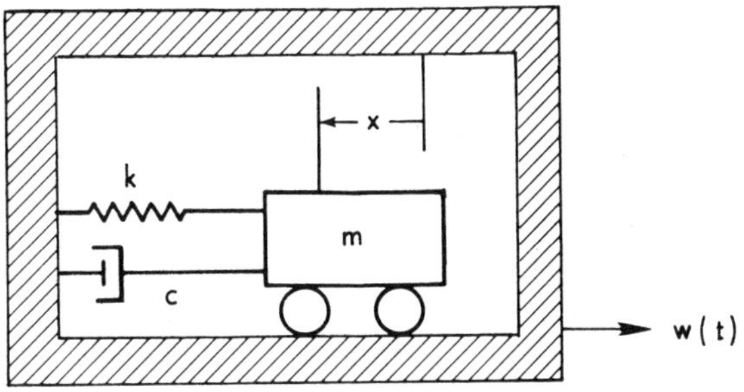

Figure 3.1-3 Second-Order Physical System

The equation of motion of the system is obtained from Newton's second law

$$\sum f_x = ma \qquad (3.1\text{-}8)$$

The forces acting are $\Sigma f_x = -kx - c\dot{x}$ corresponding to the spring and damper. Total acceleration is $a = \ddot{x} - w(t)$ so that

$$m\ddot{x} + c\dot{x} + kx = mw(t) \qquad (3.1\text{-}9)$$

If the state vector is defined as

$$\underline{x} = \begin{bmatrix} x \\ \dot{x} \end{bmatrix} \qquad (3.1\text{-}10)$$

the appropriate equation for the system dynamics is

$$\begin{bmatrix} \dot{x} \\ \ddot{x} \end{bmatrix} = \begin{bmatrix} 0 & 1 \\ -k/m & -c/m \end{bmatrix} \begin{bmatrix} x \\ \dot{x} \end{bmatrix} + \begin{bmatrix} 0 \\ w(t) \end{bmatrix} \qquad (3.1\text{-}11)$$

Example 3.1-2

Application of state-space notation when the system is described by a block diagram is illustrated using Fig. 3.1-4, which is a single-axis inertial navigation system Schuler loop error diagram (Ref. 2). The following definitions apply: φ is the platform tilt angle (rad), δv is the system velocity error (fps), δp is the system position error (ft), R is earth radius (ft), g is local gravity (fps^2), ϵ_g is gyro random drift rate (rad/sec), and ϵ_a is accelerometer uncertainty (fps^2). The system state variables are chosen as the outputs of the three integrators so that the state vector is

$$\underline{x} = \begin{bmatrix} \varphi \\ \delta v \\ \delta p \end{bmatrix} \tag{3.1-12}$$

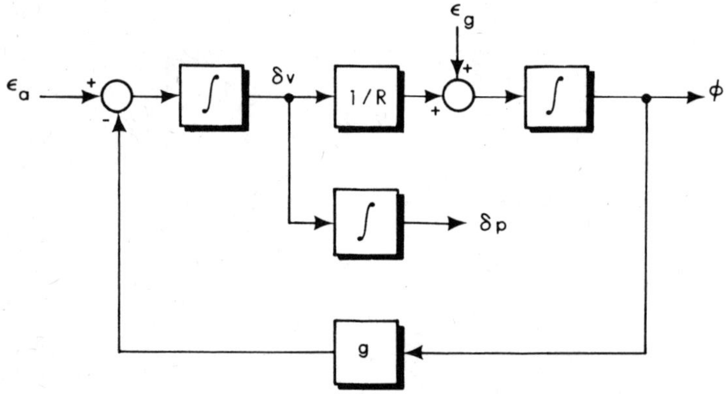

Figure 3.1-4 Single-Axis Schuler Loop Error Diagram

No control $\underline{u}$ is applied and, therefore, the system dynamics equation can be written by inspection as

$$\begin{bmatrix} \dot{\varphi} \\ \delta \dot{v} \\ \delta \dot{p} \end{bmatrix} = \begin{bmatrix} 0 & 1/R & 0 \\ -g & 0 & 0 \\ 0 & 1 & 0 \end{bmatrix} \begin{bmatrix} \varphi \\ \delta v \\ \delta p \end{bmatrix} + \begin{bmatrix} \epsilon_g \\ \epsilon_a \\ 0 \end{bmatrix} \tag{3.1-13}$$

An equivalent form for the random forcing function is

$$G\underline{w} = \begin{bmatrix} 1 & 0 \\ 0 & 1 \\ 0 & 0 \end{bmatrix} \begin{bmatrix} \epsilon_g \\ \epsilon_a \end{bmatrix} \tag{3.1-14}$$

As previously noted the only requirement is that the product $G\underline{w}$ has the same dimension as the vector $\underline{x}$.

3.2 TRANSITION MATRIX

Having provided a mathematical formulation for the description of physical systems, we next seek techniques for solution of the system dynamics equation. The first step involves solving the equation when the forcing function $\underline{w}$ and control function $\underline{u}$ are not present.

The homogeneous unforced matrix differential equation corresponding to Eq. (3.1-1) is

$$\dot{\underline{x}}(t) = F(t)\underline{x}(t) \tag{3.2-1}$$

Suppose that at some time, t_0, all but one of the outputs of the system integrators are set to zero and no inputs are present. Also, assume that the nonzero integrator output is given a magnitude of one. The behavior of the state vector for all times t, where $t \geqslant t_0$, can be expressed in terms of a time-varying "solution vector," $\underline{\varphi}_i(t,t_0)$, where the subscript refers to the integrator whose output is initially nonzero

$$\underline{\varphi}_i(t,t_0) = \begin{bmatrix} x_1(t,t_0)_i \\ x_2(t,t_0)_i \\ \vdots \\ x_n(t,t_0)_i \end{bmatrix} \tag{3.2-2}$$

If the initial condition on the ith integrator is something other than unity – a scale factor c, for example – we find from the linear behavior of the system

$$\underline{\varphi}_i(t,t_0,c) = c\underline{\varphi}_i(t,t_0) \tag{3.2-3}$$

Now, due to the superposition principle, if integrators i and j both have nonzero outputs c_i and c_j at time t_0, the system response is the sum of the individual response vectors, viz:

$$\underline{\varphi}_{i,j}(t,t_0,c_i,c_j) = c_i\underline{\varphi}_i(t,t_0) + c_j\underline{\varphi}_j(t,t_0) \tag{3.2-4}$$

But this can be written as a product of the matrix

$$\left[\underline{\varphi}_i \;\middle|\; \underline{\varphi}_j\right]$$

and the vector

$$\begin{bmatrix} c_i \\ c_j \end{bmatrix}$$

In general, every integrator can have a nonzero value at t_0; these values comprise the state $\underline{x}(t_0)$. The time history of the state is the sum of the individual effects,

$$\underline{x}(t) = \left[\ \underline{\varphi}_1(t,t_0) \;\Big|\; \underline{\varphi}_2(t,t_0) \;\Big|\; \ldots \;\Big|\; \underline{\varphi}_n(t,t_0)\ \right] \underline{x}(t_0) \tag{3.2-5}$$

which for compactness is written as

$$\underline{x}(t) = \Phi(t,t_0)\,\underline{x}(t_0) \tag{3.2-6}$$

The matrix $\Phi(t,t_0)$ is called the *transition matrix* for the system of Eq. (3.2-1). The transition matrix allows calculation of the state vector at some time t, given complete knowledge of the state vector at t_0, in the absence of forcing functions

Returning to Eq. (3.2-2), it can be seen that the solution vectors obey the differential equation

$$\frac{d\underline{\varphi}_i(t,t_0)}{dt} = F(t)\,\underline{\varphi}_i(t,t_0) \tag{3.2-7}$$

where

$$\underline{\varphi}_i(t,t) = \underline{\epsilon}_i \tag{3.2-8}$$

$$\underline{\epsilon}_1 = \begin{bmatrix} 1 \\ 0 \\ 0 \\ \cdot \\ \cdot \\ \cdot \\ 0 \end{bmatrix}, \quad \underline{\epsilon}_2 = \begin{bmatrix} 0 \\ 1 \\ 0 \\ \cdot \\ \cdot \\ \cdot \\ 0 \end{bmatrix}, \text{ etc.}$$

Similarly, the transition matrix, composed of the vectors $\underline{\varphi}_i$, obeys the equations

$$\frac{d}{dt}\Phi(t,t_0) = F(t)\,\Phi(t,t_0), \quad \Phi(t,t) = I \tag{3.2-9}$$

Figure 3.2-1 illustrates the time history of a state vector. The transition matrix $\Phi(t_1, t_0)$ describes the influence of $\underline{x}(t_0)$ on $\underline{x}(t_1)$:

$$\underline{x}(t_1) = \Phi(t_1,t_0)\underline{x}(t_0) \tag{3.2-10}$$

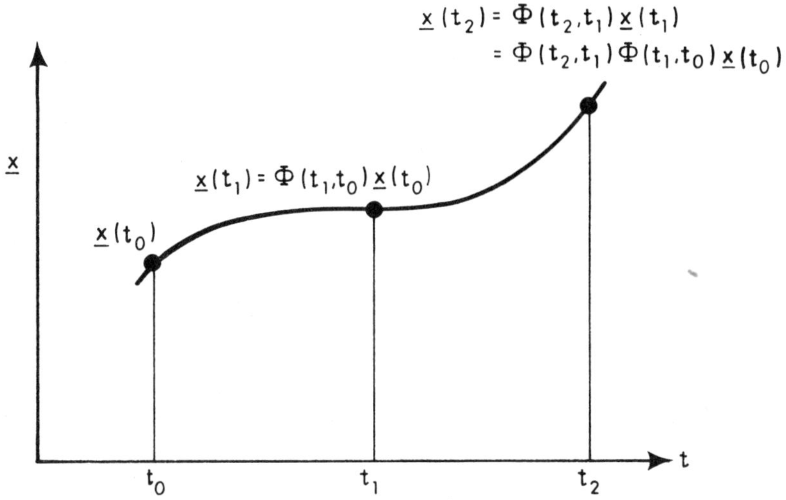

Figure 3.2-1 Conceptual Illustration of the Time Evolution of a State Vector in (n + 1)-Dimensional Space

Also,

$$\underline{x}(t_2) = \Phi(t_2,t_1)\underline{x}(t_1)$$

$$= \Phi(t_2,t_1)\Phi(t_1,t_0)\underline{x}(t_0) \qquad (3.2\text{-}11)$$

Therefore,

$$\Phi(t_2,t_0) = \Phi(t_2,t_1)\Phi(t_1,t_0) \qquad (3.2\text{-}12)$$

which is a general property of the state transition matrix, independent of the order of t_0, t_1, and t_2. Since for any t

$$\Phi(t,t) = \Phi(t,t_0)\Phi(t_0,t) = I \qquad (3.2\text{-}13)$$

premultiplying by $\Phi^{-1}(t,t_0)$ provides the useful relationship

$$\Phi^{-1}(t,t_0) = \Phi(t_0,t) \qquad (3.2\text{-}14)$$

Since the inverse of $\Phi(t,t_0)$ must exist, it necessarily follows that

$$|\Phi(t,t_o)| \neq 0 \qquad (3.2\text{-}15)$$

Other relationships involving the determinant of the transition matrix are

$$\frac{d}{dt}|\Phi(t,t_0)| = \text{trace}\,[F(t)]\;|\Phi(t,t_0)| \tag{3.2-16}$$

and

$$|\Phi(t,t_0)| = \exp\left[\int_{t_0}^{t} \text{trace}\,[F(\tau)]\;d\tau\right] \tag{3.2-17}$$

Transition Matrix for Stationary Systems – For a stationary system, the F matrix is time-invariant and the transition matrix depends only on the time interval considered, viz:

$$\Phi(t,t_0) = \Phi(t - t_0) \tag{3.2-18}$$

This is easily shown by expanding $\underline{x}(t)$ in a Taylor's series about some time, t_0,

$$\underline{x}(t) = \underline{x}(t_0) + \dot{\underline{x}}(t_0)\,(t-t_0) + \ddot{\underline{x}}(t_0)\,\frac{(t-t_0)^2}{2!} + \ldots \tag{3.2-19}$$

But from Eq. (3.2-1)

$$\dot{\underline{x}}(t_0) = F\underline{x}(t_0)$$

$$\ddot{\underline{x}}(t_0) = F\dot{\underline{x}}(t_0) = F^2\underline{x}(t_0)$$

$$\vdots$$

etc.

Substituting, the expansion becomes

$$\begin{aligned}\underline{x}(t) &= \underline{x}(t_0) + F(t-t_0)\underline{x}(t_0) + \frac{F^2(t-t_0)^2}{2!}\underline{x}(t_0) + \ldots \\ &= \left[I + F(t-t_0) + \frac{1}{2!}F^2(t-t_0)^2 + \ldots\right]\underline{x}(t_o)\end{aligned} \tag{3.2-20}$$

In Chapter 2, the matrix exponential is defined as

$$e^A = I + A + \frac{A^2}{2!} + \frac{A^3}{3!} + \ldots \tag{3.2-21}$$

Consequently, the transition matrix for the stationary system can be identified from Eq. (3.2-20) as

$$\Phi(t - t_0) = e^{F(t - t_0)} \tag{3.2-22}$$

which depends only on the stationary system dynamics F and the interval $t - t_0$. Examples of the transition matrices for several simple physical systems are illustrated below.

Example 3.2-1

Consider the circuit shown in Fig. 3.2-2, which is composed of a voltage source, v, a resistor, R, and an inductor, L. Kirchhoff's voltage law yields

$$v = iR + L \frac{di}{dt} \tag{3.2-23}$$

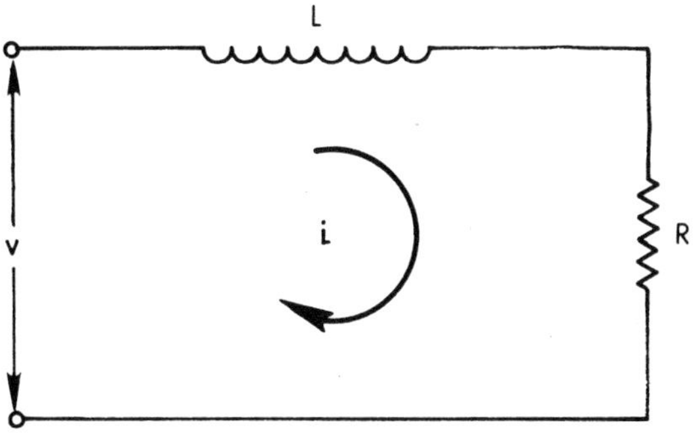

Figure 3.2-2 Elementary Electrical Circuit

We assume $i=i_0$ at $t=t_0$ and $v=0$ for all time, which yields

$$\frac{di}{dt} = -\frac{R}{L} i \tag{3.2-24}$$

The system dynamics matrix F is merely the scalar quantity $-R/L$. Elementary differential equation solution techniques yield

$$i(t) = i_0 e^{-\frac{R}{L}(t-t_0)} \tag{3.2-25}$$

as the solution to Eq. (3.2-24). From the solution we identify the transition matrix as

$$\Phi(t,t_0) = e^{-\frac{R}{L}(t-t_0)}$$

Properties of the transition matrix are readily verified. For times t_0, t_1, t_2, we write

$$\Phi(t_2,t_1) = e^{-\frac{R}{L}(t_2-t_1)} \tag{3.2-26}$$

$$\Phi(t_1,t_0) = e^{-\frac{R}{L}(t_1-t_0)} \tag{3.2-27}$$

so that

$$\begin{aligned}\Phi(t_2,t_1)\Phi(t_1,t_0) &= e^{-\frac{R}{L}(t_2-t_1)}\, e^{-\frac{R}{L}(t_1-t_0)} \\ &= e^{-\frac{R}{L}(t_2-t_0)} \\ &= \Phi(t_2,t_0)\end{aligned} \tag{3.2-28}$$

Example 3.2-2

The system shown in Fig. 3.2-3 is used as a second example of transition matrix computation. Integrator outputs x_1 and x_2 are convenient state variables; the system dynamics equation is obtained by inspection as

$$\begin{bmatrix}\dot{x}_1 \\ \dot{x}_2\end{bmatrix} = \begin{bmatrix}0 & -\omega_0 \\ \omega_0 & 0\end{bmatrix}\begin{bmatrix}x_1 \\ x_2\end{bmatrix} \tag{3.2-29}$$

Matrix multiplication yields

$$Ft = \begin{bmatrix}0 & -\omega_0 t \\ \omega_0 t & 0\end{bmatrix}; \quad F^2t^2 = \begin{bmatrix}-\omega_0^2 t^2 & 0 \\ 0 & -\omega_0^2 t^2\end{bmatrix}; \quad F^3t^3 = \begin{bmatrix}0 & \omega_0^3 t^3 \\ -\omega_0^3 t^3 & 0\end{bmatrix}; \ldots \tag{3.2-30}$$

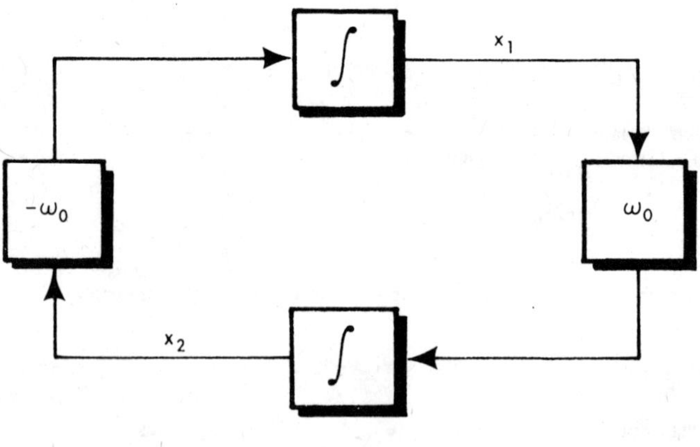

Figure 3.2-3 Second-Order Oscillatory System

so that

$$e^{Ft} = I + Ft + \frac{1}{2!} F^2 t^2 + \frac{1}{3!} F^3 t^3 \ldots$$

$$= \left[\begin{array}{c|c} 1 - \frac{1}{2}\omega_0{}^2 t^2 + \ldots & -\omega_0 t + \frac{\omega_0{}^3 t^3}{3!} - \ldots \\ \hline \omega_0 t - \frac{1}{3!}\omega_0{}^3 t^3 + \ldots & 1 - \frac{1}{2}\omega_0{}^2 t^2 + \ldots \end{array}\right] \tag{3.2-31}$$

We identify the two series in Eq. (3.2-31) as trigonometric functions:

$$\Phi(t,0) = \begin{bmatrix} \cos\omega_0 t & -\sin\omega_0 t \\ \sin\omega_0 t & \cos\omega_0 t \end{bmatrix} \tag{3.2-32}$$

Figure 3.2-3 is, in fact, a representation of a second-order oscillator; this identification is borne out by the oscillatory nature of the transition matrix.

3.3 MATRIX SUPERPOSITION INTEGRAL

Having computed the homogeneous solution, we now seek the differential equation particular solution. Consider first the linear system including forcing function inputs:

$$\dot{\underline{x}}(t) = F(t)\underline{x}(t) + L(t)\underline{u}(t) \tag{3.3-1}$$

Referring to Fig. 3.3-1, we see that the effect of the input to the ith integrator of Fig. 3.1-2 over a small interval $(\tau - \Delta\tau, \tau)$ can be represented as an impulse whose area is the ith row of the term $L(\tau)\underline{u}(\tau)$ times the interval $\Delta\tau$. This impulse produces a small change in the output of the integrator,

$$\Delta x_i(\tau) = \left(L(\tau)\underline{u}(\tau)\right)_i \Delta\tau \tag{3.3-2}$$

The change in the entire state vector can be expressed as

$$\underline{\Delta x}(\tau) = \begin{bmatrix} \Delta x_1(\tau) \\ \Delta x_2(\tau) \\ \cdot \\ \cdot \\ \cdot \\ \Delta x_n(\tau) \end{bmatrix} = L(\tau)\underline{u}(\tau)\Delta\tau \tag{3.3-3}$$

The effect of this small change in the state at any subsequent point in time is given by

$$\underline{\Delta x}(t)\ [\text{given an impulsive input } L(\tau)\underline{u}(\tau)\,\Delta\tau] = \Phi(t,\tau)L(\tau)\underline{u}(\tau)\Delta\tau \tag{3.3-4}$$

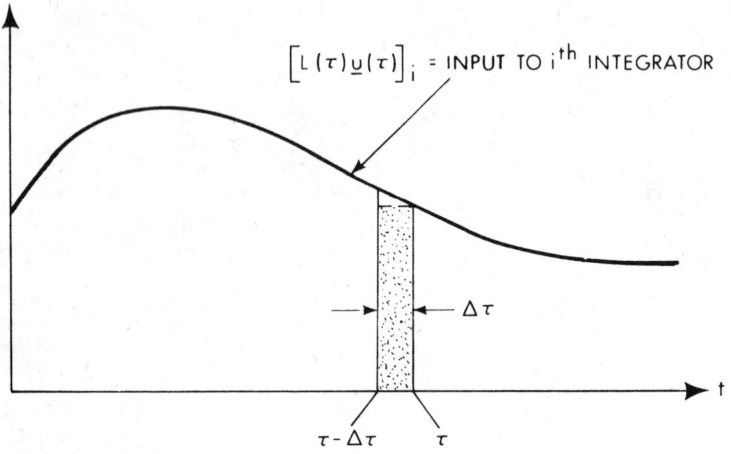

Figure 3.3-1 Representation of the Input to the ith Integrator as an Impulse

Because the system is linear, the response to an input can be viewed as the sum of individual responses to the impulses comprising the input. In the limit as $\Delta\tau \to 0$ the effect of the input on the state at some time, t, is given by

$$\underline{x}(t) = \int_{-\infty}^{t} \Phi(t,\tau)L(\tau)\underline{u}(\tau)d\tau \tag{3.3-5}$$

If the state at some time, t_0, is known, it is only necessary to observe the input after t_0, by utilizing Eq. (3.2-6):

$$\underline{x}(t) = \Phi(t,t_0)\underline{x}(t_0) + \int_{t_0}^{t} \Phi(t,\tau)L(\tau)\underline{u}(\tau)d\tau \tag{3.3-6}$$

Equation (3.3-6) is often called the *matrix superposition integral*. Of course, differentiation of Eq. (3.3-6) can be shown to result in Eq. (3.3-1). An alternate derivation of the matrix superposition integral is shown in Example 3.3-1 and its application to a physical system is illustrated in Example 3.3-2.

Example 3.3-1

The solution to the homogeneous equation is

$$\underline{x}(t) = \Phi(t,t_0)\underline{x}(t_0) \tag{3.3-7}$$

We seek a solution of a similar nature, which also includes the effect of the forcing function. The assumed form is

$$\underline{x}(t) = \Phi(t,t_0)\underline{\xi}(t) \tag{3.3-8}$$

Substitution in Eq. (3.3-1) yields

$$\frac{d}{dt}[\Phi(t,t_0)\underline{\xi}(t)] = F(t)\Phi(t,t_0)\underline{\xi}(t) + L(t)\underline{u}(t)$$

or

$$F(t)\,\Phi(t,t_0)\,\underline{\xi}(t) + \Phi(t,t_0)\,\dot{\underline{\xi}}(t) = F(t)\,\Phi(t,t_0)\,\underline{\xi}(t) + L(t)\,\underline{u}(t) \tag{3.3-9}$$

which reduces to

$$\dot{\underline{\xi}}(t) = \Phi(t_0,t)\,L(t)\,\underline{u}(t) \tag{3.3-10}$$

Solving for $\underline{\xi}(t)$, we obtain

$$\underline{\xi}(t) = \underline{\xi}(t_0) + \int_{t_0}^{t} \Phi(t_0,\tau)\,L(\tau)\,\underline{u}(\tau)\,d\tau \tag{3.3-11}$$

Substituting this result in Eq. (3.3-8) yields

$$\underline{x}(t) = \Phi(t,t_0)\,\underline{\xi}(t_0) + \int_{t_0}^{t} \Phi(t,t_0)\,\Phi(t_0,\tau)\,L(\tau)\,\underline{u}(\tau)\,d\tau$$

But $\underline{\xi}(t_0) = \underline{x}(t_0)$ so that the desired solution to Eq. (3.3-1) is

$$\underline{x}(t) = \Phi(t,t_0)\,\underline{x}(t_0) + \int_{t_0}^{t} \Phi(t,\tau)\,L(\tau)\,\underline{u}(\tau)\,d\tau \tag{3.3-12}$$

Example 3.3-2

The electrical circuit of Fig. 3.2-2 is now subject to a nonzero input voltage, v. We assume initial condition $i=i_0$ at time $t=0$. Prior to $t=0$, the voltage v is zero; for $t \geqslant 0$ it is constant at $v=V$. The system equation is ($t \geqslant 0$)

$$\frac{di}{dt} = -\frac{R}{L}i + \frac{V}{L} \tag{3.3-13}$$

Recall that the transition matrix is

$$\Phi(t,t_0) = e^{-\frac{R}{L}(t-t_0)} \tag{3.3-14}$$

or

$$\Phi(t,0) = e^{-\frac{R}{L}t}$$

Substituting in Eq. (3.3-6), we obtain

$$i(t) = i_0 e^{-\frac{R}{L}t} + \int_0^t e^{-\frac{R}{L}(t-\tau)} \frac{V}{L} d\tau$$

$$= i_0 e^{-\frac{R}{L}t} + \frac{V}{R}\left(1 - e^{-\frac{R}{L}t}\right) \tag{3.3-15}$$

The solution of the dynamics equation, when a random input is present, proceeds in an analogous fashion. Thus, corresponding to

$$\dot{\underline{x}}(t) = F(t)\,\underline{x}(t) + G(t)\,\underline{w}(t) + L(t)\,\underline{u}(t) \tag{3.3-16}$$

we directly find

$$\underline{x}(t) = \Phi(t,t_0)\underline{x}(t_0) + \int_{t_0}^{t} \Phi(t,\tau)G(\tau)\underline{w}(\tau)\,d\tau + \int_{t_0}^{t} \Phi(t,\tau)L(\tau)\underline{u}(\tau)\,d\tau \tag{3.3-17}$$

3.4 DISCRETE FORMULATION

If interest is focussed on the system state at discrete points in time, t_k, k=1,2, . . . , the resulting *difference equation* is, from Eq. (3.3-17),

$$\underline{x}_{k+1} = \Phi_k \underline{x}_k + \Gamma_k \underline{w}_k + \Lambda_k \underline{u}_k \tag{3.4-1}$$

where

$$\Phi_k = \Phi(t_{k+1}, t_k) \tag{3.4-2}$$

$$\Gamma_k \underline{w}_k = \int_{t_k}^{t_{k+1}} \Phi(t_{k+1}, \tau)\, G(\tau)\, \underline{w}(\tau)\, d\tau \tag{3.4-3}$$

$$\Lambda_k \underline{u}_k = \int_{t_k}^{t_{k+1}} \Phi(t_{k+1}, \tau)\, L(\tau)\, \underline{u}(\tau)\, d\tau \tag{3.4-4}$$

In general, Eqs. (3.4-3) and (3.4-4) provide unique definitions of only the *products* $\Gamma_k \underline{w}_k$ and $\Lambda_k \underline{u}_k$ and not the individual terms Γ_k, $\underline{w}_k$, Λ_k and $\underline{u}_k$. In the following discussion, care is taken to make this distinction clear. Notice that if $\underline{w}(t)$ is a vector of random processes, $\underline{x}_k$ and $\Gamma_k \underline{w}_k$ will be vectors of random processes. Equation (3.4-1) is illustrated in Fig. 3.4-1.*

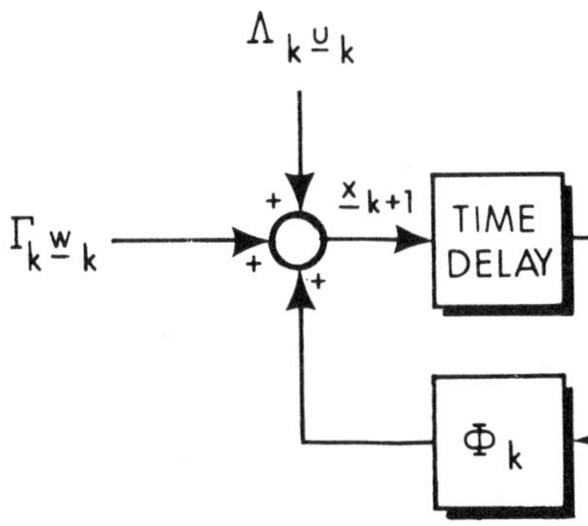

Figure 3.4-1 Illustration of Discrete Representation of Linear Dynamics Equation

It is important to note that, in subsequent discussions, we do not deal with all discrete systems, but rather, only those that can be derived from continuous systems. Without this restriction there could be extensive computational difficulty – e.g., in the discrete system whose transition matrix Φ_k is not invertible. By considering only the subset of discrete systems noted above, the invertibility of Φ_k is assured.

3.5 SYSTEM OBSERVABILITY AND CONTROLLABILITY

OBSERVABILITY

A discussion of observability requires that the concept of a *measurement* be employed. Measurements are briefly discussed in Section 2.1 with regard to least-squares estimation and further discussion is given in Chapter 4. At this point it suffices to say that measurements can indeed be made. They are denoted by $\underline{z}_k$ and are assumed linearly related to the discrete system state $\underline{x}_k$ by the observation matrix H_k, as

*In the sequel, the case of $\Gamma_k = I$ and $\underline{u}_k = \underline{0}$ is treated quite often.

$$\underline{z}_k = H_k \underline{x}_k + \underline{v}_k \tag{3.5-1}$$

where $\underline{v}_k$ is the measurement noise. Given a sequence of measurements $\underline{z}_0$, $\underline{z}_1, \ldots, \underline{z}_k$, the *observability condition* defines our ability to determine $\underline{x}_0, \underline{x}_1, \ldots, \underline{x}_k$ from the measurements.

Consider the discrete deterministic, constant n^{th}-order system* (Ref. 3)

$$\underline{x}_{k+1} = \Phi \underline{x}_k \tag{3.5-2}$$

with n scalar noise-free measurements

$$z_k = H\underline{x}_k, \quad k=0,1,2,\ldots,n-1 \tag{3.5-3}$$

so that H is a constant, n-dimensional row vector. We may write

$$\begin{aligned} z_0 &= H\underline{x}_0 \\ z_1 &= H\underline{x}_1 = H\Phi\underline{x}_0 \\ &\vdots \\ z_{n-1} &= H\underline{x}_{n-1} = H\Phi^{n-1}\underline{x}_0 \end{aligned} \tag{3.5-4}$$

Therefore,

$$\begin{bmatrix} z_0 \\ z_1 \\ \vdots \\ z_{n-1} \end{bmatrix} = \begin{bmatrix} H \\ H\Phi \\ \vdots \\ H\Phi^{n-1} \end{bmatrix} \underline{x}_0 = \Xi^T \underline{x}_0 \tag{3.5-5}$$

If $\underline{x}_0$ is to be determined uniquely, the matrix Ξ^T (or equivalently Ξ) must have an inverse – i.e., be nonsingular. Thus, the observability condition is that the matrix

$$\Xi = \left[H^T \;\middle|\; \Phi^T H^T \;\middle|\; \ldots \;\middle|\; (\Phi^T)^{n-1} H^T \right] \tag{3.5-6}$$

be of rank n. The same condition will apply when measurements are a *vector*, $\underline{z}$; extension to *continuous* systems is straightforward.

More complete statements of observability and the observability condition for continuous systems follow:

---

*A time-invariant discrete system is denoted by $\Phi_k = \Phi$, $\Lambda_k = \Lambda$, etc.

- A system is observable at time $t_1 > t_0$ if it is possible to determine the state $\underline{x}(t_0)$ by observing $\underline{z}(t)$ in the interval (t_0, t_1). If *all* states $\underline{x}(t)$ corresponding to all $\underline{z}(t)$ are observable, the system is *completely observable.*

- A continuous deterministic n^{th}-order, constant coefficient, linear dynamic system is observable if, and only if, the matrix

$$\Xi = \left[H^T \;\vdots\; F^T H^T \;\vdots\; \left(F^T\right)^2 H^T \;\vdots\; \cdots \;\vdots\; \left(F^T\right)^{n-1} H^T \right] \tag{3.5-7}$$

has rank n.

Example 3.5-1

Consider the third-order system in Fig. 3.5-1 described by:

$$\begin{bmatrix} \dot{x}_1 \\ \dot{x}_2 \\ \dot{x}_3 \end{bmatrix} = \begin{bmatrix} 0 & 0 & 0 \\ 0 & 0 & 0 \\ 1 & 1 & 0 \end{bmatrix} \begin{bmatrix} x_1 \\ x_2 \\ x_3 \end{bmatrix} + \begin{bmatrix} w_1 \\ w_2 \\ 0 \end{bmatrix} \tag{3.5-8}$$

If measurements can be made only at the output of the final integrator, then

$$z = x_3 \tag{3.5-9}$$

so that

$$H = [0 \quad 0 \quad 1] \tag{3.5-10}$$

We compute

$$H^T = \begin{bmatrix} 0 \\ 0 \\ 1 \end{bmatrix}; \quad F^T H^T = \begin{bmatrix} 1 \\ 1 \\ 0 \end{bmatrix}; \quad F^T F^T H^T = \begin{bmatrix} 0 \\ 0 \\ 0 \end{bmatrix}$$

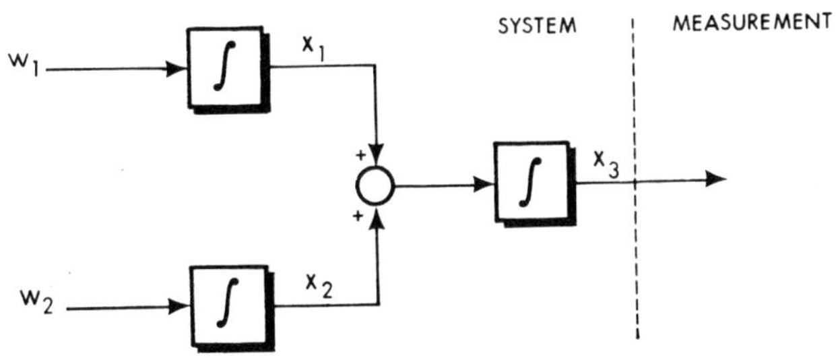

Figure 3.5-1 Third-Order System with Output Observation Only

and form the matrix

$$\Xi = \begin{bmatrix} 0 & 1 & 0 \\ 0 & 1 & 0 \\ 1 & 0 & 0 \end{bmatrix} \tag{3.5-11}$$

A square n × n matrix has rank n if it has a nonzero determinant. The determinant of Ξ is zero so that the matrix has rank less than 3; thus, the system is not observable. The physical interpretation of this result is that it is impossible to distinguish between the spectrally identical states x_1 and x_2 when the only available measurement is their sum.

CONTROLLABILITY

We now concern ourselves with determining conditions such that it is possible to control the state of a deterministic linear dynamic system – i.e., to select an input so that the state takes any desired value after n stages. A simple derivation of controllability for the case of a discrete system driven by a scalar control is analogous to the derivation of observability; it is left as an exercise for the reader. The condition is that the matrix

$$\Theta = \left[\Lambda \,\middle|\, \Phi\Lambda \,\middle|\, \ldots \,\middle|\, \Phi^{n-1}\Lambda\right] \tag{3.5-12}$$

be of rank n. We proceed directly to the statements of controllability and the controllability condition for a continuous system:

- A system is controllable at time $t_1 > t_0$ if there exists a control $\underline{u}(t)$ such that any arbitrary state $\underline{x}(t_0) = \underline{\delta}$ can be driven to another arbitrary state $\underline{x}(t_1) = \underline{\gamma}$.
- A continuous deterministic n^{th}-order, constant coefficient, linear dynamic system is controllable if, and only if, the matrix

$$\Theta = \left[L \,\middle|\, FL \,\middle|\, F^2L \,\middle|\, \ldots \,\middle|\, F^{n-1}L\right] \tag{3.5-13}$$

has rank n.

Example 3.5-2

Consider the system in Fig. 3.5-2. We wish to find conditions on α and β such that the system is controllable. The system is second-order and described by:

$$\begin{bmatrix} \dot{x}_1 \\ \dot{x}_2 \end{bmatrix} = \begin{bmatrix} -\alpha & 0 \\ 0 & -\beta \end{bmatrix} \begin{bmatrix} x_1 \\ x_2 \end{bmatrix} + \begin{bmatrix} 1 \\ 1 \end{bmatrix} u \tag{3.5-14}$$

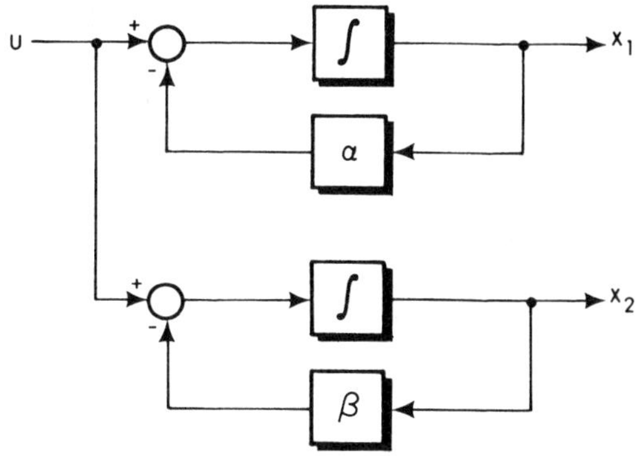

Figure 3.5-2 Parallel First-Order Systems with a Common Control Input

so that

$$F = \begin{bmatrix} -\alpha & 0 \\ 0 & -\beta \end{bmatrix} ; \; L = \begin{bmatrix} 1 \\ 1 \end{bmatrix}$$

yielding

$$\Theta = \begin{bmatrix} 1 & -\alpha \\ 1 & -\beta \end{bmatrix} \tag{3.5-15}$$

To determine the rank of Θ, we compute its determinant

$$| \Theta | = -\beta + \alpha \tag{3.5-16}$$

The controllability condition requires a nonzero determinant for Θ

$$0 \neq -\beta + a$$

or

$$\alpha \neq \beta \tag{3.5-17}$$

The interpretation of this result is quite straightforward. If α and β are equal, the two first-order systems are identical and there is no way that an input u could, by itself, produce different values of x_1 and x_2. The controllability condition requires that x_1 and x_2 can be driven to *any* arbitrary values: with $\alpha = \beta$ the condition cannot be met and, therefore, the system is uncontrollable. It is important to note that the test applied to the matrix Θ only *establishes* controllability; it does not provide a means to determine the input u required to control the system. This latter task can be quite difficult.

Nonuniqueness of Model – Having introduced the concept of a measurement, we have completed the basic state-space structure of a linear system. For a continuous linear system, the general state-space model is

$$\underline{\dot{x}}(t) = F(t)\underline{x}(t) + G(t)\,\underline{w}(t) \tag{3.5-18}$$

$$\underline{z}(t) = H(t)\,\underline{x}(t) + \underline{v}(t) \tag{3.5-19}$$

For a discrete linear system, the model is

$$\underline{x}_{k+1} = \Phi_k \underline{x}_k + \Gamma_k \underline{w}_k \tag{3.5-20}$$

$$\underline{z}_k = H_k \underline{x}_k + \underline{v}_k \tag{3.5-21}$$

These models are not unique; given the pertinent system input and output quantities – i.e., $\underline{x}(0)$, $\underline{w}(t)$ and $\underline{z}(t)$ in the continuous case – there are many different sets of F(t), G(t) and H(t) which will yield the same overall input-output behavior. As noted in Ref. 4, choosing a particular set of F, G and H corresponds to the choice of a coordinate system. This choice can have considerable impact in numerical analyses as well as affecting system observability and controllability as described above.

3.6 COVARIANCE MATRIX

In the following discussions both the system state and forcing function are vectors whose elements are *random variables*. The state vector obeys a relationship of the form of Eq. (3.1-1); while the forcing function $\underline{w}$ is, in general, assumed to be uncorrelated in time (*white noise*). In the discrete formulation, the forcing function $\underline{w}_k$ will be assumed uncorrelated from observation time to observation time (*white sequence*). For the remainder of this chapter, we restrict consideration to those systems for which the control input $\underline{u}$ is zero.

It is generally assumed that random variables have zero ensemble average values – i.e., they are unbiased. The fact that a variable is *known* to be biased implies knowledge of the bias value. Consequently, a new quantity with the bias removed (i.e., subtracted) can be defined. This does not suggest that certain *constant* state variables cannot be considered; as long as their distribution over the ensemble of possible values is unbiased, constant state variables are admissible. It should be pointed out that if the state at some time t_0 is unbiased, the state will remain unbiased. This can be illustrated by taking the ensemble average of both sides of Eq. (3.4-1) where $\underline{u}_k = \underline{0}$:

$$\begin{aligned} E[\underline{x}_{k+1}] &= E[\Phi_k \underline{x}_k + \Gamma_k \underline{w}_k] \\ &= \Phi_k E[\underline{x}_k] + \Gamma_k E[\underline{w}_k] \\ &= \underline{0} \end{aligned} \tag{3.6-1}$$

The random state and forcing function vectors are frequently described in terms of their *covariance matrices*. The cross-covariance matrix of two vectors $\underline{r}$ and $\underline{s}$ is defined in terms of the outer products:

$$\begin{aligned}\text{cross-covariance of } \underline{r} \text{ and } \underline{s} &= E[(\underline{r} - E[\underline{r}])(\underline{s} - E[\underline{s}])^T] \\ &= E[\underline{r}\underline{s}^T] - E[\underline{r}]\, E[\underline{s}^T]\end{aligned} \tag{3.6-2}$$

When $\underline{r} = \underline{s}$, Eq. (3.6-2) defines the *covariance* of $\underline{r}$; it is simply a matrix whose elements are the second moments of the random components $r_1, r_2, \ldots, r_n$. In the sequel, we define the error $\underline{\tilde{x}}$ in the estimate of a state vector to be the difference between the estimated ($\underline{\hat{x}}$) and actual ($\underline{x}$) values:

$$\underline{\tilde{x}} \triangleq \underline{\hat{x}} - \underline{x} \tag{3.6-3}$$

The covariance of $\underline{\tilde{x}}$, designated P, is then given by

$$P = E[\underline{\tilde{x}}\underline{\tilde{x}}^T] \tag{3.6-4}$$

It provides a statistical measure of the uncertainty in $\underline{x}$. The notation permits us to discuss the properties of the covariance matrix independently of the mean value of the state.

Some features of the covariance matrix can be seen by treating the error in knowledge of two random system state variables,

$$\underline{\tilde{x}} = \begin{bmatrix} \tilde{x}_1 \\ \tilde{x}_2 \end{bmatrix} \tag{3.6-5}$$

The covariance matrix of $\underline{\tilde{x}}$ is

$$\begin{aligned} P &= E\left\{\begin{bmatrix} \tilde{x}_1^2 & \tilde{x}_1\tilde{x}_2 \\ \tilde{x}_1\tilde{x}_2 & \tilde{x}_2^2 \end{bmatrix}\right\} \\ &= \begin{bmatrix} E[\tilde{x}_1^2] & E[\tilde{x}_1\tilde{x}_2] \\ E[\tilde{x}_1\tilde{x}_2] & E[\tilde{x}_2^2] \end{bmatrix} \end{aligned} \tag{3.6-6}$$

Notice that the covariance matrix of an n-state vector is an n × n *symmetric* matrix; this fact will be used repeatedly in subsequent chapters. The diagonal elements of this covariance matrix are the mean square errors in knowledge of the state variables. Also, the trace of P is the mean square length of the vector $\underline{\tilde{x}}$. The off-diagonal terms of P are indicators of cross-correlation between the elements of $\underline{\tilde{x}}$. Specifically, they are related to the linear correlation coefficient $\rho(\tilde{x}_1, \tilde{x}_2)$ by

$$\rho(\tilde{x}_1, \tilde{x}_2) = \frac{E[\tilde{x}_1 \tilde{x}_2]}{\sigma_{\tilde{x}_1} \sigma_{\tilde{x}_2}} \tag{3.6-7}$$

where σ indicates standard deviation.

The random forcing functions are also described in terms of their covariance matrices. In the *continuous* formulation, the covariance matrix for the white noise random forcing function, $G\underline{w}$, is given by

$$E[(G(t)\underline{w}(t))(G(\tau)\underline{w}(\tau))^T] = G(t)Q(t)G^T(t)\,\delta(t-\tau) \tag{3.6-8}$$

where the operator δ is the Dirac delta function. The corresponding covariance matrix of the uncorrelated random sequence $\Gamma\underline{w}_k$ in the *discrete* formulation is

$$E[(\Gamma_k\underline{w}_k)(\Gamma_\ell\underline{w}_\ell)^T] = \begin{cases} \Gamma_k Q_k \Gamma_k^T, & k = \ell \\ 0, & k \neq \ell \end{cases} \tag{3.6-9}$$

The noise covariance matrix is computed using the definition of $\Gamma_k\underline{w}_k$ given in Eq. (3.4-3), viz:

$$\Gamma_k\underline{w}_k = \int_{t_k}^{t_{k+1}} \Phi(t_{k+1},\tau)G(\tau)\underline{w}(\tau)\,d\tau \tag{3.6-10}$$

yielding

$$\Gamma_k Q_k \Gamma_k^T = \tag{3.6-11}$$

$$E\left\{\int_{t_k}^{t_{k+1}} \int_{t_k}^{t_{k+1}} \Phi(t_{k+1},\tau)G(\tau)\underline{w}(\tau)\underline{w}^T(\alpha)G^T(\alpha)\Phi^T(t_{k+1},\alpha)\,d\tau\,d\alpha\right\}$$

The expectation operator may be taken inside the integral since, loosely speaking, the expectation of a sum is the sum of the expectations and an integral may be thought of as an infinite series. We obtain

$$\Gamma_k Q_k \Gamma_k^T = \int_{t_k}^{t_{k+1}} \int_{t_k}^{t_{k+1}} \Phi(t_{k+1},\tau)G(\tau)E[\underline{w}(\tau)\underline{w}^T(\alpha)]\,G^T(\alpha)\Phi(t_{k+1},\alpha)d\tau\,d\alpha$$

$$= \int_{t_k}^{t_{k+1}} \int_{t_k}^{t_{k+1}} \Phi(t_{k+1},\tau)G(\tau)Q(\tau)\delta(\tau-\alpha)G^T(\alpha)\Phi(t_{k+1},\alpha)d\tau\,d\alpha$$

(3.6-12)

The properties of the Dirac delta function, $\delta(\tau - \alpha)$, are used to perform the integration over α, yielding

$$\Gamma_k Q_k \Gamma_k{}^T = \int_{t_k}^{t_{k+1}} \Phi(t_{k+1}, \tau)\, G(\tau)\, Q(\tau)\, G^T(\tau)\, \Phi^T(t_{k+1}, \tau)\, d\tau \quad (3.6\text{-}13)$$

which is the result sought.

There is an important distinction to be made between the matrices Q(t) and Q_k. The former is a *spectral density matrix*, whereas the latter is a *covariance matrix*. A spectral density matrix may be converted to a covariance matrix through multiplication by the Dirac delta function, $\delta(t - \tau)$; since the delta function has units of 1/time, it follows that the units of the two matrices Q(t) and Q_k are different. This subject is again discussed in Chapter 4, when the continuous optimal filter is derived from the discrete formulation.

The above discussion concerns covariance matrices formed between vectors of equal dimension n. The matrices generated are square and n × n. Covariance matrices can also be formed between two vectors of unlike dimension; an example of this situation arises in Chapter 7.

3.7 PROPAGATION OF ERRORS

Consider the problem of estimating the state of a dynamic system in which the state vector $\underline{x}$ is known at some time t_k with an uncertainty expressed by the error covariance matrix

$$P_k = E[\tilde{\underline{x}}_k \tilde{\underline{x}}_k{}^T] \quad (3.7\text{-}1)$$

where the error vector, $\tilde{\underline{x}}_k$, is the difference between the true state, $\underline{x}_k$, and the estimate, $\hat{\underline{x}}_k$;

$$\tilde{\underline{x}}_k = \hat{\underline{x}}_k - \underline{x}_k \quad (3.7\text{-}2)$$

It is desired to obtain an estimate at a later point in time, t_{k+1}, which will have an *unbiased* error, $\tilde{\underline{x}}_{k+1}$. To form the estimate (i.e., the predictable portion of $\hat{\underline{x}}_{k+1}$ given $\hat{\underline{x}}_k$) the known state transition matrix Φ_k of Eq. (3.4-1) is used, resulting in

$$\hat{\underline{x}}_{k+1} = \Phi_k \hat{\underline{x}}_k \quad (3.7\text{-}3)$$

To show that the error in the estimate at t_{k+1} is unbiased, subtract Eq. (3.4-1) from Eq. (3.7-3) to obtain

$$\tilde{\underline{x}}_{k+1} = \Phi_k \tilde{\underline{x}}_k - \Gamma_k \underline{w}_k \quad (3.7\text{-}4)$$

The expected value of the error is

$$E[\tilde{\underline{x}}_{k+1}] = \Phi_k E[\tilde{\underline{x}}_k] - \Gamma_k E[\underline{w}_k] = \underline{0} \qquad (3.7\text{-}5)$$

Thus, under the assumptions that $\tilde{\underline{x}}_k$ and $\underline{w}_k$ are unbiased, it can be seen that Eq. (3.7-3) permits extrapolation of the state vector estimate without introducing a bias.

If a *known* input was provided to the system during the interval (t_k, t_{k+1}) this would appear as the additional term, $\Lambda_k \underline{u}_k$, in Eq. (3.4-1),

$$\underline{x}_{k+1} = \Phi_k \underline{x}_k + \Gamma_k \underline{w}_k + \Lambda_k \underline{u}_k \qquad (3.7\text{-}6)$$

Since the input is known, an identical quantity is added to the estimate of Eq. (3.7-3) and thus, the estimation error would be unchanged.

Equation (3.7-4) can be used to develop a relationship for projecting the error covariance matrix P from time t_k to t_{k+1}. The error covariance matrix P_{k+1} is

$$P_{k+1} = E[\tilde{\underline{x}}_{k+1} \tilde{\underline{x}}_{k+1}^T] \qquad (3.7\text{-}7)$$

From Eq. (3.7-4)

$$\begin{aligned} \tilde{\underline{x}}_{k+1} \tilde{\underline{x}}_{k+1}^T &= (\Phi_k \tilde{\underline{x}}_k - \Gamma_k \underline{w}_k)(\Phi_k \tilde{\underline{x}}_k - \Gamma_k \underline{w}_k)^T \\ &= \Phi_k \tilde{\underline{x}}_k \tilde{\underline{x}}_k^T \Phi_k^T - \Phi_k \tilde{\underline{x}}_k \underline{w}_k^T \Gamma_k^T - \Gamma_k \underline{w}_k \tilde{\underline{x}}_k^T \Phi_k^T + \Gamma_k \underline{w}_k \underline{w}_k^T \Gamma_k^T \end{aligned} \qquad (3.7\text{-}8)$$

Taking the expected value of the terms of Eq. (3.7-8) and using the fact that the estimation error at t_k and the noise $\Gamma_k \underline{w}_k$ are uncorrelated (a consequence of the fact that $\Gamma_k \underline{w}_k$ is a white sequence), namely

$$E[\tilde{\underline{x}}_k (\Gamma_k \underline{w}_k)^T] = 0 \qquad (3.7\text{-}9)$$

the equation for projecting the error covariance is found to be

$$P_{k+1} = \Phi_k P_k \Phi_k^T + \Gamma_k Q_k \Gamma_k^T \qquad (3.7\text{-}10)$$

From Eq. (3.7-10) it can be seen that the size of the random system disturbance (i.e., the "size" of $\Gamma_k Q_k \Gamma_k^T$) has a direct bearing on the magnitude of the error covariance at any point in time. Less obvious is the effect of dynamic system stability, as reflected in the transition matrix, on covariance behavior. In broad terms, a very stable system* will cause the first term in Eq. (3.7-10) to be smaller than the covariance P_k. No restrictions regarding the stability of the system were made in the development above and the error covariance of an unstable system will grow unbounded in the absence of

*That is, one whose F-matrix only has eigenvalues with large, negative real parts.

measurements taken on the state. A system with neutral stability* will also exhibit an unbounded error growth if appropriate process noise is present.

The expression for covariance propagation in the continuous system formulation is obtained through use of limiting arguments applied to Eq. (3.7-10). The noise covariance is:

$$\Gamma_k Q_k \Gamma_k{}^T = \int_{t_k}^{t_{k+1}} \Phi(t_{k+1},\tau)\, G(\tau)\, Q(\tau)\, G^T(\tau)\, \Phi^T(t_{k+1},\tau) d\tau \qquad (3.7\text{-}11)$$

For $t_{k+1} - t_k = \Delta t \rightarrow 0$, this is replaced by

$$\Gamma_k Q_k \Gamma_k{}^T \rightarrow GQG^T \, \Delta t \qquad (3.7\text{-}12)$$

where terms of order Δt^2 have been dropped. The differential equation for the transition matrix is

$$\dot{\Phi}(t) = F(t)\, \Phi(t)$$

or

$$\frac{\Phi(t + \Delta t) - \Phi(t)}{\Delta t} = F(t)\, \Phi(t) \qquad (3.7\text{-}13)$$

as $\Delta t \rightarrow 0$. Rearranging terms we deduce that, for $\Delta t \rightarrow 0$,

$$\Phi_k \rightarrow I + F\Delta t \qquad (3.7\text{-}14)$$

Substituting in Eq. (3.7-10), we have

$$\begin{aligned} P_{k+1} &= (I + F\Delta t)\, P_k\, (I + F\Delta t)^T + GQG^T \, \Delta t \\ &= P_k + (FP_k + P_k F^T + GQG^T)\, \Delta t + FP_k F^T \, \Delta t^2 \end{aligned} \qquad (3.7\text{-}15)$$

This equation can be rearranged in the form

$$\frac{P_{k+1} - P_k}{\Delta t} = FP_k + P_k F^T + GQG^T + FP_k F^T \, \Delta t \qquad (3.7\text{-}16)$$

As $\Delta t \rightarrow 0$, the equation becomes

$$\dot{P}(t) = F(t)P(t) + P(t)F^T(t) + G(t)Q(t)\, G^T(t) \qquad (3.7\text{-}17)$$

*That is, one whose F-matrix has some purely imaginary eigenvalue pairs.

which is the continuous form of the covariance propagation equation. This is the so-called *linear variance equation*.

3.8 MODELING AND STATE VECTOR AUGMENTATION

The error propagation equations obtained in Section 3.7 are developed under the assumption that the system random disturbances ($\underline{w}(t)$ or $\underline{w}_k$) are not correlated in time. Suppose, however, that important time correlation does exist. The characterization of system disturbances which have significant time correlation may be accomplished by means of "state vector augmentation." That is, the dimension of the state vector is increased by including the correlated disturbances as well as a description of system dynamic behavior in appropriate rows of an enlarged F (or Φ) matrix. In this manner, correlated random noises are taken to be state variables of a fictitious linear dynamic system which is itself excited by white noise. This *model* serves two purposes; it provides proper autocorrelation characteristics through specification of the linear system and the strength of the driving noise and, in addition, the random nature of the signal follows from the random excitation. Most correlated system disturbances can be described to a good approximation by a combination of one or more of the several types of models described in this section. The problem of correlated disturbances is treated for both continuous and discrete formulations.

State Vector Augmentation – The augmentation of the state vector to account for correlated disturbances is done using Eq. (3.1-1):

$$\dot{\underline{x}} = F\underline{x} + G\underline{w} \tag{3.8-1}$$

Suppose $\underline{w}$ is composed of correlated quantities $\underline{w}_1$ and uncorrelated quantities $\underline{w}_2$,

$$\underline{w} = \underline{w}_1 + \underline{w}_2 \tag{3.8-2}$$

If $\underline{w}_1$ can be modeled by a differential equation

$$\dot{\underline{w}}_1 = F_w\underline{w}_1 + \underline{w}_3 \tag{3.8-3}$$

where $\underline{w}_3$ is a vector composed of uncorrelated noises, then the augmented state vector $\underline{x}'$ is given by

$$\underline{x}' = \begin{bmatrix} \underline{x} \\ \underline{w}_1 \end{bmatrix} \tag{3.8-4}$$

and the augmented state differential equation, driven only by uncorrelated disturbances, is

$$\underline{\dot{x}}' = \begin{bmatrix} \dot{\underline{x}} \\ \dot{\underline{w}}_1 \end{bmatrix} = \begin{bmatrix} F & G \\ O & F_w \end{bmatrix} \begin{bmatrix} \underline{x} \\ \underline{w}_1 \end{bmatrix} + \begin{bmatrix} G & O \\ O & I \end{bmatrix} \begin{bmatrix} \underline{w}_2 \\ \underline{w}_3 \end{bmatrix} \tag{3.8-5}$$

We now consider a number of specific correlation models for system disturbances, in each instance scalar descriptions are presented.

Random Constant – The random constant is a non-dynamic quantity with a fixed, albeit random, amplitude. The continuous random constant is described by the state vector differential equation

$$\dot{x} = 0 \tag{3.8-6}$$

The corresponding discrete process is described by

$$x_{k+1} = x_k \tag{3.8-7}$$

The random constant can be thought of as the output of an integrator which has no input but has a random initial condition [see Fig. 3.8-1(a)].

Random Walk – The random walk process results when uncorrelated signals are integrated. It derives its name from the example of a man who takes fixed-length steps in arbitrary directions. In the limit, when the number of steps is large and the individual steps are short in length, the distance travelled in a particular direction resembles the random walk process. The state variable differential equation for the random walk process is

$$\dot{x} = w \tag{3.8-8}$$

where $E[w(t)w(\tau)] = q(t)\ \delta(t-\tau)$. A block diagram representation of this equation is shown in Fig. 3.8-1(b). The equivalent discrete process is

$$x_{k+1} = x_k + w_k \tag{3.8-9}$$

where the noise covariance is [Eq. (3.6-13)] $q_k = q(t_{k+1}-t_k)$. The scalar version of the continuous linear variance equation

$$\dot{p} = 2fp + g^2 q \tag{3.8-10}$$

is used to determine the time behavior of the random walk variable. For f=0, g=1, we obtain

$$\dot{p} = q$$

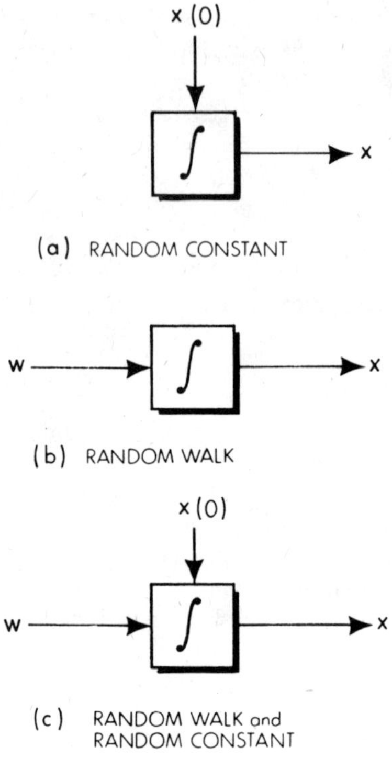

Figure 3.8-1 Block Diagrams for Random Constant and Random Walk Processes

so that

$$E[x^2] = p = qt \tag{3.8-11}$$

A combination of the random walk and random constant can be represented by the use of only one state variable. This is illustrated in Fig. 3.8-1(c).

Random Ramp – Frequently, random errors which exhibit a definite time-growing behavior are present. The random ramp, a function which grows *linearly* with time, can often be used to describe them. The growth rate of the random ramp is a random quantity with a given probability density. Two state elements are necessary to describe the random ramp:

$$\dot{x}_1 = x_2$$
$$\dot{x}_2 = 0 \tag{3.8-12}$$

The state x_1 is the random ramp process; x_2 is an auxiliary variable whose initial condition provides the slope of the ramp. This initial condition is exhibited in the form of a mean square slope, $E[x_2^2(0)]$. From the solution of Eqs. (3.8-12) the mean square value of x_1 is seen to grow parabolically with time, viz:

$$E[x_1^2(t)] = E[x_2^2(0)]\,t^2 \tag{3.8-13}$$

A combination of a random ramp, random walk and random constant can be represented by the use of only two state variables as illustrated in Fig. 3.8-2. The equivalent discrete version of the random ramp is defined by the two variables

$$x_{1_{k+1}} = x_{1_k} + (t_{k+1} - t_k)x_{2_k}$$
$$x_{2_{k+1}} = x_{2_k} \tag{3.8-14}$$

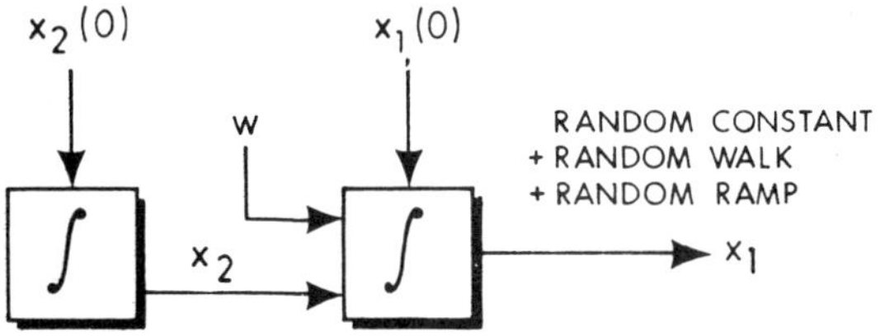

Figure 3.8-2 Generation of Three Random Characteristics by the Addition of Only Two State Variables

Exponentially Correlated Random Variable – A random quantity whose autocorrelation function is a decreasing exponential

$$\varphi_{xx}(\tau) = \sigma^2 e^{-\beta|\tau|} \tag{3.8-15}$$

is frequently a useful representation of random system disturbances. Recall from Chapter 2 that this autocorrelation function is representative of a first-order gauss-markov process. This quantity is often used to provide an *approximation* for a band-limited signal, whose spectral density is flat over a finite bandwidth. The exponentially correlated random variable is generated by passing an uncorrelated signal through a linear first-order feedback system as shown in Fig. 3.8-3. The differential equation of the state variable is

$$\dot{x} = -\beta x + w \tag{3.8-16}$$

The mean square value of the exponentially correlated random variable is constant if the mean square initial condition on the integrator is taken as $q/2\beta$. The mean square value of x is found from the scalar covariance equation, Eq. (3.8-10). Since q is constant, $\dot{p}=0$ in steady state, so that $(f = -\beta, g = 1)$

$$0 = -2p\beta + q \tag{3.8-17}$$

yielding

$$E[x^2] = p = \frac{q}{2\beta} \tag{3.8-18}$$

The discrete version of the exponentially correlated random variable is described by

$$x_{k+1} = e^{-\beta(t_{k+1}-t_k)} x_k + w_k \tag{3.8-19}$$

Using Eq. (3.7-11), we readily find the covariance of w_k as

$$q_k = \frac{q}{2\beta}\left[1 - e^{-2\beta(t_{k+1} - t_k)}\right] \tag{3.8-20}$$

Periodic Random Quantities – Random variables which exhibit periodic behavior also arise in physical systems. A useful autocorrelation function model is given by

$$\varphi_{x_1x_1}(\tau) = \sigma^2 e^{-\beta|\tau|} \cos\omega|\tau| \tag{3.8-21}$$

where the values of β and ω are chosen on the basis of the physics of the situation or to fit empirical autocorrelation data. Two state variables are necessary to represent a random variable with this autocorrelation function. One pair of quantities which provides this relation obeys the following differential equations (Ref.5)

$$\dot{x}_1 = x_2 + w$$

$$\dot{x}_2 = -\alpha^2 x_1 - 2\beta x_2 + (\alpha-2\beta) w \tag{3.8-22}$$

where

$$\alpha = (\beta^2 + \omega^2)^{1/2} \tag{3.8-23}$$

The spectral density of the white noise w is $2\beta\sigma^2\delta(\tau)$.

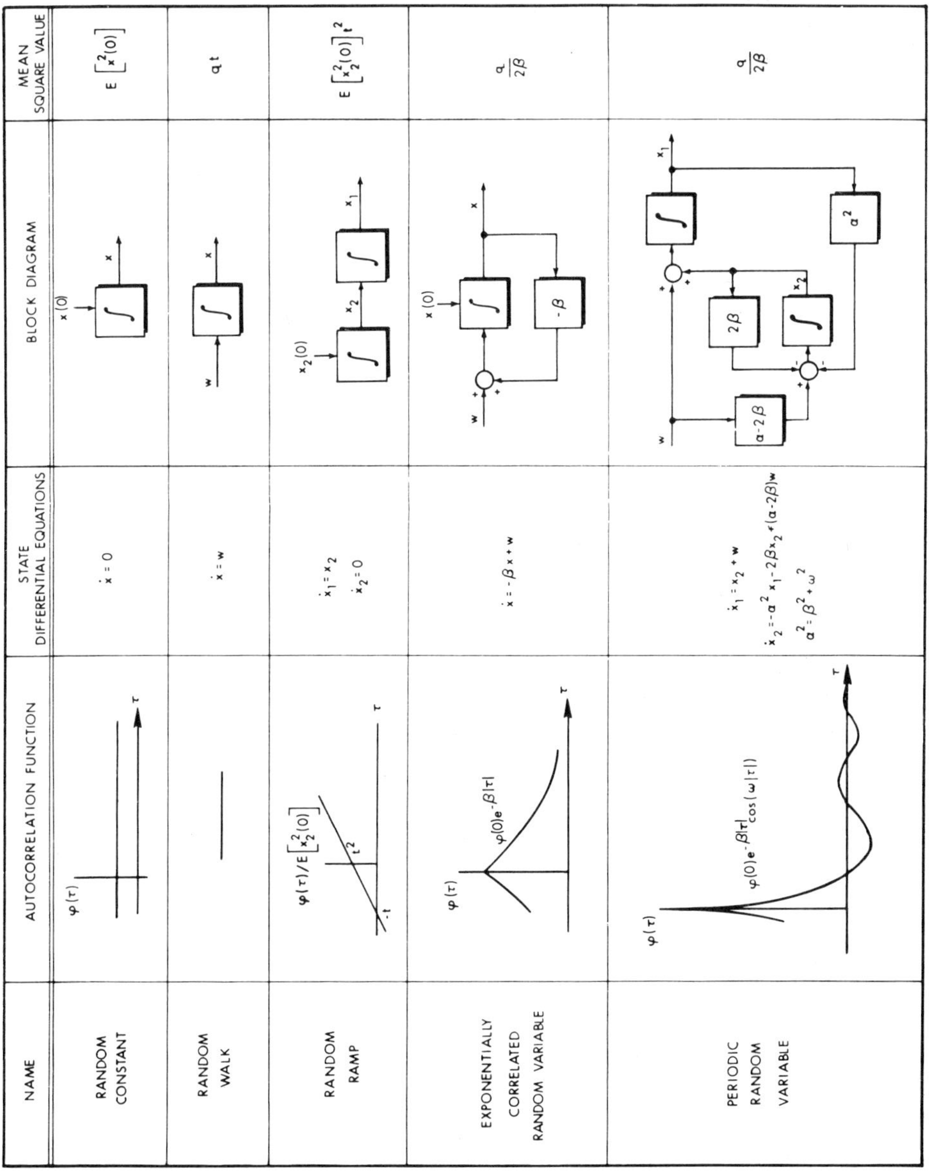

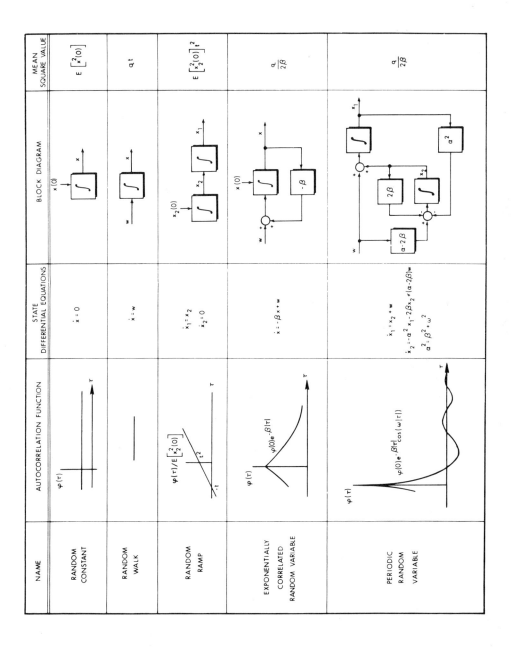

Figure 3.8-3 Continuous Time Models for Random Variables

A summary of the continuous models described above is given in Fig. 3.8-3. The discrete models which have been presented are summarized in Fig. 3.8-4. Occasionally, other more complex random error models arise. For example, Ref. 6 discusses a time and distance correlated error whose autocorrelation function is given by

$$\varphi(\tau,d) = \sigma^2 e^{-|\tau|/T} e^{-|d|/D} \tag{3.8-24}$$

where T and D are the first-order correlation time and first-order correlation distance, respectively. This correlation also occasionally appears when considering geophysical phenomena (e.g., Ref. 7). However, the majority of system measurement errors and disturbances can be described by some combination of the random variable relations summarized in Fig. 3.8-3.

3.9 EMPIRICAL MODEL IDENTIFICATION

Subsequent discussions assume that any random process under study has been modeled as a linear system driven by gaussian white noise. In this section, methods for determining the best linear-gaussian model to fit an observed sequence of data are discussed. As illustrated in Fig. 3.9-1, a model for the process is produced by combining empirical data and prior knowledge of underlying physical mechanisms. The derived model and the data are then used in an estimation process as described in succeeding chapters.

It is assumed that the available data are a sequence of scalar observations,

$$z_1, z_2, \ldots, z_N$$

A time-invariant, discrete linear system of the form

$$\underline{x}_{k+1} = \Phi \underline{x}_k + S\, r_k \tag{3.9-1}$$

$$z_k = H\, \underline{x}_k + r_k \tag{3.9-2}$$

is used to fit the data. It can be shown that this type of model is general enough to fit any sequence of observations generated by a stable, time-invariant linear system driven by gaussian noise. Here r_k, referred to as the residual at time t_k, is the difference between the observation at time t_k and the output the model would produce based on inputs up to but not including time t_k. The residuals reflect the degree to which the model fits the data.

Model identification typically proceeds in the following way. Once data are acquired, a preliminary model for the random process is chosen. An optimal smoothing algorithm (see Chapter 5) is used to determine the initial state and the inputs most likely to have produced the observed data. Next, residuals are examined to determine how well the model fits the data. A new model is selected if the residuals do not display the desired characteristics, and the procedure is repeated.

NAME	STATE DIFFERENCE EQUATION	BLOCK DIAGRAM	RELATION of DISCRETE q_k to CONTINUOUS q
RANDOM CONSTANT	$x_{k+1} = x_k$	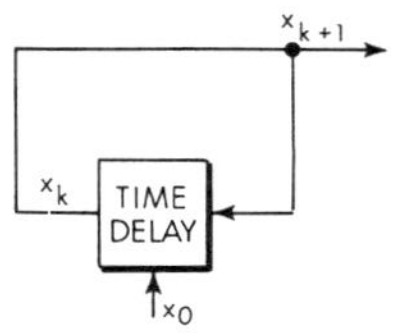	$q_k = 0$
RANDOM WALK	$x_{k+1} = x_k + w_k$	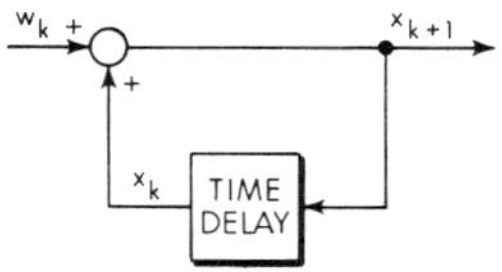	$q_k = q(t_{k+1} - t_k)$
RANDOM RAMP	$x_{1_{k+1}} = x_{1_k} + (t_{k+1} - t_k) x_{2_k}$ $x_{2_{k+1}} = x_{2_k}$	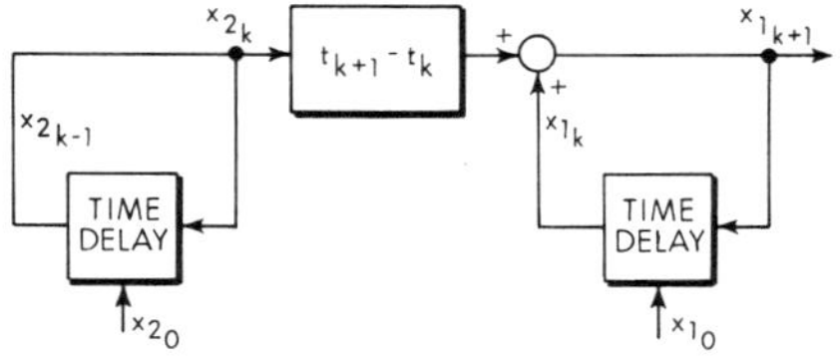	$q_k = 0$
EXPONENTIALLY CORRELATED RANDOM VARIABLE	$x_{k+1} = e^{-\beta(t_{k+1} - t_k)} x_k + w_k$	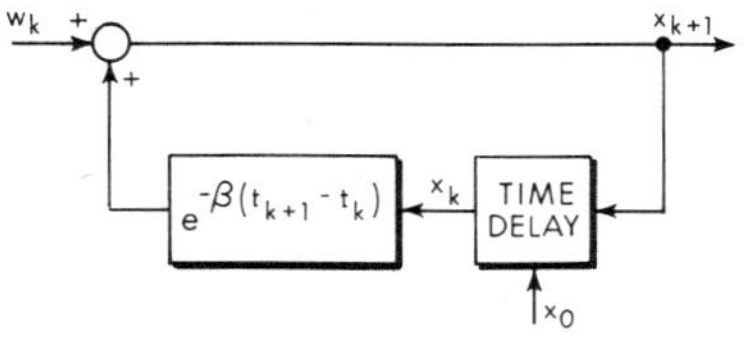	$q_k = \frac{q}{2\beta}\left[1 - e^{-2\beta(t_{k+1} - t_k)}\right]$

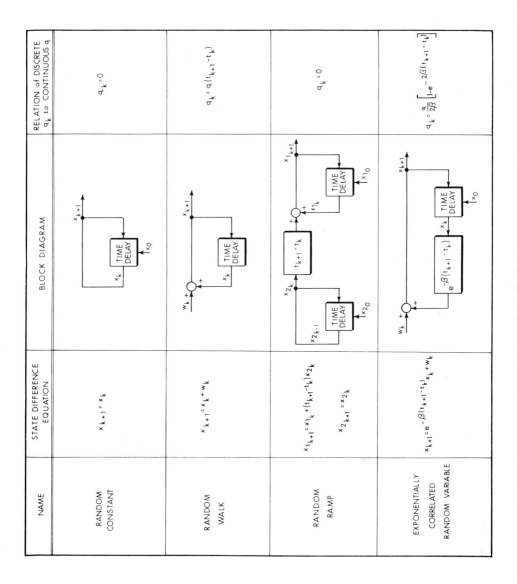

Figure 3.8-4 Discrete Version of Error Models

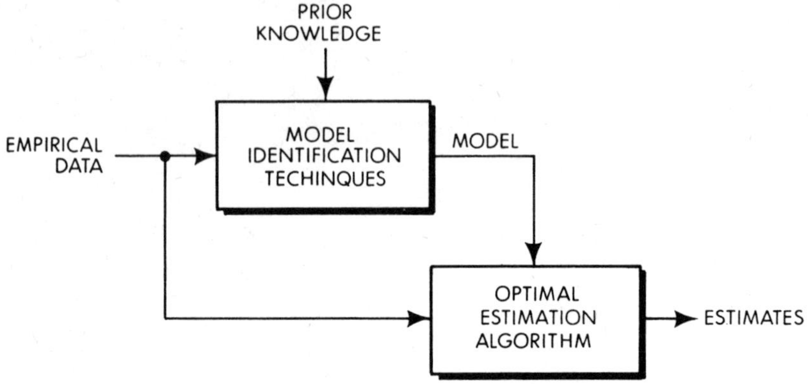

Figure 3.9-1 Empirical Model Identification

AUTOCORRELATION TECHNIQUES

Since certain of the models considered here can produce stationary gaussian random processes, it is natural to give primary consideration to the sample mean and sample autocorrelation function for the data when attempting model identification. Given a string of data z_i, i = 1,2, . . . , N, taken at constant intervals, Δt, the sample mean and sample autocorrelation function are determined as

$$m = \frac{1}{N} \sum_{i=1}^{N} z_i \tag{3.9-3}$$

$$\hat{\varphi}_{zz}(\ell \Delta t) = \frac{1}{N - \ell} \sum_{i=1}^{N-\ell} (z_i - m)(z_{i+\ell} - m)^T \quad , \quad \ell = 0, 1, \ldots, N-1 \tag{3.9-4}$$

A preliminary model for the process can be obtained by choosing one whose mean and autocorrelation function closely approximate the sample mean and sample autocorrelation function. Although seemingly straightforward, the procedure is often complicated by the fact that data might not be taken over a sufficient period of time or at sufficiently frequent intervals. It is, of course, necessary to apply physical insight to the problem in order to correctly interpret calculated statistical quantities. The following example provides an illustration.

Example 3.9-1

Deflections of the vertical are angular deviations of the true gravity vector from the direction postulated by the regular model called the reference ellipsoid (Ref. 8). Consider the data sample shown in Fig. 3.9-2; it consists of measurements of one component of the deflection of the vertical, ξ, taken at 12.5 nm intervals across the 35th parallel in the United States (Ref. 9).

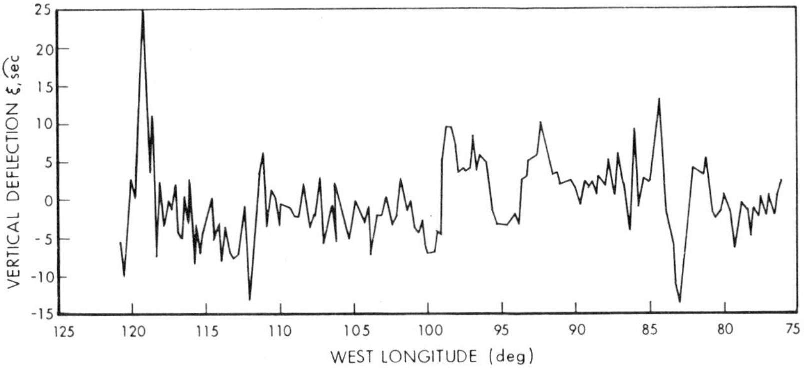

Figure 3.9-2 Meridian Component, Vertical Deflections of Gravity – 35th Parallel, United States (Ref. 9)

We are concerned with spatial rather than temporal correlation so that the shift parameter in the autocorrelation function is distance. The sample autocorrelation function calculated according to Eq. (3.9-4) is shown in Fig. 3.9-3. The formula is evaluated only out to 12 shifts, i.e., 150 nm; beyond that point two effects cloud the validity of the results. First, as the number of shifts ℓ increases, the number of terms $N - \ell$ in the summation decreases, and confidence in the derived values is reduced. Secondly, calculated autocorrelation values at shifts greater than 150 nm are sufficiently small that any measurement noise present tends to dominate.

Vertical deflection autocorrelation values are seen to fall off exponentially. A first-order model least-squares curve-fit results in the *empirical autocorrelation function* shown in Fig. 3.9-3. It has the form

$$\hat{\varphi}_{\xi\xi}(d) = \sigma_\xi^2\, e^{-|d|/D} + m_\xi^2 \tag{3.9-5}$$

where the standard deviation is determined as

$$\sigma_\xi = 5.2\ \widehat{\text{sec}}$$

the mean is found to be

$$m_\xi = 0.2\ \widehat{\text{sec}}$$

the correlation distance is

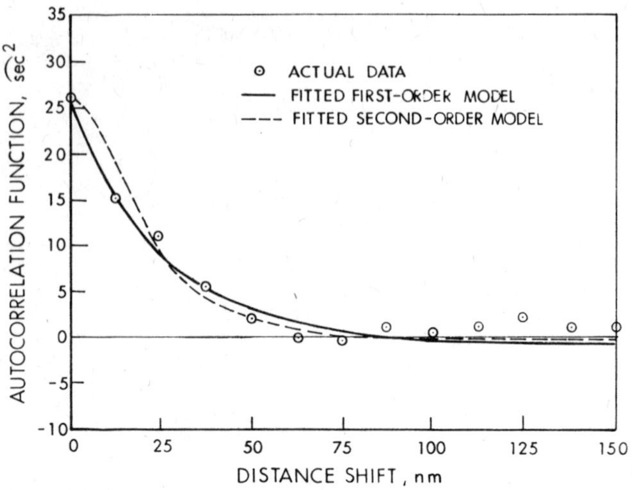

Figure 3.9-3 Vertical Deflection Sample Autocorrelation Function – 35th Parallel U.S.

D = 25.1 nm

and d is the distance shift parameter. As described in Ref. 10, the fitted function from which Eq. (3.9-5) is obtained is derived using techniques which compensate for finite data. Thus, it is not unnatural that the function is negative at large distance shifts. In this example the correlation distance D is approximately twice the data interval. The structure of the actual autocorrelation function cannot be accurately determined for small distance shifts – e.g., less than 12.5 nm.

An exponential (first-order gauss-markov) autocorrelation function describes a process whose correlation falls off exponentially at all distance shifts, including d = 0. Equivalently, the derivative of the autocorrelation function is nonzero at zero distance shift,

$$\left.\frac{\partial}{\partial d}\,\hat{\varphi}_{\xi\xi}(d)\right|_{d=0} = -\frac{1}{D}\,\sigma_\xi^2 \neq 0 \qquad (3.9\text{-}6)$$

This result is not *physically* appealing since vertical deflections are caused by nonuniform mass distributions in the earth, which are not likely to change instantaneously in distance as would be implied by a nonzero derivative. Thus, consideration is given to the second-order gauss-markov process, whose autocorrelation function does have a zero derivative at zero shift. The form of the function is

$$\hat{\varphi}_{\xi\xi}(d) = \sigma_\xi^2 \left(1 + \frac{|d|}{D}\right) e^{-|d|/D} + m_\xi^2 \qquad (3.9\text{-}7)$$

and by least-squares curve fitting the characteristic distance is determined to be

D = 11.6 nm

The empirically determined quantities σ_ξ and m_ξ are unchanged. Details of applying the second-order gauss-markov model to gravity phenomena are given in Ref. 11. Comparisons of the first- and second-order processes with the empirical autocorrelation values are given in Fig. 3.9-3.

Finite data length implies an inability to determine the form of an autocorrelation function at large shifts, whereas finite data spacing implies an inability to determine the form at small shifts. Together, the two correspond to a limited quantity of data, thus resulting in classical statistical limitations in estimating autocorrelation function parameters. These factors are treated analytically in Refs. 12 and 13; brief summaries are given here.

Consider a zero mean random process whose statistics are described by the autocorrelation function

$$\varphi_{zz}(\tau) = \sigma_z^2 \, e^{-|\tau|/T_0} \tag{3.9-8}$$

A sequence of measurements taken at intervals Δt over a period T is used to obtain an empirical autocorrelation function

$$\hat{\varphi}_{zz}(\ell\Delta t) = \frac{1}{N-\ell} \sum_{i=1}^{N-\ell} z_i z_{i+\ell} \quad , \quad \ell = 0, 1, 2, \ldots, N-1 \tag{3.9-9}$$

where $N = T/\Delta t$. In Fig. 3.9-4, *expected* values of the normalized empirical autocorrelation function are plotted for various values of T/T_0. The case of $T/T_0 \to \infty$ corresponds to the actual function, Eq. (3.9-8). Clearly, for small values of T/T_0, empirically derived autocorrelation functions may be quite different from the function corresponding to the underlying physical model.

Data limitations, expressed as a finite value for N, lead to an uncertainty in the estimation of autocorrelation function parameters. If the actual process variance is σ_z^2 and the empirically derived estimate is $\hat{\sigma}_z^2$, the uncertainty in the estimate is expressed by

$$E\,[(\hat{\sigma}_z^2 - \sigma_z^2)^2] = \frac{\sigma_z^4}{N} \tag{3.9-10}$$

If the actual process correlation time is T_0 and the empirically derived estimate is $\hat{T}_0$, the uncertainty in the estimate is expressed by

$$E\,[(\hat{T}_0 - T_0)^2] = T_0^2 \, \frac{e^{2\Delta t/T_0} - 1}{N(\Delta t/T_0)^2} \tag{3.9-11}$$

This relationship is illustrated in Fig. 3.9-5; the ideal situation corresponds to T large and Δt small, relative to T_0.

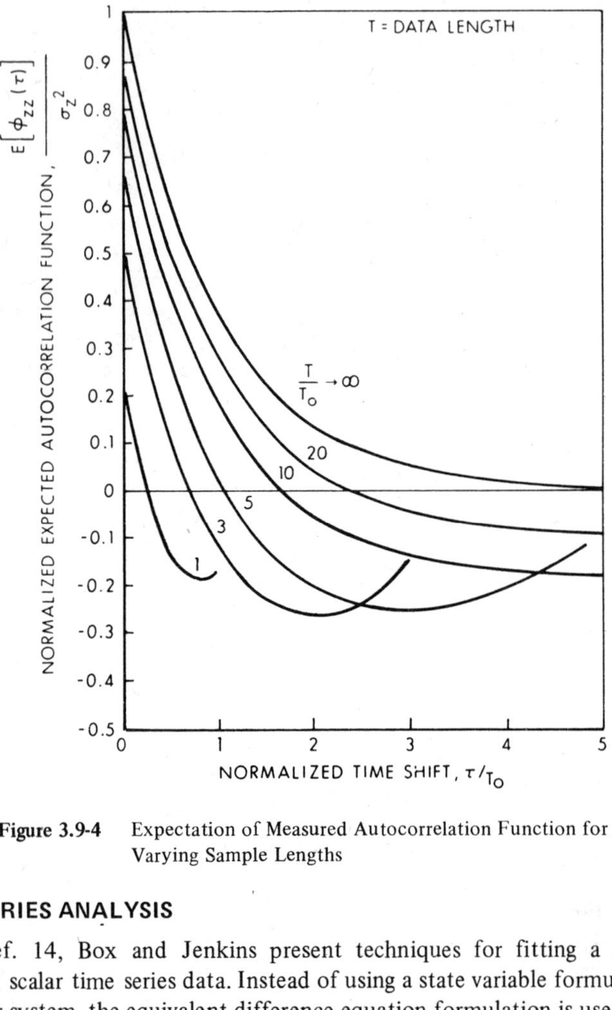

Figure 3.9-4 Expectation of Measured Autocorrelation Function for Varying Sample Lengths

TIME SERIES ANALYSIS

In Ref. 14, Box and Jenkins present techniques for fitting a model to empirical scalar time series data. Instead of using a state variable formulation for the linear system, the equivalent difference equation formulation is used,

$$z_k = \sum_{i=1}^{p} b_i\, z_{k-i} + r_k - \sum_{i=1}^{q} c_i\, r_{k-i} \tag{3.9-12}$$

Here z_k is the observation of the time series at time t_k and r_k, called the *residual* at time t_k, is an uncorrelated gaussian random variable. The summation limits p and q as well as the parameters b_i and c_i are adjusted to fit the data. It is assumed that the data have been detrended; that is, a linear system has been found which adequately models the mean value of the data. For example, the random bias and random ramp models discussed previously can be used. The detrended data sequence is assumed to have stationary statistics.

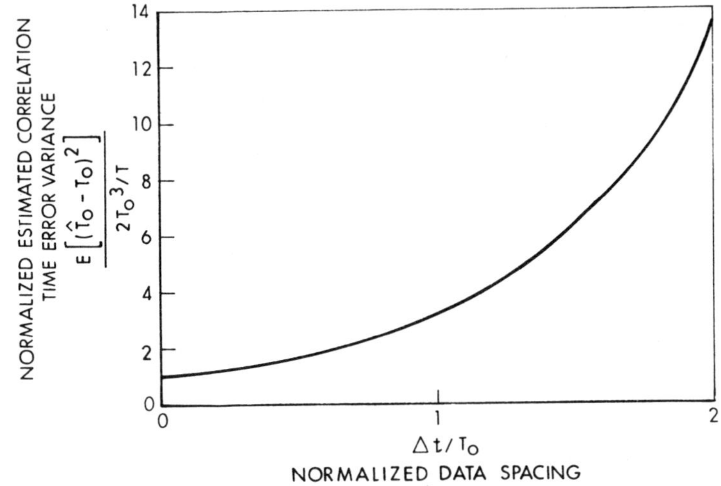

Figure 3.9-5 Variance of Estimated Correlation Time of Exponential Autocorrelation Function

A state variable representation of this model has the states

$$\underline{x}_k = \begin{bmatrix} r_{k-q+1} \\ r_{k-q+2} \\ \cdot \\ \cdot \\ \cdot \\ r_{k-2} \\ r_{k-1} \\ \hline z_{k-p+1} \\ z_{k-p+2} \\ \cdot \\ \cdot \\ \cdot \\ z_{k-1} \\ \hline \tilde{z}_k(-) \end{bmatrix} \tag{3.9-13}$$

Here

$$\tilde{z}_k(-) = z_k - r_k \tag{3.9-14}$$

represents the prediction of z_k based on the model and on knowledge of the infinite number of z's prior to time t_k. It is easy to see that

$$\begin{aligned}\overset{\infty}{z}_{k+1}(-) &= b_1 z_k - c_1 r_k + \sum_{i=2}^{p} b_i z_{k+1-i} - \sum_{i=2}^{q} c_i r_{k+1-i} \\ &= b_1 \overset{\infty}{z}_k(-) + (b_1 - c_1) r_k + \sum_{i=2}^{p} b_i z_{k+1-i} - \sum_{i=2}^{q} c_i r_{k+1-i}\end{aligned} \tag{3.9-15}$$

Therefore the state variable model is

$$\underline{x}_{k+1} = \left[\begin{array}{cc:cc} \begin{matrix}0\\0\\ \vdots\\0\end{matrix} & I & & 0 \\ 0 & 0 \;\cdots\; 0 & & \\ \hdashline & & \begin{matrix}0\\0\\ \vdots\\0\end{matrix} & I \\ & 0 & & \\ \hdashline -c_q & -c_{q-1} \;\cdots\; -c_2 & b_p & b_{p-1} \;\cdots\; b_1 \end{array}\right] \underline{x}_k + \left[\begin{array}{c} 0\\0\\ \vdots\\0\\1\\ \hdashline 0\\0\\ \vdots\\0\\1\\ \hdashline b_1 - c_1 \end{array}\right] r_k \tag{3.9-16}$$

To gain insight into how the characteristics of the time series model are affected by the coefficients of the difference equation, the general form of Eq. (3.9-12) is specialized to several particular cases. The characteristics of these forms are reflected in their autocorrelation functions. Therefore, by studying these specialized forms, it is easier to identify a model that matches the observed time series and also has a minimum number of parameters to be determined.

Autoregressive (AR) Process – If it is assumed that the present observation is a linear combination of past observations plus a gaussian random variable, Eq. (3.9-12) becomes

$$z_k = \sum_{i=1}^{p} b_i z_{k-i} + r_k \tag{3.9-17}$$

This type of time series is called an autoregressive (AR) process. Note that the residual r_k is the only portion of the measurement z_k which cannot be predicted from previous measurements. It is assumed that the coefficients of this difference equation have been chosen so that the linear system is stable, thus making the autoregressive process stationary. The characteristic nature of this process is reflected in its autocorrelation function. Multiplying both sides of Eq. (3.9-17) by delayed z_k and taking the ensemble expectation of the result, we find

$$\varphi_{zz}(k) = \sum_{i=1}^{p} b_i \varphi_{zz}(k-i) \tag{3.9-18}$$

Thus, the autocorrelation function of an AR process obeys the homogeneous difference equation for the process. For a stationary AR process, the solution to the homogeneous difference equation is given by linear combinations of damped sinusoids and exponentials.

It is desirable to express the difference equation coefficients in terms of the autocorrelation function values; estimates of the coefficients can then be obtained by using the sample autocorrelation function. Evaluation of Eq. (3.9-19) for p successive shifts results in a set of p linear equations which can be written as

$$\begin{bmatrix} \varphi_{zz}(0) & \varphi_{zz}(1) & \cdots & \varphi_{zz}(p-1) \\ \varphi_{zz}(1) & \varphi_{zz}(0) & \cdots & \varphi_{zz}(p-2) \\ \cdot & \cdot & & \cdot \\ \cdot & \cdot & & \cdot \\ \cdot & \cdot & & \cdot \\ \varphi_{zz}(p-1) & \varphi_{zz}(p-2) & \cdots & \varphi_{zz}(0) \end{bmatrix} \begin{bmatrix} b_1 \\ b_2 \\ \cdot \\ \cdot \\ \cdot \\ b_p \end{bmatrix} = \begin{bmatrix} \varphi_{zz}(1) \\ \varphi_{zz}(2) \\ \cdot \\ \cdot \\ \cdot \\ \varphi_{zz}(p) \end{bmatrix} \tag{3.9-19}$$

This set of linear equations, called the Yule-Walker equations, can be solved for the correlation matrix values and correlation vector using estimates of the autocorrelation function. The state variable representation of an autoregressive process has only the p states,

$$\underline{x}_k = \begin{bmatrix} z_{k-p+1} \\ z_{k-p+2} \\ \cdot \\ \cdot \\ \cdot \\ z_{k-2} \\ z_{k-1} \\ \overset{\infty}{z}_k(-) \end{bmatrix} \tag{3.9-20}$$

Moving Average (MA) Process – If the time series is assumed to be generated by a finite linear combination of past and present inputs only, the process is called a moving average (MA) process. Under this assumption, the time series is generated by the difference equation

$$z_k = r_k - \sum_{i=1}^{q} c_i r_{k-i} \tag{3.9-21}$$

This model always produces a stationary process. It is assumed that the coefficients are chosen so that model is also invertible – i.e., the input sequence can be completely determined from knowledge of the observed output sequence. Under the assumption that the covariance of the input sequence is

$$\begin{aligned} E[r_k r_\ell] &= \sigma_r^2 \qquad \text{for } \ell = k \\ &= 0 \qquad \text{for } \ell \neq k \end{aligned} \tag{3.9-22}$$

the corresponding autocorrelation function of the observations is

$$\begin{aligned} \varphi_{zz}(k) &= \left(-c_k + \sum_{i=1}^{q-k} c_i c_{k+i}\right) \sigma_r^2 \qquad \text{for } k \leqslant q \\ &= 0 \qquad \text{for } k > q \end{aligned} \tag{3.9-23}$$

Thus, the autocorrelation function for an MA process has a finite number of non-zero values, and cuts off at the order of the process.

Mixed Autoregressive Moving Average (ARMA) Processes – A more general stationary time series can be generated by combining the AR and MA processes to get a mixed autoregressive moving average (ARMA) process. The difference equation model for the process takes the general form of Eq. (3.9-12), but the allowable range of coefficients is restricted so that the process is stationary and the model is invertible. The process autocorrelation function is identical to the pure AR process after (q – p) shifts. Thus, the autocorrelation function is given by

$$\varphi_{zz}(k) = \sum_{i=1}^{p} b_i \varphi_{zz}(k-i) \quad \text{for } k > q \tag{3.9-24}$$

with p initial conditions $\varphi_{zz}(q), \varphi_{zz}(q-1), \ldots, \varphi_{zz}(q-p+1)$. If $q < p$, the autocorrelation function will consist of damped exponentials and sinusoids determined by the difference equation coefficients and initial conditions. If $p \leqslant q$, there will be $(q-p+1)$ initial values which do not follow the general pattern. An example is given by the first-order mixed ARMA process,

$$z_k = b_1 z_{k-1} + r_k - c_1 r_{k-1}$$

The autocorrelation function for this process is

$$\varphi_{zz}(0) = \frac{1 + c_1{}^2 - 2 c_1 b_1}{1 - b_1{}^2} \sigma_r{}^2$$

$$\varphi_{zz}(1) = b_1 \varphi_{zz}(0) - c_1 \sigma_r{}^2$$

$$= \frac{(1 - b_1 c_1)(b_1 - c_1)}{1 - b_1{}^2} \sigma_r{}^2$$

$$\varphi_{zz}(k) = b_1 \varphi_{zz}(k-1) \quad \text{for } k \geqslant 2$$

Note that $\varphi_{zz}(0)$ and $\varphi_{zz}(1)$ depend upon both the autoregressive and moving average parameters. The autocorrelation function is exponential except for $\varphi_{zz}(0)$, which does not follow the exponential pattern.

Autoregressive Integrated Moving Average (ARIMA) Processes – If the measurements do not exhibit stationary statistics, the AR, MA and ARMA models cannot be used directly. In certain situations this difficulty can be overcome by differencing the data. For example, suppose the differences

$$d_k = z_k - z_{k-1} \tag{3.9-25}$$

are found to have stationary statistics. Then an ARMA process can be used to model these differences, and the measurements can be modeled as the sum of the differences plus an initial condition. Such a process is called an autoregressive integrated moving average (ARIMA) process; the term "integrated" refers to the summation of the differences. Note that the random walk process discussed in Section 3.8 is the simplest form of an ARIMA process.

Fast Fourier Transforms – In 1965 Cooley and Tukey (Ref. 15) described a computationally efficient algorithm for obtaining Fourier coefficients. The fast Fourier transform (FFT) is a method for computing the discrete Fourier transform of a time series of discrete data samples. Such time series result when digital analysis techniques are used for analyzing a continuous waveform. The time series will represent completely the continuous waveform, provided the waveform is frequency band-limited and the samples are taken at a rate at least twice the highest frequency present. The discrete Fourier transform of the time series is closely related to the Fourier integral transform of the continuous waveform.

The FFT has applicability in the generation of statistical error models from series of test data. The algorithm can be modified to compute the *autocorrelation function* of a one-dimensional real sequence or the *cross-correlation function* and *convolution* of two one-dimensional real sequences. It can also be used to estimate the *power spectral density* of a one-dimensional real continuous waveform from a sequence of evenly-spaced samples. The considerable efficiency of the FFT, relative to conventional analysis techniques, and the availability of outputs from which statistical error models are readily obtained, suggest that the FFT will be of considerable utility in practical applications of linear system techniques. Computational aspects of applying the FFT as noted above are discussed in Ref. 16.

REFERENCES

1. DeRusso, D.M., Roy, R.J., and Close, C.M., *State Variables for Engineers*, John Wiley & Sons, Inc., New York, 1966.
2. Leondes, C.T., ed., *Guidance and Control of Aerospace Vehicles*, McGraw-Hill Book Co., Inc., New York, 1963.
3. Lee, R.C.K., *Optimal Estimation, Identification, and Control*, M.I.T. Press, Cambridge, Mass., 1964.
4. Schweppe, F.C., *Uncertain Dynamic Systems*, Prentice-Hall, Englewood Cliffs, N.J., 1973.
5. Fitzgerald, R.J., "Filtering Horizon – Sensor Measurements for Orbital Navigation," *AIAA Guidance and Control Conference Proceedings*, August 1966, pp. 500-509.
6. Wilcox, J.C., "Self-Contained Orbital Navigation Systems with Correlated Measurement Errors," *AIAA/ION Guidance and Control Conference Proceedings*, August 1965, pp. 231-247.
7. D'Appolito, J.A. and Kasper, J.F., Jr., "Predicted Performance of an Integrated OMEGA/Inertial Navigation System," *Proc. National Aerospace Electronics Conference*, May 1971.
8. Heiskanen, W.A. and Moritz, H., *Physical Geodesy*, W.H. Freeman, San Francisco, 1967.
9. Rice, D.A., "A Geoidal Section in the United States," *Bull. Geod.*, Vol. 65, 1962.
10. Levine, S.A. and Gelb, A., "Effect of Deflections of the Vertical on the Performance of a Terrestrial Inertial Navigation System," *J. Spacecraft and Rockets*, Vol. 6, No. 9, September 1969.
11. Kasper, J.F., Jr., "A Second-Order Markov Gravity Anomaly Model," *J. Geophys. Res.*, Vol. 76, No. 32, November 1971.
12. Bendat, J.S. and Piersol, A.G., *Measurement and Analysis of Random Data*, John Wiley & Sons, Inc., New York, 1966.
13. Weinstock, H., "The Description of Stationary Random Rate Processes," M.I.T. Instrumentation Laboratory, E-1377, July 1963.
14. Box, G.E.P., and Jenkins, G.M., *Time Series Analysis, Forecasting and Control*, Holden-Day, San Francisco, 1970.
15. Cooley, J.W. and Tukey, J.W., "An Algorithm for the Machine Calculation of Complex Fourier Series," *Math. of Comput.*, Vol. 19, pp. 297-301, April 1965.
16. Cochran, W.T., *et. al.*, "What is the Fast Fourier Transform?", *Proc. of the IEEE*, Vol. 55, No. 10, pp. 1664-1674, October 1967.

PROBLEMS

Problem 3-1

Show that

$$P(t) = \Phi(t,t_0)P(t_0)\,\Phi^T(t,t_0) + \int_{t_0}^{t} \Phi(t,\tau)G(\tau)Q(\tau)G^T(\tau)\,\Phi^T(t,\tau)\,d\tau$$

is the *solution* of the linear variance equation.

Problem 3-2

Use the solution given in Problem 1 to show that the solution to

$$FP + PF^T = -Q$$

is

$$P = \int_0^{\infty} e^{F^T t} Q e^{Ft} dt$$

where F and Q are constant matrices.

Problem 3-3

Extensive analysis of certain geophysical data yields the following temporal autocorrelation function

$$\hat{\varphi}(\tau) = \sigma^2 \left(\alpha_1{}^2 + \alpha_2{}^2 \cos \omega\tau + \alpha_3{}^2\, e^{-\beta|\tau|}\right)$$

where ω is earth angular rate ($2\pi/24\ hr^{-1}$) and τ is time shift. Derive a set of state vector equations to describe this random process.

Problem 3-4

For the system shown in Fig. 3-1, where the autocorrelation function is

$$\varphi_{x_2x_2}(\tau) = \sigma^2 e^{-\beta|\tau|}$$

show that

$$P(t) = \frac{\sigma^2}{\beta^2}\left[\begin{array}{c|c} 2(\beta t - 1 + e^{-\beta t}) & \beta(1-e^{-\beta t}) \\ \hline \beta(1-e^{-\beta t}) & \beta^2 \end{array}\right]$$

for the state vector taken as $\underline{x} = [x_1\ x_2]^T$.

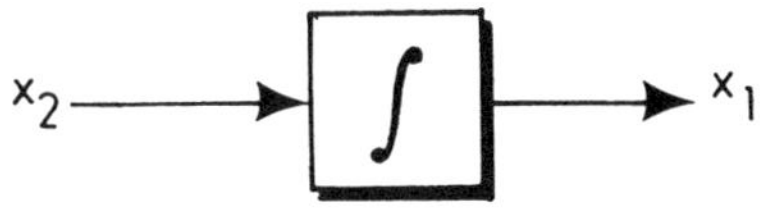

Figure 3-1

Problem 3-5

Consider the single-axis error model for an inertial navigation system shown in Fig. 3-2.

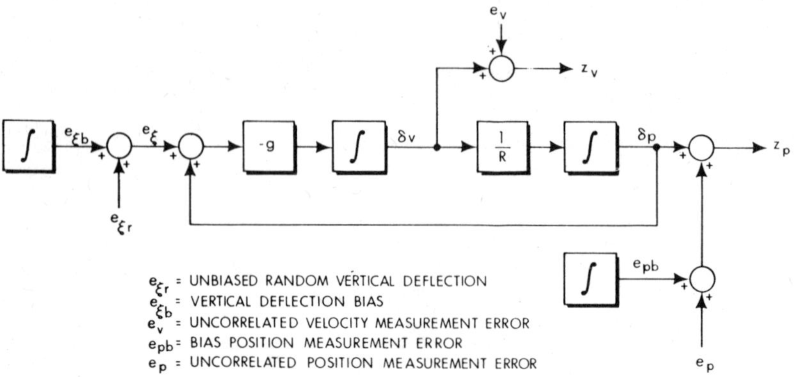

Figure 3-2

In this simplified model, gyro drift rate errors are ignored. It is desired to estimate the vertical deflection process, e_ξ. Consider the system state vector consisting of the four states:

$$\underline{x}^T = [e_{pb} \;\; \delta p \;\; \delta v \;\; e_{\xi b}]$$

(a) For position measurements only (z_v not available), set up the observability matrix and determine whether this system is observable. (b) Now assume we have both position and velocity measurements. Is this system observable? (c) Now assume $e_{pb} = 0$ and can be eliminated from the state error model. Is this system observable with position measurements only?

Problem 3-6

For the linear system whose dynamics are described by

$$F = \begin{bmatrix} 0 & 1 & 0 \\ 0 & 0 & 1 \\ 0 & 0 & -\alpha \end{bmatrix}$$

Show that for small αT, the state transition matrix is

$$\Phi(T) = \begin{bmatrix} 1 & T & \frac{T^2}{2} \\ 0 & 1 & T \\ 0 & 0 & 1 \end{bmatrix}$$

Problem 3-7

For the constant discrete n^{th}-order deterministic system driven by a scalar control

$$\underline{x}_{k+1} = \Phi \underline{x}_k + \underline{\lambda} u_k$$

Show that the controllability criterion is that the rank of

$$\Theta = \left[\underline{\lambda} \;\middle|\; \Phi\underline{\lambda} \;\middle|\; \cdots \;\middle|\; \Phi^{n-1}\underline{\lambda} \right]$$

be n where $\underline{x}_0$, $\underline{x}_n$ and the u_i are given. (*Hint*: Describe the vector $\underline{x}_n - \Phi^n \underline{x}_0$ in terms of $u_0, \ldots, u_{n-1}$ and powers of Φ.)

Problem 3-8

Show that for the stationary linear system described in Fig. 3-3, the state transition matrix is ($\Delta t \geqslant 0$)

$$\Phi(\Delta t) = \left[\begin{array}{c|c} e^{-\Delta t/T_1} & \dfrac{T_1 T_2}{T_1 - T_2}\left(e^{-\Delta t/T_1} - e^{-\Delta t/T_2}\right) \\ \hline 0 & e^{-\Delta t/T_2} \end{array}\right]$$

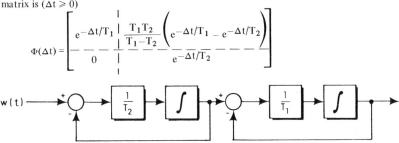

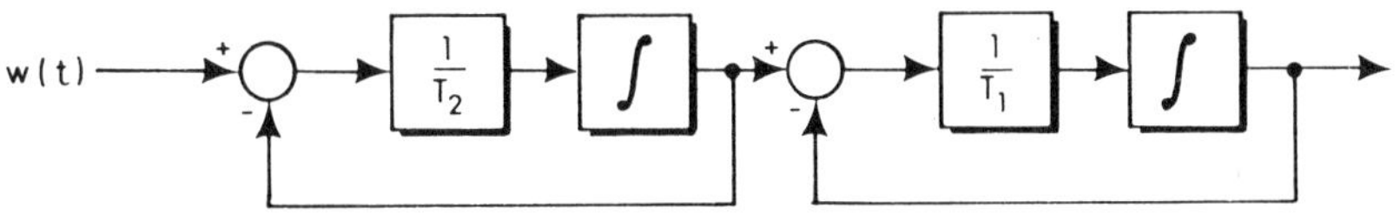

Figure 3-3

Problem 3-9

For the system illustrated in Fig. 3-4, show that

$$E\left[x_1^2(t)\right] = q_1 t$$

$$E\left[x_2^2(t)\right] = \frac{q_1 t^3}{3} + q_2 t$$

$$E\left[x_3^2(t)\right] = \frac{q_1 t^5}{20} + \frac{q_2 t^3}{3} + q_3 t$$

$$\vdots$$

$$E\left[x_n^2(t)\right] = \sum_{i=1}^{n} \frac{q_{n+1-i}\, t^{2i-1}}{(2i-1)(i-1)!(i-1)!}$$

where the white noise inputs w_i have spectral densities $q_i \delta(t)$.

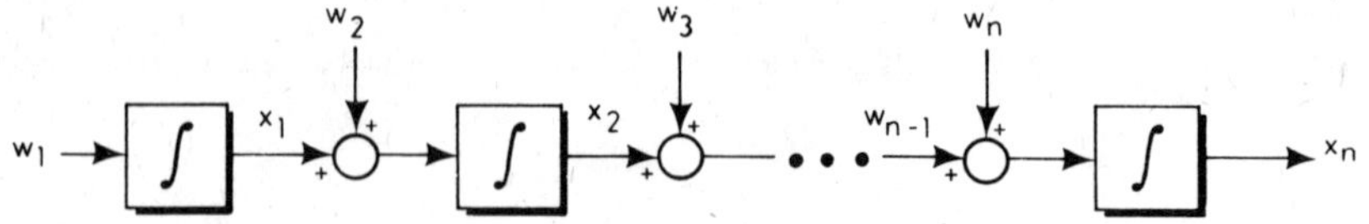

Figure 3-4

Problem 3-10

For the second-order system shown in Fig. 3-5, use the linear variance equation to obtain

$$E\left[x_1^2(t)\right] = \frac{q}{4\zeta\omega^3}$$

$$E\left[x_2^2(t)\right] = \frac{q}{4\zeta\omega}$$

where the white noise input w has spectral density $q\delta(t)$, in the steady state.

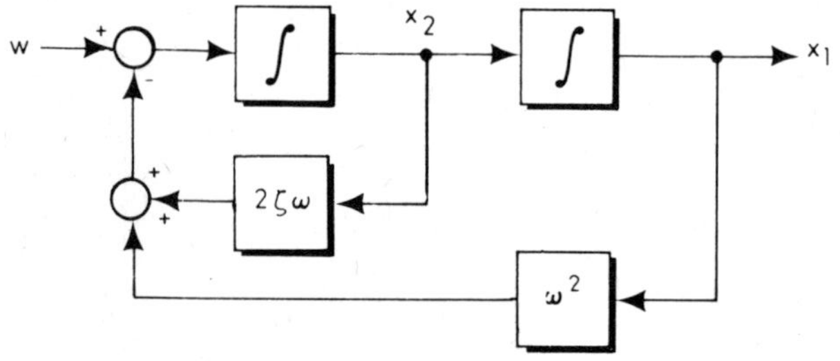

Figure 3-5

Problem 3-11

For the first-order system shown in Fig. 3-6, choose the gain K to minimize the mean square error between the command input r(t) and the system output c(t). Define a state vector $\underline{x}^T = [c \;\; r]$ and obtain the steady-state solution of the linear variance equation, P_{ss}. Define the quantity $e(t) = c(t) - r(t)$ and compute its steady-state mean square value as

$$\sigma_e^2 = [1 \;\; -1] \, P_{ss} \begin{bmatrix} 1 \\ -1 \end{bmatrix}$$

For $\sigma^2 = \beta = 1.0$ and $N = 0.5$, show that $K = 1.0$.

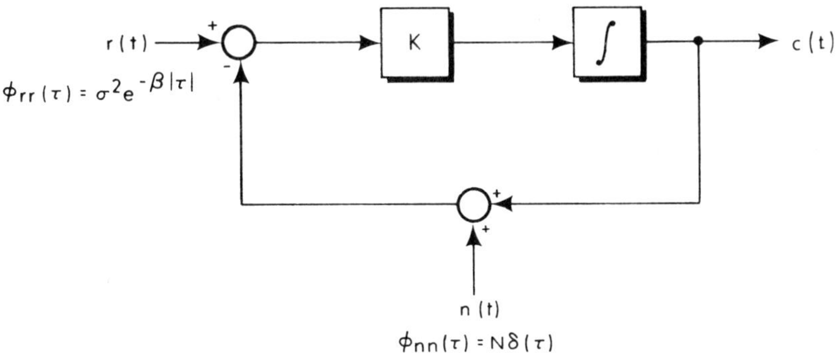

Figure 3-6

4. OPTIMAL LINEAR FILTERING

The preceding chapters discuss a number of properties of random processes and develop state-vector models of randomly excited linear systems. Now, we are ready to take up the principal topic of the book – namely, the estimation of a state vector from measurement data corrupted by noise. Optimal estimates that minimize the estimation error, in a well-defined statistical sense, are of particular interest. This chapter is devoted to the subject of optimal filtering for linear systems. *Filtering* refers to estimating the state vector at the current time, based upon all past measurements. *Prediction* refers to estimating the state at a future time; we shall see that prediction and filtering are closely related. Chapter 5 discusses *smoothing*, which means estimating the value of the state at some prior time, based on all measurements taken up to the current time.

Our approach to development of the optimal linear filter is to argue the *form* it should take, to specify a suitable *criterion of optimality* and to proceed directly to optimization of the assumed form. Before embarking upon this course, however, we *briefly* provide some background on the subjects of desirable *characteristics* of estimators in general, *alternative approaches* to optimal estimation and motivation for optimal *linear* estimators.

An estimate, $\hat{\underline{x}}$, is the computed value of a quantity, $\underline{x}$, based upon a set of measurements, $\underline{z}$. An *unbiased* estimate is one whose expected value is the same as that of the quantity being estimated. A *minimum variance* (unbiased) estimate has the property that its error variance is less than or equal to that of any other unbiased estimate. A *consistent* estimate is one which converges to the true value of $\underline{x}$, as the number of measurements increases. Thus, we shall look for unbiased, minimum variance, consistent estimators.

Let us assume that the set of ℓ measurements, $\underline{z}$, can be expressed as a linear combination of the n elements of a constant vector $\underline{x}$ plus a random, additive measurement error, $\underline{v}$. That is, the measurement process is modeled as

$$\underline{z} = H\underline{x} + \underline{v} \tag{4.0-1}$$

where $\underline{z}$ is an $\ell \times 1$ vector, $\underline{x}$ is an $n \times 1$ vector, H is an $\ell \times n$ matrix and $\underline{v}$ is an $\ell \times 1$ vector. For $\ell > n$ the measurement set contains redundant information. In *least-squares* estimation, one chooses as $\hat{\underline{x}}$ that value which minimizes the sum of squares of the deviations, $z_i - \hat{z}_i$; i.e., minimizes the quantity

$$J = (\underline{z} - H\hat{\underline{x}})^T (\underline{z} - H\hat{\underline{x}}) \tag{4.0-2}$$

The resulting least-squares estimate ($\ell \geqslant n$), found by setting $\partial J/\partial \hat{\underline{x}} = \underline{0}$ (Sec. 2.1), is

$$\hat{\underline{x}} = (H^T H)^{-1} H^T \underline{z} \tag{4.0-3}$$

If, instead, one seeks to minimize the weighted sum of squares of deviations,

$$J = (\underline{z} - H\hat{\underline{x}})^T R^{-1} (\underline{z} - H\hat{\underline{x}}) \tag{4.0-4}$$

where R^{-1} is an $\ell \times \ell$ symmetric, positive definite weighting matrix, the *weighted-least-squares* estimate

$$\hat{\underline{x}} = (H^T R^{-1} H)^{-1} H^T R^{-1} \underline{z} \tag{4.0-5}$$

is obtained. These results have no direct probabilistic interpretation; they were derived through deterministic argument only. Consequently, the least-squares estimates may be preferred to other estimates when there is no basis for assigning probability density functions to $\underline{x}$ and $\underline{z}$. Alternatively, one may use the *maximum likelihood* philosophy, which is to take as $\hat{\underline{x}}$ that value which maximizes the probability of the measurements $\underline{z}$ that actually occurred, taking into account known statistical properties of $\underline{v}$. There is still no statistical model assumed for the variable $\underline{x}$. In the simple example above, the conditional probability density function for $\underline{z}$, conditioned on a given value for $\underline{x}$, is just the density for $\underline{v}$ centered around $H\underline{x}$. With $\underline{v}$ taken as a zero mean, gaussian distributed observation with covariance matrix R, we have

$$p(\underline{z}|\underline{x}) = \frac{1}{(2\pi)^{\ell/2}|R|^{1/2}} \exp\left[-\frac{1}{2}(\underline{z} - H\underline{x})^T R^{-1} (\underline{z} - H\underline{x})\right] \tag{4.0-6}$$

To maximize $p(\underline{z}|\underline{x})$ we minimize the exponent in brackets. This is equivalent to minimizing the cost function in Eq. (4.0-4), although now a probabilistic basis for choosing R exists. The result, of course, is as given in Eq. (4.0-5). Still another approach is *Bayesian* estimation, where statistical models are available for both $\underline{x}$ and $\underline{z}$, and one seeks the *a posteriori* conditional density function, $p(\underline{x}|\underline{z})$, since it contains *all* the statistical information of interest. In general, $p(\underline{x}|\underline{z})$ is evaluated as (Bayes' theorem)

$$p(\underline{x}|\underline{z}) = \frac{p(\underline{z}|\underline{x})\, p(\underline{x})}{p(\underline{z})} \tag{4.0-7}$$

where $p(\underline{x})$ is the *a priori* probability density function of $\underline{x}$, and $p(\underline{z})$ is the probability density function of the measurements. Depending upon the criterion of optimality, one can compute $\hat{\underline{x}}$ from $p(\underline{x}|\underline{z})$. For example, if the object is to maximize the probability that $\hat{\underline{x}} = \underline{x}$, the solution is $\hat{\underline{x}}$ = mode* of $p(\underline{x}|\underline{z})$. When the *a priori* density function $p(\underline{x})$ is uniform (which implies no knowledge of $\underline{x}$ between its allowable limits), this estimate is equal to the maximum likelihood estimate. If the object is to find a generalized *minimum variance* Bayes' estimate, that is, to minimize the cost functional

$$J = \int_{-\infty}^{\infty}\int_{-\infty}^{\infty} \cdots \int_{-\infty}^{\infty} (\hat{\underline{x}} - \underline{x})^T S\, (\hat{\underline{x}} - \underline{x})\, p(\underline{x}|\underline{z})\, dx_1 dx_2 \ldots dx_n \tag{4.0-8}$$

where S is an arbitrary, positive semidefinite matrix, we simply set $\partial J/\partial \hat{\underline{x}} = \underline{0}$ to find, *independent of S*, that

$$\hat{\underline{x}} = \int_{-\infty}^{\infty}\int_{-\infty}^{\infty} \cdots \int_{-\infty}^{\infty} \underline{x}\, p(\underline{x}|\underline{z})\, dx_1 dx_2 \ldots dx_n = E[\underline{x}|\underline{z}] \tag{4.0-9}$$

which is the *conditional mean* estimate. Equation (4.0-8) has the characteristic structure

$$J = \int_{-\infty}^{\infty}\int_{-\infty}^{\infty} \cdots \int_{-\infty}^{\infty} L(\tilde{\underline{x}})\, p(\underline{x}|\underline{z})\, dx_1 dx_2 \ldots dx_n \tag{4.0-10}$$

where $L(\tilde{\underline{x}})$ is a scalar "loss function" of the estimation error

$$\tilde{\underline{x}} = \hat{\underline{x}} - \underline{x} \tag{4.0-11}$$

The result given in Eq. (4.0-9) holds for a wide variety of loss functions in addition to that used in Eq. (4.0-8), with some mild restrictions on the form of $p(\underline{x}|\underline{z})$; in all these cases the minimum variance Bayes' estimate (conditional mean) is also *the* optimal Bayes' estimate (Ref. 1). Assuming gaussian distributions for $\underline{x}$ and $\underline{v}$, the result of evaluating $E[\underline{x}|\underline{z}]$ in Eq. (4.0-9) is

*The *peak* value of $p(\underline{x}|\underline{z})$.

$$\hat{\underline{x}} = (P_0^{-1} + H^T R^{-1} H)^{-1} H^T R^{-1} \underline{z} \tag{4.0-12}$$

where P_0 is the *a priori* covariance matrix of $\underline{x}$.

In comparing the various estimation methods just discussed, we note that if there is little or no *a priori* information, P_0^{-1} is very small and Eq. (4.0-12) becomes Eq. (4.0-5). And if we argue that all measurement errors are uncorrelated (i.e., R is a diagonal matrix) and all errors have equal variance (i.e., $R = \sigma^2 I$), Eq. (4.0-5) reduces to Eq. (4.0-3). In his important work, Kalman (Ref. 2) formulated and solved the Wiener problem for gauss-markov sequences through use of state-space representation and the viewpoint of conditional distributions and expectations. His results also reduce to those given above. Therefore, the important conclusion is reached that for gaussian random variables, *identical results are obtained by all these methods as long as the assumptions are the same in each case* (Ref. 3). This property motivates us to consider primarily Bayesian minimum variance estimators.

Now we observe that $\hat{\underline{x}}$ in Eq. (4.0-12) is a *linear* operation on the measurement data. Furthermore, it is proven elsewhere (Ref. 4) that, for a gaussian time-varying signal, the optimal (minimum mean square error) predictor is a linear predictor. Additionally, as a practical fact, most often all we know about the characterization of a given random process is its autocorrelation function. But there always exists a gaussian random process possessing the same autocorrelation function; we therefore might as well assume that the given random process is itself gaussian. That is, the two processes are indistinguishable from the standpoint of the amount of knowledge postulated. On the basis of these observations we are led to consider the optimal estimator as a linear operator in most applications.

Henceforth, unless stated otherwise, the term *optimal estimator* refers to one which minimizes the mean square estimation error. Next, we consider the recursive form of the linear estimator, which applies to gaussian random sequences. The discrete time problem is considered first; the continuous time equivalent is then obtained through a simple limiting procedure. Intuitive concepts, situations of special interest and examples comprise the remainder of the chapter.

4.1 RECURSIVE FILTERS

A recursive filter is one in which there is no need to store past measurements for the purpose of computing present estimates. This concept is best demonstrated by the following example.

Example 4.1-1

Consider the problem of estimating a scalar nonrandom* constant, x, based on k noise-corrupted measurements, z_i, where $z_i = x + v_i$ ($i = 1, 2, \ldots k$). Here v_i represents the

*x is unknown and has no defined statistical properties.

measurement noise, which we assume to be a white sequence. An unbiased, minimum variance estimate $\hat{x}_k$ results from *averaging* the measurements (this can be shown from Eq. (4.0-3)); thus, we choose

$$\hat{x}_k = \frac{1}{k} \sum_{i=1}^{k} z_i \tag{4.1-1}$$

When an additional measurement becomes available we have, as the new estimate

$$\hat{x}_{k+1} = \frac{1}{k+1} \sum_{i=1}^{k+1} z_i \tag{4.1-2}$$

This expression can be manipulated to evidence the prior estimate, viz:

$$\hat{x}_{k+1} = \frac{k}{k+1} \left(\frac{1}{k} \sum_{i=1}^{k} z_i \right) + \frac{1}{k+1} z_{k+1} = \frac{k}{k+1} \hat{x}_k + \frac{1}{k+1} z_{k+1} \tag{4.1-3}$$

Hence, by employing Eq. (4.1-3) rather than Eq. (4.1-2) to compute $\hat{x}_{k+1}$, the need to store past measurements is eliminated – all previous information is embodied in the prior estimate (plus the measurement index, k) – and we have a *recursive*, linear estimator. Note that Eq. (4.1-3) can be written in the alternative recursive form

$$\hat{x}_{k+1} = \hat{x}_k + \frac{1}{k+1} (z_{k+1} - \hat{x}_k)$$

in which the new estimate is given by the prior estimate plus an appropriately weighted difference between the new measurement and its expected value, given by the prior estimate. The quantity $z_{k+1} - \hat{x}_k$ is called the measurement *residual*.

In the example we dealt with scalar quantities; generalization of the concept to vector quantities proceeds directly. Consider a discrete system whose state at time t_k is denoted by $\underline{x}(t_k)$ or simply $\underline{x}_k$, where $\underline{w}_k$ is a zero mean, white sequence of covariance Q_k,

$$\underline{x}_k = \Phi_{k-1} \underline{x}_{k-1} + \underline{w}_{k-1} \tag{4.1-4}$$

Measurements are taken as linear combinations of the system state variables, corrupted by uncorrelated noise. The measurement equation is written in vector-matrix notation as

$$\underline{z}_k = H_k \underline{x}_k + \underline{v}_k \tag{4.1-5}$$

where $\underline{z}_k$ is the set of ℓ measurements at time t_k, namely, $z_{1k}, z_{2k}, \ldots, z_{\ell k}$, arranged in vector form

$$\underline{z}_k = \begin{bmatrix} z_{1k} \\ z_{2k} \\ \vdots \\ z_{\ell k} \end{bmatrix} \tag{4.1-6}$$

H_k is the measurement matrix at time t_k; it describes the linear combinations of state variables which comprise $\underline{z}_k$ in the absence of noise. The dimension of the measurement matrix is $\ell \times n$, corresponding to ℓ-dimensioned measurements of an n-dimensioned state. The term $\underline{v}_k$ is a vector of random noise quantities (zero mean, covariance R_k) corrupting the measurements.

Given a *prior* estimate of the system state at time t_k, denoted $\hat{\underline{x}}_k(-)$, we seek an updated estimate, $\hat{\underline{x}}_k(+)$, based on use of the measurement, $\underline{z}_k$. In order to avoid a growing memory filter, this estimate is *sought* in the linear, recursive form*

$$\hat{\underline{x}}_k(+) = K_k'\hat{\underline{x}}_k(-) + K_k\underline{z}_k \tag{4.1-7}$$

where K_k' and K_k are time-varying weighting matrices, as yet unspecified. Although the following derivation is for an assumed recursive, single-stage filter, the result has been shown to be the solution for a more general problem. If $\underline{w}_k$, $\underline{v}_k$ are gaussian, the filter we will find is *the* optimal multi-stage filter; a nonlinear filter cannot do better (Ref. 2). In other cases we will simply have determined the optimal *linear* filter.

4.2 DISCRETE KALMAN FILTER

It is possible to derive the Kalman filter by optimizing the assumed form of the linear estimator. An equation for the *estimation error* after incorporation of the measurement can be obtained from Eq. (4.1-7) through substitution of the measurement equation [Eq. (4.1-5)] and the defining relations (tilde denotes estimation error)

$$\begin{aligned} \hat{\underline{x}}_k(+) &= \underline{x}_k + \tilde{\underline{x}}_k(+) \\ \hat{\underline{x}}_k(-) &= \underline{x}_k + \tilde{\underline{x}}_k(-) \end{aligned} \tag{4.2-1}$$

The result is

$$\tilde{\underline{x}}_k(+) = [K_k' + K_kH_k - I]\underline{x}_k + K_k'\tilde{\underline{x}}_k(-) + K_k\underline{v}_k \tag{4.2-2}$$

By definition $E[\underline{v}_k] = \underline{0}$. Also, if $E[\tilde{\underline{x}}_k(-)] = \underline{0}$, this estimator will be *unbiased* (i.e., $E[\tilde{\underline{x}}_k(+)] = \underline{0}$) for any given state vector $\underline{x}_k$ only if the term in square brackets is zero. Thus, we *require*

*Throughout the text, (−) and (+) are used to denote the times immediately *before* and immediately *after* a discrete measurement, respectively.

$$K'_k = I - K_k H_k \tag{4.2-3}$$

and the estimator takes the form

$$\hat{\underline{x}}_k(+) = (I - K_k H_k)\hat{\underline{x}}_k(-) + K_k \underline{z}_k \tag{4.2-4}$$

or alternatively,

$$\hat{\underline{x}}_k(+) = \hat{\underline{x}}_k(-) + K_k \left[\underline{z}_k - H_k \hat{\underline{x}}_k(-)\right] \tag{4.2-5}$$

The corresponding estimation error is, from Eqs. (4.1-5), (4.2-1) and (4.2-5),

$$\tilde{\underline{x}}_k(+) = (I - K_k H_k)\, \tilde{\underline{x}}_k(-) + K_k \underline{v}_k \tag{4.2-6}$$

Error Covariance Update – Using Eq. (4.2-6) the expression for the change in the error covariance matrix when a measurement is employed can be derived. From the definition

$$P_k(+) = E\left[\tilde{\underline{x}}_k(+)\, \tilde{\underline{x}}_k(+)^T\right] \tag{4.2-7}$$

Eq. (4.2-6) gives

$$\begin{aligned} P_k(+) = E\Big\{&(I - K_k H_k)\tilde{\underline{x}}_k(-)\left[\tilde{\underline{x}}_k(-)^T(I - K_k H_k)^T + \underline{v}_k^T K_k^T\right] \\ &+ K_k \underline{v}_k \left[\tilde{\underline{x}}_k(-)^T(I - K_k H_k)^T + \underline{v}_k^T K_k^T\right]\Big\} \end{aligned} \tag{4.2-8}$$

By definition,

$$E\left[\tilde{\underline{x}}_k(-)\tilde{\underline{x}}_k(-)^T\right] = P_k(-) \tag{4.2-9}$$

$$E\left[\underline{v}_k \underline{v}_k^T\right] = R_k \tag{4.2-10}$$

and, as a result of measurement errors being uncorrelated,

$$E\left[\tilde{\underline{x}}_k(-)\, \underline{v}_k^T\right] = E\left[\underline{v}_k\, \tilde{\underline{x}}_k(-)^T\right] = 0 \tag{4.2-11}$$

Thus

$$P_k(+) = (I - K_k H_k)\, P_k(-)\, (I - K_k H_k)^T + K_k R_k K_k^T \tag{4.2-12}$$

Optimum Choice of K_k – The criterion for choosing K_k is to minimize a weighted scalar sum of the diagonal elements of the error covariance matrix $P_k(+)$. Thus, for the cost function we choose

$$J_k = E\left[\tilde{\underline{x}}_k(+)^T\, S\, \tilde{\underline{x}}_k(+)\right] \tag{4.2-13}$$

where S is any positive semidefinite matrix. As demonstrated in Eq. (4.0-9), the optimal estimate is independent of S; hence, we may as well choose S = I, yielding

$$J_k = \text{trace}\,[P_k(+)] \tag{4.2-14}$$

This is equivalent to minimizing the *length* of the estimation error vector. To find the value of K_k which provides a minimum, it is necessary to take the partial derivative of J_k with respect to K_k and equate it to zero. Use is made of the relation for the partial derivative of the trace of the product of two matrices A and B (with B symmetric),

$$\frac{\partial}{\partial A}\,[\text{trace}\,(ABA^T)] = 2AB$$

From Eqs. (4.2-12) and (4.2-13) the result is

$$-2\,(I - K_kH_k)\,P_k\,(-)\,H_k{}^T + 2\,K_kR_k = 0$$

Solving for K_k,

$$K_k = P_k\,(-)\,H_k{}^T\,[H_kP_k\,(-)\,H_k{}^T + R_k]^{-1} \tag{4.2-15}$$

which is referred to as the *Kalman gain matrix*. Examination of the Hessian of J_k reveals that this value of K_k does indeed *minimize* J_k [Eq. (2.1-79)].

Substitution of Eq. (4.2-15) into Eq. (4.2-12) gives, after some manipulation,

$$P_k(+) = P_k(-) - P_k(-)\,H_k{}^T\,[H_kP_k(-)\,H_k{}^T + R_k]^{-1}\,H_kP_k(-) \tag{4.2-16a}$$

$$= [I - K_kH_k]\,P_k(-) \tag{4.2-16b}$$

which is the optimized value of the updated estimation error covariance matrix.

Thus far we have described the discontinuous state estimate and error covariance matrix behavior *across* a measurement. The extrapolation of these quantities *between* measurements is (Section 3.7)

$$\hat{\underline{x}}_k\,(-) = \Phi_{k-1}\,\hat{\underline{x}}_{k-1}(+) \tag{4.2-17}$$

$$P_k\,(-) = \Phi_{k-1}P_{k-1}(+)\,\Phi_{k-1}{}^T + Q_{k-1} \tag{4.2-18}$$

See Fig. 4.2-1 for a "timing diagram" of the various quantities involved in the discrete optimal filter equations. The equations of the discrete Kalman filter are summarized in Table 4.2-1.

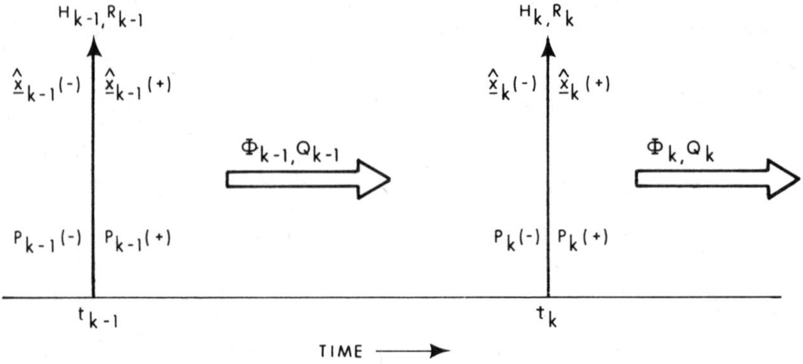

Figure 4.2-1 Discrete Kalman Filter Timing Diagram

Figure 4.2-2 illustrates these equations in block diagram form. The Kalman filter to be *implemented* appears outside the dashed-line box. That appearing inside the box is simply a mathematical abstraction – a *model* of what we think the system and measurement processes are. Of course, the Kalman filter we implement is based upon this model. Chapters 7 and 8 dwell on the practical consequences of this fact. It is often said that the Kalman filter generates its own *error analysis*. Clearly, this refers to the computation of P_k, which provides an indication of the accuracy of the estimate. Again, Chapters 7 and 8 explore the practical meaning of P_k in view of modeling errors and other unavoidable factors.

In the linear, discrete Kalman filter, calculations at the covariance level ultimately serve to provide K_k, which is then used in the calculation of mean values (i.e., the estimate $\hat{\underline{x}}_k$). There is no feedback from the state equations to

TABLE 4.2-1 SUMMARY OF DISCRETE KALMAN FILTER EQUATIONS

System Model	$\underline{x}_k = \Phi_{k-1}\underline{x}_{k-1} + \underline{w}_{k-1}, \quad \underline{w}_k \sim N(\underline{0}, Q_k)$
Measurement Model	$\underline{z}_k = H_k\underline{x}_k + \underline{v}_k, \quad \underline{v}_k \sim N(\underline{0}, R_k)$
Initial Conditions	$E[\underline{x}(0)] = \hat{\underline{x}}_0, \; E[(\underline{x}(0) - \hat{\underline{x}}_0)(\underline{x}(0) - \hat{\underline{x}}_0)^T] = P_0$
Other Assumptions	$E[\underline{w}_k\underline{v}_j^T] = 0$ for all j, k
State Estimate Extrapolation	$\hat{\underline{x}}_k(-) = \Phi_{k-1}\hat{\underline{x}}_{k-1}(+)$
Error Covariance Extrapolation	$P_k(-) = \Phi_{k-1}P_{k-1}(+)\,\Phi_{k-1}^T + Q_{k-1}$
State Estimate Update	$\hat{\underline{x}}_k(+) = \hat{\underline{x}}_k(-) + K_k[\underline{z}_k - H_k\hat{\underline{x}}_k(-)]$
Error Covariance Update	$P_k(+) = [I - K_kH_k]\,P_k(-)$
Kalman Gain Matrix	$K_k = P_k(-)\,H_k^T[H_kP_k(-)H_k^T + R_k]^{-1}$

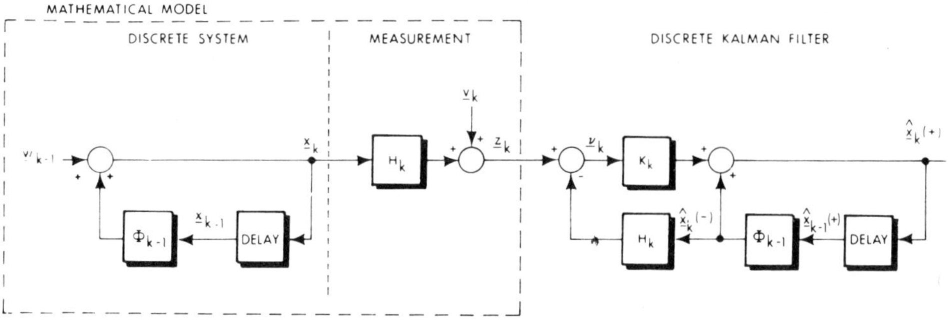

Figure 4.2-2 System Model and Discrete Kalman Filter

the covariance equations. This is illustrated in Fig. 4.2-3, which is essentially a simplified computer flow diagram of the discrete Kalman filter.

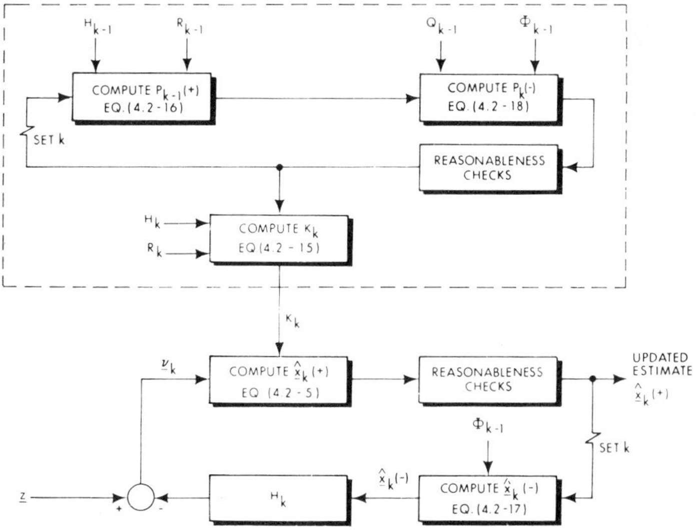

Figure 4.2-3 Discrete Kalman Filter Information Flow Diagram

A Simpler Form for K_k – There is a matrix inversion relationship which states that, for $P_k(+)$ as given in Eq. (4.2-16a), $P_k^{-1}(+)$ is expressible as:

$$P_k^{-1}(+) = P_k^{-1}(-) + H_k^T R_k^{-1} H_k \tag{4.2-19}$$

This relationship can easily be verified by showing that $P_k(+)\,P_k^{-1}(+) = I$. We use this result to manipulate K_k as follows,

$$K_k = [P_k(+)\,P_k^{-1}(+)]\,P_k(-)\,H_k^T\,[H_kP_k(-)H_k^T + R_k]^{-1}$$
$$= P_k(+)[P_k^{-1}(-) + H_k^TR_k^{-1}H_k]\,P_k(-)H_k^T\,[H_kP_k(-)H_k^T + R_k]^{-1}$$

Expanding and collecting terms yields

$$K_k = P_k(+)H_k^T\,[I + R_k^{-1}H_kP_k(-)H_k^T]\,[H_kP_k(-)H_k^T + R_k]^{-1}$$
$$= P_k(+)H_k^T\,R_k^{-1} \tag{4.2-20}$$

which is the simpler form sought.

A Property of the Optimal Estimator – We may verify by direct calculation that

$$E[\underline{\hat{x}}_k(+)\underline{\tilde{x}}_k(+)^T] = 0 \tag{4.2-21}$$

That is, the optimal estimate and its error are uncorrelated (orthogonal). The state and its estimate at time $t_1(+)$ are

$$\underline{x}_1 = \Phi_0\underline{x}_0 + \underline{w}_0 \tag{4.2-22}$$

and

$$\underline{\hat{x}}_1(+) = \Phi_0\underline{\hat{x}}_0 + K_1\,[\underline{z}_1 - H_1\,\Phi_0\underline{\hat{x}}_0]$$
$$= \Phi_0\underline{\hat{x}}_0 + K_1\,[-H_1\,\Phi_0\underline{\tilde{x}}_0 + H_1\underline{w}_0 + \underline{v}_1] \tag{4.2-23}$$

respectively, where the measurement equation, $\underline{z}_1 = H_1\underline{x}_1 + \underline{v}_1$ has been employed. Subtracting Eq. (4.2-22) from Eq. (4.2-23) yields an equation in $\underline{\tilde{x}}_1$ – i.e.,

$$\underline{\tilde{x}}_1(+) = (\Phi_0 - K_1H_1\Phi_0)\underline{\tilde{x}}_0 + (K_1H_1 - I)\underline{w}_0 + K_1\underline{v}_1 \tag{4.2-24}$$

Since $E[\underline{\hat{x}}_0\underline{\tilde{x}}_0^T] = 0$, we directly calculate the quantity of interest as

$$E[\underline{\hat{x}}_1(+)\underline{\tilde{x}}_1(+)^T] = E\Big\{[\Phi_0\underline{\hat{x}}_0 + K_1(-H_1\Phi_0\underline{\tilde{x}}_0 + H_1\underline{w}_0 + \underline{v}_1)]\,\cdot$$
$$[\underline{\tilde{x}}_0^T(\Phi_0 - K_1H_1\Phi_0)^T + \underline{w}_0^T(K_1H_1 - I)^T + \underline{v}_1^TK_1^T]\Big\}$$
$$= -K_1H_1\Phi_0P_0(\Phi_0^T - \Phi_0^TH_1^TK_1^T) + K_1H_1Q_0(H_1^TK_1^T - I)$$
$$+ K_1R_1K_1^T$$

$$
\begin{aligned}
E[\hat{\underline{x}}_1(+)\tilde{\underline{x}}_1(+)^T] &= -K_1H_1(\Phi_0P_0\Phi_0{}^T + Q_0) + K_1H_1(\Phi_0P_0\Phi_0{}^T + Q_0)H_1{}^TK_1{}^T \\
&\quad + K_1R_1K_1{}^T \\
&= -K_1H_1P_1(-)[I - K_1{}^TH_1{}^T] + K_1R_1K_1{}^T \\
&= K_1(-H_1P_1(+)^T + R_1K_1{}^T) \\
&= 0
\end{aligned}
$$

where use has been made of Eq. (4.2-20). We may now repeat the process for $E[\hat{\underline{x}}_2(+)\tilde{\underline{x}}_2(+)^T]$ and, by induction, verify Eq. (4.2-21).

Kalman Filter Examples – Several applications of optimal filtering are presented in this section. Since analytic results are sought, only relatively low-order systems are considered. However, it is well to note that, in practice, Kalman optimal filters for 50th order systems (and even higher) have been considered, and 10th and 20th order Kalman filters are relatively common. With increasing availability of small, powerful digital computers, the main factors which limit the size of implemented Kalman filters are no longer computer limitations *per se*, but rather *modeling errors* and associated *sensitivity problems.* These are discussed in Chapters 7 and 8.

Example 4.2-1

Estimate the value of a constant x, given discrete measurements of x corrupted by an uncorrelated gaussian noise sequence with zero mean and variance r_0.

The scalar equations describing this situation are

$$x_{k+1} = x_k \qquad \text{System}$$

$$z_k = x_k + v_k \qquad \text{Measurement}$$

where

$$v_k \sim N(0, r_0)$$

that is, v_k is an uncorrelated gaussian noise sequence with zero mean and variance r_0. For this problem, $\varphi_0(t_k,t_j) = h = 1$ and $q_k = 0$ yielding the variance propagation equation

$$p_{k+1}(-) = p_k(+)$$

and single-stage optimal update equation

$$
\begin{aligned}
p_{k+1}(+) &= p_{k+1}(-) - p_{k+1}(-)\,[p_{k+1}(-) + r_0]^{-1}p_{k+1}(-) \\
&= \frac{p_{k+1}(-)}{1 + \dfrac{p_{k+1}(-)}{r_0}}
\end{aligned}
$$

$$p_{k+1}(+) = \frac{p_k(+)}{1 + \dfrac{p_k(+)}{r_0}}$$

This difference equation can be solved, starting with $p_0(+) = p_0$, by noting that

$$p_1(+) = \frac{p_0}{1 + \dfrac{p_0}{r_0}}$$

$$p_2(+) = \frac{p_1(+)}{1 + \dfrac{p_1(+)}{r_0}} = \frac{p_0}{1 + \dfrac{2p_0}{r_0}}$$

$$\vdots$$

$$p_k(+) = \frac{p_0}{1 + \dfrac{p_0}{r_0} k}$$

The discrete optimal filter for this example problem is

$$\hat{x}_{k+1} = \hat{x}_k + \frac{\dfrac{p_0}{r_0}}{1 + \dfrac{p_0}{r_0} k} [z_k - \hat{x}_k]$$

For sufficiently large k, $\hat{x}_{k+1} = \hat{x}_k = \hat{x}$ as new measurements provide essentially no new information.

Example 4.2-2

Consider two unknown but *correlated* constants, x_1 and x_2. We wish to determine the improvement in knowledge of x_1 which is possible through processing a single noisy measurement of x_2.

The vector and matrix quantities of interest in this example are:

$$\underline{x} = \begin{bmatrix} x_1 \\ x_2 \end{bmatrix}, \quad H = [0\ 1], \quad P(-) = \begin{bmatrix} p_{11}(-) & p_{12}(-) \\ p_{12}(-) & p_{22}(-) \end{bmatrix} = \begin{bmatrix} \sigma_1^2 & \sigma_{12}^2 \\ \sigma_{12}^2 & \sigma_2^2 \end{bmatrix}$$

where P(–) is the covariance matrix describing the uncertainty in $\underline{x}$ before the measurement, that is, σ_1^2 is the initial mean square error in knowledge of x_1, σ_2^2 is the initial mean square error in knowledge of x_2, and σ_{12}^2 measures the corresponding cross-correlation. Computing the *updated* covariance matrix, P(+), according to Eq. (4.2-16), yields

$$P(+) = \begin{bmatrix} p_{11}(+) & p_{12}(+) \\ p_{12}(+) & p_{22}(+) \end{bmatrix} = \left[\begin{array}{c|c} \sigma_1^2 \left(\dfrac{\sigma_2^2(1-\rho^2) + r_2}{\sigma_2^2 + r_2} \right) & \sigma_{12}^2 \left(\dfrac{r_2}{\sigma_2^2 + r_2} \right) \\ \hline \sigma_{12}^2 \left(\dfrac{r_2}{\sigma_2^2 + r_2} \right) & \sigma_2^2 \left(\dfrac{r_2}{\sigma_2^2 + r_2} \right) \end{array}\right]$$

where r_2 denotes the measurement noise covariance, and ρ is the correlation coefficient defined by [Eq. (3.6-7)]

$$\rho = \frac{\sigma_{12}^2}{\sigma_1 \sigma_2}$$

A few limiting cases are worth examining. First, in the case where the measurement is perfect (i.e., $r_2 = 0$), the final uncertainty in the estimate of x_2, $p_{22}(+)$, is, of course, zero. Also, when $\rho = 0$, the final uncertainty in the estimate of x_1 is, as expected, equal to the initial uncertainty; nothing can be learned from the measurement in this case. Finally, in the case where $\rho = \pm 1$, the final uncertainty in the estimate of x_1 is given by

$$p_{11}(+) = \sigma_1^2 \left(\frac{1}{1 + \sigma_2^2/r_2} \right)$$

and the amount of information gained [i.e., the reduction in $p_{11}(+)$] depends upon the ratio of initial mean square error in knowledge of x_2 to the mean square error in measurement of x_2. All of these results are intuitively satisfying.

Example 4.2-3

Omega is a world-wide navigation system, utilizing phase comparison of 10.2 kHz continuous-wave radio signals. The user employs a propagation correction, designed to account for *globally* predictable phenomena (diurnal effects, earth conductivity variations, etc.), to bring theoretical phase values into agreement with observed phase measurements. The residual Omega phase errors are known to exhibit correlation over large distances (e.g., > 1000 nm). Design a data processor that, operating in a limited geographic area, can process error data gathered at two measurement sites and infer best estimates of phase error at other locations nearby.

In the absence of other information, assume that the phase errors are well-modeled as a zero-mean markov process in space, with variance σ_φ^2. Further assume that the phase error process is isotropic – i.e., it possesses the same statistics in all directions, and that the measurement error is very small.

We may now proceed in the usual way. First, denote by φ_2 and φ_3 the two phase error measurements, and by φ_1 the phase error to be estimated. Then we can form $\underline{x}$ and $\underline{z}$ as

$$\underline{x} = \begin{bmatrix} \varphi_1 \\ \varphi_2 \\ \varphi_3 \end{bmatrix}, \quad \underline{z} = \begin{bmatrix} \varphi_2 \\ \varphi_3 \end{bmatrix}$$

whence it follows that

$$H = \begin{bmatrix} 0 & 1 & 0 \\ 0 & 0 & 1 \end{bmatrix}$$

Also, by definition

$$P(0) = \sigma_\varphi^2 \left[\begin{array}{c|c|c} 1 & e^{-r_{12}/d} & e^{-r_{13}/d} \\ \hline e^{-r_{12}/d} & 1 & e^{-r_{23}/d} \\ \hline e^{-r_{13}/d} & e^{-r_{23}/d} & 1 \end{array} \right]$$

where r_{ij} is the distance between sites i and j, and d is the correlation distance of the phase error random process. The best estimator (in the sense of minimum mean square error) is given by

$$\hat{\underline{x}}(+) = K\underline{z}$$

since $\hat{\underline{x}}(-) = \underline{0}$. K is given by

$$K = P(0)\, H^T \left(HP(0)\, H^T\right)^{-1}$$

and the following result is obtained:

$$\hat{\varphi}_1 = \frac{1}{1 - e^{-2r_{23}/d}} \left\{ \begin{array}{l} \varphi_2 \left[e^{-r_{12}/d} - e^{-(r_{13} + r_{23})/d}\right] \\ \quad + \varphi_3 \left[e^{-r_{13}/d} - e^{-(r_{12} + r_{23})/d}\right] \end{array} \right\}$$

$$\hat{\varphi}_2 = \varphi_2$$

$$\hat{\varphi}_3 = \varphi_3$$

The first equation shows that $\hat{\varphi}_1$ is computed as a weighted sum of the perfect measurements. The last two equations are, of course, a direct consequence of the assumed perfect measurements.

This technique has been successfully employed with actual Omega data; see Fig. 4.2-4.

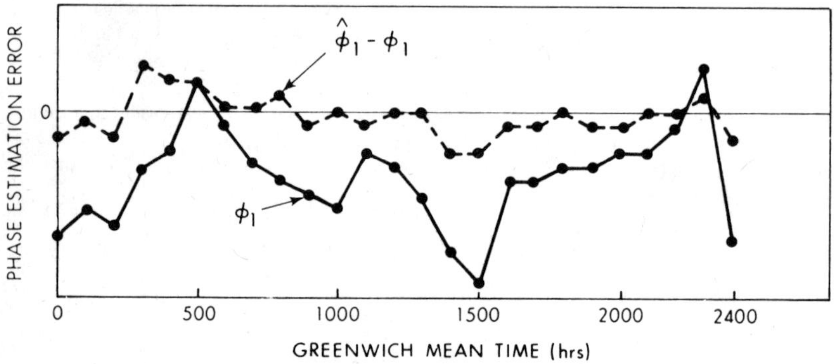

Figure 4.2-4 Phase Estimation Error – Kings Point, N.Y. (Ref. 14)

Example 4.2-4

Study the design of an optimal data processing scheme by which an Inertial Navigation System (INS), in a balloon launched ionospheric probe, can be initialized in early flight using radio position updates.

In view of the intended application, it is appropriate to choose a model of INS error dynamics valid only over a several minute period. Taking single-axis errors into consideration, we may write

$$\delta p(t) \approx \delta p(0) + \delta v(0)t + \delta a(0)\,\frac{t^2}{2}$$

where $\delta p(0)$, $\delta v(0)$ and $\delta a(0)$ represent initial values of position, velocity and acceleration errors, and higher-order terms have been dropped. Thus, we may write, as the state equations for this system

$$\delta \dot{p} = \delta v$$

$$\delta \dot{v} = \delta a$$

$$\delta \dot{a} = 0$$

or, equivalently,

$$\begin{bmatrix} \delta \dot{p} \\ \delta \dot{v} \\ \delta \dot{a} \end{bmatrix} = \underbrace{\begin{bmatrix} 0 & 1 & 0 \\ 0 & 0 & 1 \\ 0 & 0 & 0 \end{bmatrix}}_{F} \underbrace{\begin{bmatrix} \delta p \\ \delta v \\ \delta a \end{bmatrix}}_{\underline{x}(t)}$$

The radio position error is indicated by $e_p(t)$, see Fig. 4.2-5.

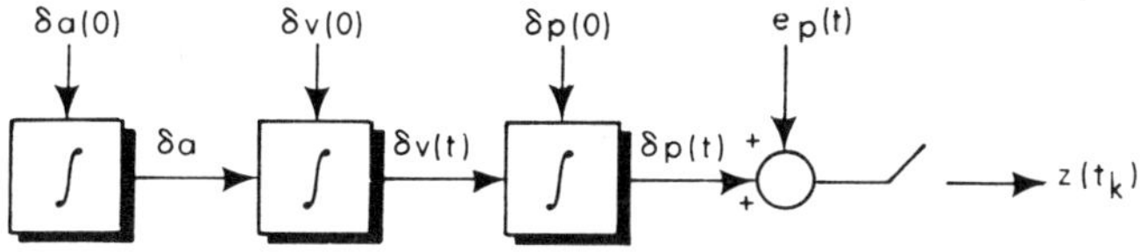

Figure 4.2-5 System and Discrete Measurement Model

The transition matrix for this time-invariant system is readily computed as

$$\Phi(t,0) = e^{Ft}$$

$$= I + Ft + F^2 \frac{t^2}{2!} + F^3 \frac{t^3}{3!} + \ldots$$

$$= \begin{bmatrix} 1 & 0 & 0 \\ 0 & 1 & 0 \\ 0 & 0 & 1 \end{bmatrix} + \begin{bmatrix} 0 & 1 & 0 \\ 0 & 0 & 1 \\ 0 & 0 & 0 \end{bmatrix} t + \begin{bmatrix} 0 & 0 & 1 \\ 0 & 0 & 0 \\ 0 & 0 & 0 \end{bmatrix} \frac{t^2}{2} + 0$$

$$= \begin{bmatrix} 1 & t & t^2/2 \\ 0 & 1 & t \\ 0 & 0 & 1 \end{bmatrix}$$

For any initial estimate $\hat{\underline{x}}(0)$ of the system state vector, we may compute the expected value of $\hat{\underline{x}}(T^-)$, just prior to the first fix, as follows:

$$\hat{\underline{x}}(T^-) = \Phi(T,0)\hat{\underline{x}}(0)$$

$$= \begin{bmatrix} \delta\hat{p}(0) + \delta\hat{v}(0)T + \delta\hat{a}(0)T^2/2 \\ \delta\hat{v}(0) + \delta\hat{a}(0)T \\ \delta\hat{a}(0) \end{bmatrix}$$

Also, corresponding to an initially diagonal covariance matrix P(0), the value of $P(T^-)$ is computed according to (overbar denotes ensemble expectation)

$$P(T^-) = \Phi(T,0)\, P(0)\, \Phi^T(T,0)$$

$$= \left[\begin{array}{c|c|c} \overline{\delta p^2(0)} + \overline{\delta v^2(0)}T^2 + \overline{\delta a^2(0)}T^4/4 & \overline{\delta v^2(0)}T + \overline{\delta a^2(0)}T^3/2 & \overline{\delta a^2(0)}T^2/2 \\ \hline & \overline{\delta v^2(0)} + \overline{\delta a^2(0)}T^2 & \overline{\delta a^2(0)}T \\ \hline & & \overline{\delta a^2(0)} \end{array}\right]$$

which is *symmetric*, but *not diagonal.* That is, position error is now correlated with velocity and acceleration errors.

The scalar position measurement is obtained by forming the difference between radio-indicated and inertially-indicated positions, viz:

$$z = p_{ind}(\text{radio}) - p_{ind}(\text{INS})$$

$$= p_t + e_p - (p_t + \delta p)$$

$$= -\delta p + e_p$$

where the indicated quantities, p_{ind}, have been described in terms of their true value, p_t, plus error components. This is equivalent to writing

$$z(T) = \underbrace{[-1 \quad 0 \quad 0]}_{H} \begin{bmatrix} \delta p(T) \\ \delta v(T) \\ \delta a(T) \end{bmatrix} + \underbrace{e_p(T)}_{v}$$

which enables identification of H and the measurement error. Let us assume that the radio position errors are stationary and uncorrelated from fix to fix. The effect of a fix at time T is computed according to ($\sigma_p^2 = \overline{e_p^2}$):

$$P(T^+) = P(T^-) - P(T^-)H^T[HP(T^-)H^T + R]^{-1}HP(T^-)$$

$$= P(T^-) - \frac{1}{p_{11}(T^-) + \sigma_p^2} \left[\begin{array}{c|c|c} p_{11}^2(T^-) & p_{11}(T^-)p_{12}(T^-) & p_{11}(T^-)p_{13}(T^-) \\ \hline & p_{12}^2(T^-) & p_{12}(T^-)p_{13}(T^-) \\ \hline & & p_{13}^2(T^-) \end{array}\right]$$

The upper left corner element of $P(T^+)$ is, for example,

$$p_{11}(T^+) = p_{11}(T^-) - \frac{p_{11}^2(T^-)}{p_{11}(T^-) + \sigma_p^2} \tag{4.2-25}$$

$$\approx \sigma_p^2$$

under the assumption $p_{11}(T^-) \gg \sigma_p^2$. Thus, the first position fix reduces the INS position error to approximately the fix error (actually, somewhat *below* it). Similar calculations show that the velocity and acceleration errors are essentially unchanged. The *next* position fix reduces the INS velocity error. The Kalman gain at time t_k for this system is

$$K_k = P_k(+)\, H_k^T R_k^{-1}$$

$$= \begin{bmatrix} p_{11}(t_k^+) & p_{12}(t_k^+) & p_{13}(t_k^+) \\ & p_{22}(t_k^+) & p_{23}(t_k^+) \\ & & p_{33}(t_k^+) \end{bmatrix} \begin{bmatrix} -1 \\ 0 \\ 0 \end{bmatrix} \frac{1}{\sigma_p^2}$$

$$= \begin{bmatrix} k_1(t_k) \\ k_2(t_k) \\ k_3(t_k) \end{bmatrix}$$

In the optimal filter, illustrated in Fig. 4.2-6, the sampler is understood to represent an impulse modulator. Note the model of the system imbedded in the Kalman filter.

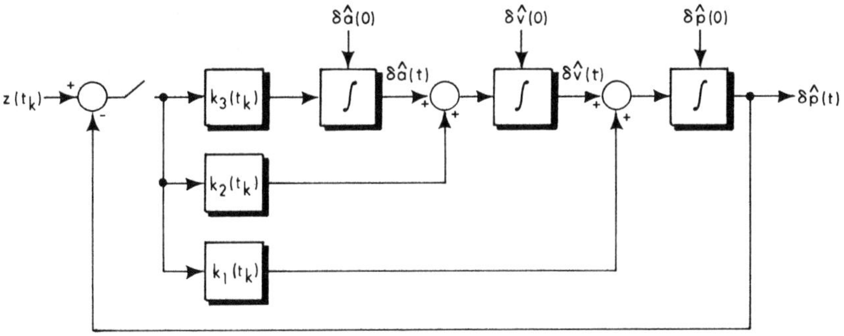

Figure 4.2-6 Optimal Filter Configuration

Figure 4.2-7 shows the result of a computer simulation of the complete system. The first fix occurs at t = 0, the second at t = 0.5 min and the third at t = 1.0 min. In this system the initial errors were $p_{11}(0) = (1\text{ nm})^2$, $p_{22}(0) = (7.3\text{ kts})^2$, $p_{33}(0) = 0$ and $\sigma_p = 30$ ft. Higher order terms were included but, as indicated earlier, these are not of primary importance here. Thus, the results are largely as predicted by the analysis herein.

4.3 CONTINUOUS KALMAN FILTER

The transition from the discrete to the continuous formulation of the Kalman filter is readily accomplished. First, in order to go from the discrete system and measurement models [Eqs. (4.1-4, 5)] to the continuous models*

*Throughout the remainder of the text, time dependence of all continuous time quantities will often be suppressed for notational convenience. For example, $\underline{x}(t)$ and Q(t) will be denoted by $\underline{x}$ and Q, etc.

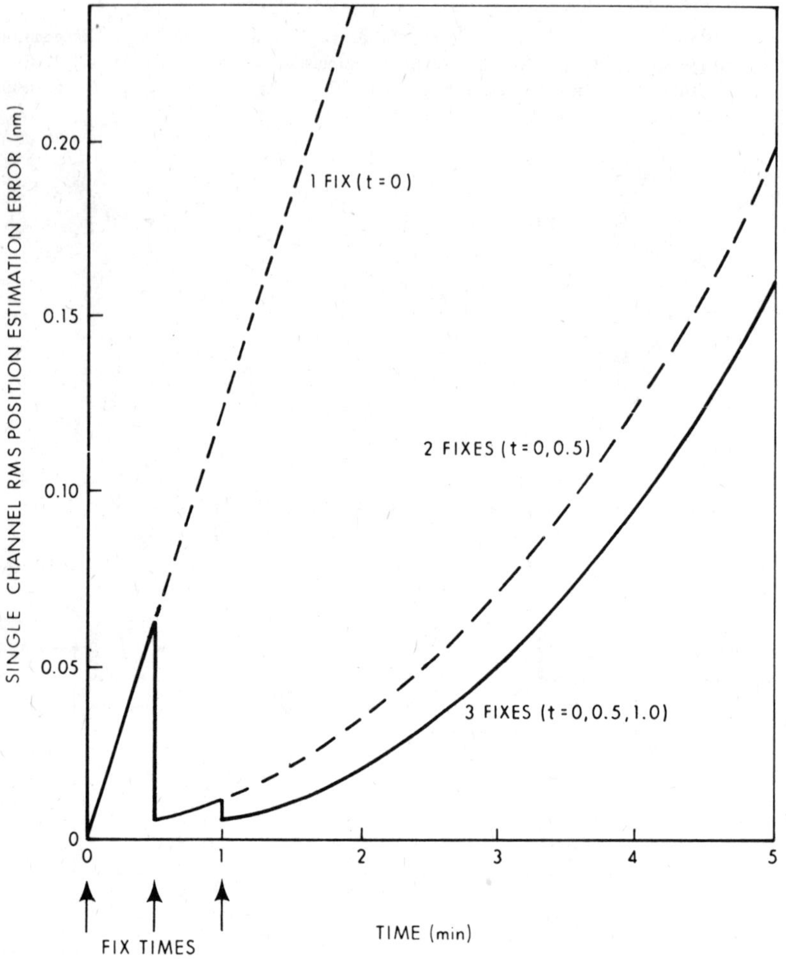

Figure 4.2-7 Optimal Use of Position Fixes in an Ionospheric Probe Inertial Navigation System

$$\dot{\underline{x}} = F\underline{x} + G\underline{w} \tag{4.3-1}$$

$$\underline{z} = H\underline{x} + \underline{v} \tag{4.3-2}$$

where $\underline{w}$, $\underline{v}$ are zero mean white noise processes with spectral density matrices Q and R, respectively, it is necessary to observe the following equivalences, valid in the limit as $t_k - t_{k-1} = \Delta t \to 0$:

$$\Phi_k \to I + F\Delta t$$

$$Q_k \rightarrow GQG^T \Delta t$$

$$R_k \rightarrow R/\Delta t$$

The first two of these relationships were derived in Section 3.7. What now remains is to establish the equivalence between the discrete white noise sequence $\underline{v}_k$ and the (non-physically realizable) continuous white noise process $\underline{v}$. Note that, whereas $R_k = E[\underline{v}_k\underline{v}_k^T]$ is a covariance matrix, R(t) defined by $E[\underline{v}(t)\underline{v}^T(\tau)] = R(t)\delta(t-\tau)$ is a spectral density matrix (the Dirac function $\delta(t-\tau)$ has units of 1/time). The covariance matrix $R(t)\delta(t-\tau)$ has infinite-valued elements. The discrete white noise sequence can be made to approximate the continuous white noise process by shrinking the pulse lengths (Δt) and increasing their amplitude, such that $R_k \rightarrow R/\Delta t$. That is, in the limit as $\Delta t \rightarrow 0$, the discrete noise sequence tends to one of infinite-valued pulses of zero duration, such that the area under the "impulse" autocorrelation function is $R_k\Delta t$, equal to the area R under the continuous white noise impulse autocorrelation function.

Using these expressions, our approach is simply one of writing the appropriate difference equations and observing their behavior in the limit as $\Delta t \rightarrow 0$. The notation will be kept as simple as possible and, in keeping with the rest of the presentation, the derivations are heuristic and to the point. The reader interested in a more rigorous derivation is referred to Kalman and Bucy (Ref. 5). For present purposes it shall be assumed that R is non-singular – i.e., R^{-1} exists. In addition, it is assumed that $\underline{w}$ and $\underline{v}$ are uncorrelated.

CONTINUOUS PROPAGATION OF COVARIANCE

In discrete form, the state error covariance matrix was shown to propagate according to

$$P_{k+1}(-) = \Phi_k P_k(+)\Phi_k^T + Q_k \tag{4.3-3}$$

This is now rewritten as

$$P_{k+1}(-) = [I + F\Delta t]P_k(+)[I + F\Delta t]^T + GQG^T \Delta t \tag{4.3-4}$$

Expansion yields

$$P_{k+1}(-) = P_k(+) + [FP_k(+) + P_k(+)F^T + GQG^T]\Delta t + O(\Delta t^2)$$

where $O(\Delta t^2)$ denotes "terms of the order Δt^2." As a consequence of optimal use of a measurement, $P_k(+)$ can be expressed as [Eq. (4.2-16b)]

$$P_k(+) = [I - K_k H_k]P_k(-)$$

Inserting this expression into the equation for $P_{k+1}(-)$ and rearranging terms yields

$$\frac{P_{k+1}(-) - P_k(-)}{\Delta t} = FP_k(-) + P_k(-)F^T + GQG^T - \frac{1}{\Delta t} K_k H_k P_k(-)$$
$$-FK_k H_k P_k(-) - K_k H_k P_k(-) F^T + O(\Delta t) \qquad (4.3\text{-}5)$$

Examining the term $1/\Delta t\ K_k$, we note that [Eq. (4.2-15)]

$$\frac{1}{\Delta t} K_k = \frac{1}{\Delta t} P_k(-) H_k^T [H_k P_k(-) H_k^T + R_k]^{-1}$$
$$= P_k(-) H_k^T [H_k P_k(-) H_k^T \Delta t + R_k \Delta t]^{-1} \qquad (4.3\text{-}6)$$
$$= P_k(-) H_k^T [H_k P_k(-) H_k^T \Delta t + R]^{-1}$$

Thus, in the limit as $\Delta t \to 0$ we get

$$\lim_{\Delta t \to 0} \frac{1}{\Delta t} K_k = PH^T R^{-1} \qquad (4.3\text{-}7)$$

and, simultaneously,

$$\dot{P} = FP + PF^T + GQG^T - PH^T R^{-1} HP \qquad (4.3\text{-}8)$$

Tracing the development of the terms on the right-side of this equation, it is clear that $FP + PF^T$ results from behavior of the homogeneous (unforced) system without measurements, GQG^T accounts for the *increase* of uncertainty due to process noise (this term is positive semidefinite) and $-PH^T R^{-1} HP$ accounts for the *decrease* of uncertainty as a result of measurements. Equation (4.3-8) is nonlinear in P; it is referred to as the *matrix Riccati equation.* In the absence of measurements we get

$$\dot{P} = FP + PF^T + GQG^T \qquad (4.3\text{-}9)$$

which is the *linear variance equation* previously derived in Section 3.7.

CONTINUOUS KALMAN FILTER

The discrete form of the Kalman filter may be written as [Eq. (4.2-5)]

$$\hat{\underline{x}}_k(+) = \Phi_{k-1} \hat{\underline{x}}_{k-1}(+) + K_k [\underline{z}_k - H_k \Phi_{k-1} \hat{\underline{x}}_{k-1}(+)] \qquad (4.3\text{-}10)$$

where we have made the substitution

$$\hat{\underline{x}}_k(-) = \Phi_{k-1} \hat{\underline{x}}_{k-1}(+)$$

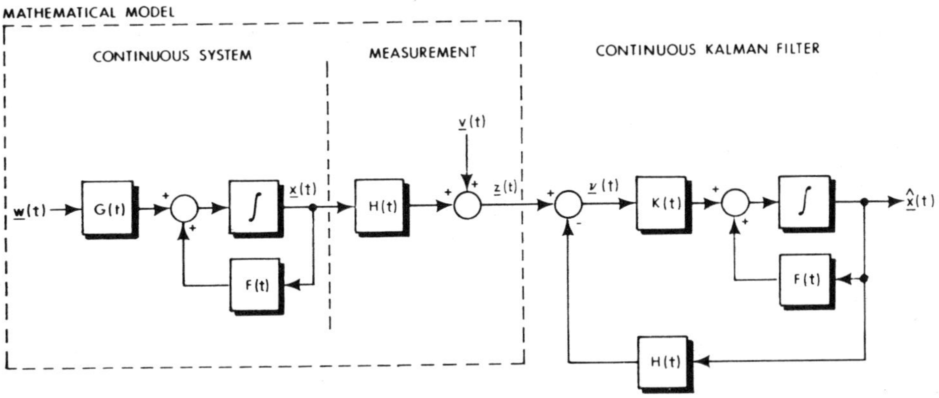

Figure 4.3-1 System Model and Continuous Kalman Filter

Replacing Φ_{k-1} by $I + F\Delta t$ and K_k by $PH^TR^{-1}\ \Delta t$ and rearranging terms yields

$$\frac{\hat{\underline{x}}_k(+) - \hat{\underline{x}}_{k-1}(+)}{\Delta t} = F\hat{\underline{x}}_{k-1}(+) + PH^TR^{-1}\left[\underline{z}_k - H_k\hat{\underline{x}}_{k-1}(+)\right] + O(\Delta t) \tag{4.3-11}$$

In the limit as $\Delta t \to 0$ this becomes

$$\dot{\hat{\underline{x}}} = F\hat{\underline{x}} + PH^TR^{-1}\left[\underline{z} - H\hat{\underline{x}}\right] \tag{4.3-12}$$

which is the continuous Kalman filter, and in which P is computed according to Eq. (4.3-8); see Fig. 4.3-1. The continuous Kalman filter equations are summarized in Table 4.3-1.

TABLE 4.3-1 SUMMARY OF CONTINUOUS KALMAN FILTER EQUATIONS (WHITE MEASUREMENT NOISE)

System Model	$\dot{\underline{x}}(t) = F(t)\underline{x}(t) + G(t)\underline{w}(t), \quad \underline{w}(t) \sim N(\underline{0}, Q(t))$
Measurement Model	$\underline{z}(t) = H(t)\underline{x}(t) + \underline{v}(t), \quad \underline{v}(t) \sim N(\underline{0}, R(t))$
Initial Conditions	$E[\underline{x}(0)] = \hat{\underline{x}}_0, \quad E[(\underline{x}(0) - \hat{\underline{x}}_0)(\underline{x}(0) - \hat{\underline{x}}_0)^T] = P_0$
Other Assumptions	$R^{-1}(t)$ exists
State Estimate	$\dot{\hat{\underline{x}}}(t) = F(t)\hat{\underline{x}}(t) + K(t)[\underline{z}(t) - H(t)\hat{\underline{x}}(t)], \ \hat{\underline{x}}(0) = \hat{\underline{x}}_0$
Error Covariance Propagation	$\dot{P}(t) = F(t)P(t) + P(t)F^T(t) + G(t)Q(t)G^T(t)$ $-K(t)R(t)K^T(t), \quad P(0) = P_0$
Kalman Gain Matrix	$K(t) = P(t)H^T(t)R^{-1}(t)$ when $E[\underline{w}(t)\underline{v}^T(\tau)] = 0$ $= [P(t)H^T(t) + G(t)C(t)]R^{-1}(t)$ when $E[\underline{w}(t)\underline{v}^T(\tau)] = C(t)\delta(t-\tau)$

The differential equation for $\underline{\tilde{x}}$ can be obtained by subtracting Eq. (4.3-1) from Eq. (4.3-12) and employing Eq. (4.3-2). Recalling that $\underline{\hat{x}} = \underline{x} + \underline{\tilde{x}}$, this results in $(K = PH^T R^{-1})$

$$\dot{\underline{\tilde{x}}} = (F - KH)\underline{\tilde{x}} - G\underline{w} + K\underline{v} \tag{4.3-13}$$

Analogous to the case of the discrete estimator, it can be shown that

$$E[\underline{\hat{x}}(t)\underline{\tilde{x}}(t)^T] = 0 \tag{4.3-14}$$

CORRELATED PROCESS AND MEASUREMENT NOISES

It is useful to obtain results for the case where the process and measurement noises are correlated, viz:

$$E[\underline{w}(t)\underline{v}^T(\tau)] = C(t)\,\delta(t - \tau) \tag{4.3-15}$$

One approach is to convert this new problem to an *equivalent* problem with no correlation between process and measurement noises. To do this, add zero to the right-side of Eq. (4.3-1), in the form*

$$\dot{\underline{x}} = F\underline{x} + G\underline{w} + D(\underline{z} - H\underline{x} - \underline{v}) \tag{4.3-16}$$

where $D = GCR^{-1}$. The filtering problem now under consideration is

$$\dot{\underline{x}} = (F - DH)\,\underline{x} + D\underline{z} + G\underline{w} - D\underline{v} \tag{4.3-17}$$

$$\underline{z} = H\underline{x} + \underline{v}$$

where in Eq. (4.3-17) $D\underline{z}$ is treated as a known input (Section 4.4) and $(G\underline{w} - D\underline{v})$ is treated as a process noise. By the choice of D specified above, $E[(G\underline{w} - D\underline{v})\underline{v}^T] = 0$, and thus the measurement and process noises in this equivalent problem are indeed uncorrelated. Applying previously derived results to this reformulated problem, it is easily verified that the solution is as given in Table 4.3-1, with the Kalman gain matrix specified by

$$K = [PH^T + GC]R^{-1} \tag{4.3-18}$$

This method can also be used to derive results for the corresponding discrete time problem.

Example 4.3-1

Estimate the value of a constant x, given a continuous measurement of x corrupted by a gaussian white noise process with zero mean and spectral density r.

---

*Method attributed to Y.C. Ho.

The scalar equations describing this situation are

$$\dot{x} = 0 \qquad \text{System}$$

$$z = x + v \qquad \text{Measurement}$$

where

$$v \sim N(0, r)$$

Figure 4.3-2 depicts the system and measurement models.

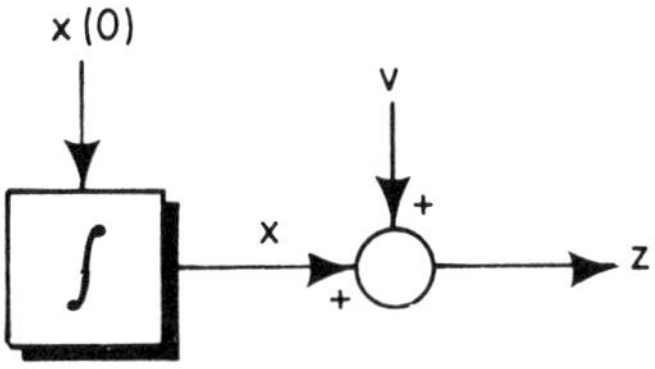

Figure 4.3-2 System and Measurement Models

For this problem, f = g = q = 0 and h = 1; thus yielding the scalar Riccati equation

$$\dot{p} = -p^2/r$$

Integrating this equation by separating variables, we get

$$\int_{p_0}^{p} \frac{dp}{p^2} = -\frac{1}{r}\int_0^t dt$$

or

$$p = \frac{p_0}{1 + \frac{p_0}{r}t}$$

Note that through the definition, $r_0 = r/T$, we see that this result is *identical* to that obtained in Example 4.2-1 at the instants t = kT (k = 0,1,2, . . .), as it should be.

The continuous Kalman gain is

$$k = \frac{p}{r} = \frac{\frac{p_0}{r}}{1 + \frac{p_0}{r}t}$$

Figure 4.3-3 illustrates the filter which, acting on z, produces an optimal estimate of x, viz:

$$\dot{\hat{x}}(t) = \frac{\frac{p_0}{r}}{1 + \frac{p_0}{r}t}\,[z - \hat{x}(t)]$$

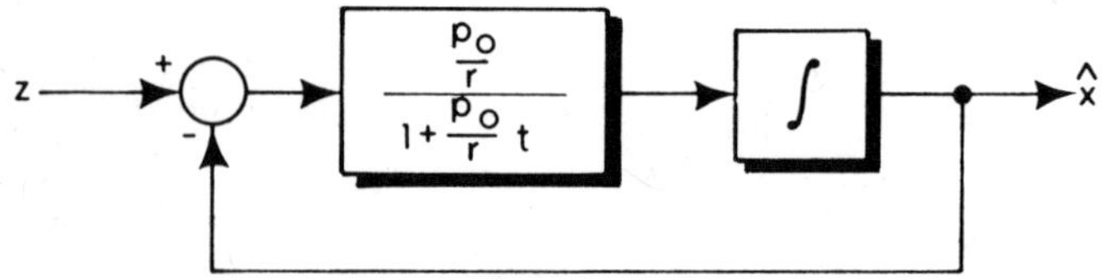

Figure 4.3-3 Optimal Filter Design

Notice that, as $t \to \infty$, $k(t) \to 0$ and thus $\hat{x}(t)$ tends to a constant, as indeed it should.

Example 4.3-2

A spacecraft is falling radially away from the earth at an almost constant speed, and is subject to small, random, high-frequency disturbance accelerations of spectral density q. Determine the accuracy to which vehicle velocity can be estimated, using ground-based doppler radar with an error of spectral density r. Assume poor initial condition information.

Let x denote the deviation from the predicted nominal spacecraft velocity, using available gravitational models. We then have

$$\dot{x} = w\,, \qquad w \sim N(0,q)$$

$$z = x+v, \qquad v \sim N(0,r)$$

and consequently ($f = 0$, $g = h = 1$)

$$\dot{p} = q - p^2/r, \qquad p(0) = p_0$$

Employing the identity

$$\int \frac{dp}{\alpha^2 - p^2} = \frac{1}{2\alpha} \ln\left(\frac{\alpha + p}{\alpha - p}\right)$$

we find directly ($\alpha \stackrel{\Delta}{=} \sqrt{rq}$, $\beta \stackrel{\Delta}{=} \sqrt{q/r}$):

$$p(t) = \alpha\left(\frac{p_0 \cosh \beta t + \alpha \sinh \beta t}{p_0 \sinh \beta t + \alpha \cosh \beta t}\right)$$

In the case of poor initial condition information, p_0 is large and thus p(t) reduces to

$$p(t) \approx \alpha \coth \beta t$$

In the steady state, $p(t) \to \alpha$ regardless of initial conditions; see Fig. 4.3-4. The Kalman filter for this problem looks like that illustrated in Fig. 4.3-3, with k(t) now given by

$$k(t) = p(t)/r$$

$$= \beta \coth \beta t$$

In the limit of large t, $k(t) \to \beta$ and not zero as in the previous example. New measurements are always processed in this case – a direct consequence of the noise term driving the system model.

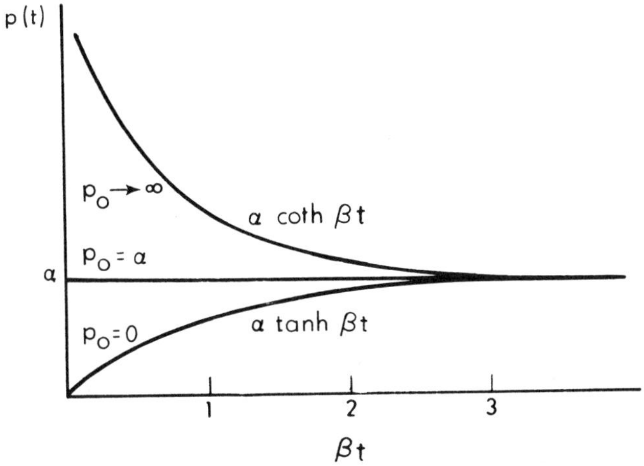

Figure 4.3-4 Tracking Error Versus Time

4.4 INTUITIVE CONCEPTS

Covariance Matrix – Inspection of the equations describing the behavior of the error covariance matrix reveals several observations which confirm our intuition about the operation of the filter. The effect of system disturbances on error covariance growth is the same as that observed when measurements were not available. The larger the statistical parameters of the disturbances as reflected in the "size" of the Q-matrix, and the more pronounced the effect of the disturbances as reflected in the "size" of the G-matrix, the more rapidly the error covariance increases.

The effect of measurement noise on the error covariance of the discrete filter is observed best in the expression

$$P_k^{-1}(+) = P_k^{-1}(-) + H_k^T R_k^{-1} H_k \tag{4.4-1}$$

Large measurement noise (R_k^{-1} is small) provides only a small increase in the inverse of the error covariance (a small *decrease* in the error covariance) when the measurement is used; the associated measurements contribute little to reduction in estimation errors. On the other hand, small measurement errors (large R_k^{-1}) cause the error covariance to decrease considerably whenever a measurement is utilized. When measurement noise is absent, Eq. (4.2-16a) must be used because R_k^{-1} does not exist.

The effect of measurement noise on the ability of the continuous Kalman filter to provide accurate estimates of the state, appears in the fourth term on the right side of Eq. (4.3-8). If noise occurs in every element of the measurement, R and R^{-1} are positive definite matrices. The term

$$P H^T R^{-1} HP \tag{4.4-2}$$

is also positive definite and the negative of this will always cause a decrease in the "size" of a nonzero error covariance matrix P. The magnitude of this term is inversely proportional to statistical parameters of the measurement noise. Larger measurement noise will cause the error covariance to diminish less rapidly or to increase, depending on the system dynamics, disturbances and the initial value of P. Smaller noise will cause the filter estimates to converge on the true values more rapidly. The effects of system disturbances and measurement noises of different magnitudes can be described graphically by considering the standard deviation of the error in the estimate of a representative state variable. This is presented in Fig. 4.4-1 for a hypothetical system which reaches statistical "steady state."

Kalman Gain Matrix – The optimality of the Kalman filter is contained in its structure and in the specification of the gain matrices. There is an intuitive logic behind the equations for the Kalman gain matrix. It can be seen from the forms

$$K_k = P_k(+) H_k^T R_k^{-1} \quad \text{or} \quad K(t) = P(t)H^T(t) R^{-1}(t) \tag{4.4-3}$$

To better observe the meaning of the expressions, assume that H is the identity matrix. In this case, both P and R^{-1} are nxn matrices. If R^{-1} is a diagonal matrix (no cross-correlation between noise terms), K results from multiplying each column of the error covariance matrix by the appropriate inverse of mean square measurement noise. Each element of the filter gain matrix is essentially the ratio between statistical measures of the uncertainty in the state estimate and the uncertainty in a measurement.

Thus, the gain matrix is "proportional" to the uncertainty in the estimate, and "inversely proportional" to the measurement noise. If measurement noise is large and state estimate errors are small, the quantity $\underline{\nu}$ in Figs. 4.2-2 and 4.3-1 is due chiefly to the noise and only small changes in the state estimates should be made. On the other hand, small measurement noise and large uncertainty in the state estimates suggest that $\underline{\nu}$ contains considerable information about errors in the estimates. Therefore, the difference between the actual and the predicted measurement will be used as a basis for strong corrections to the estimates. Hence, the filter gain matrix is specified in a way which agrees with an intuitive approach to improving the estimate.

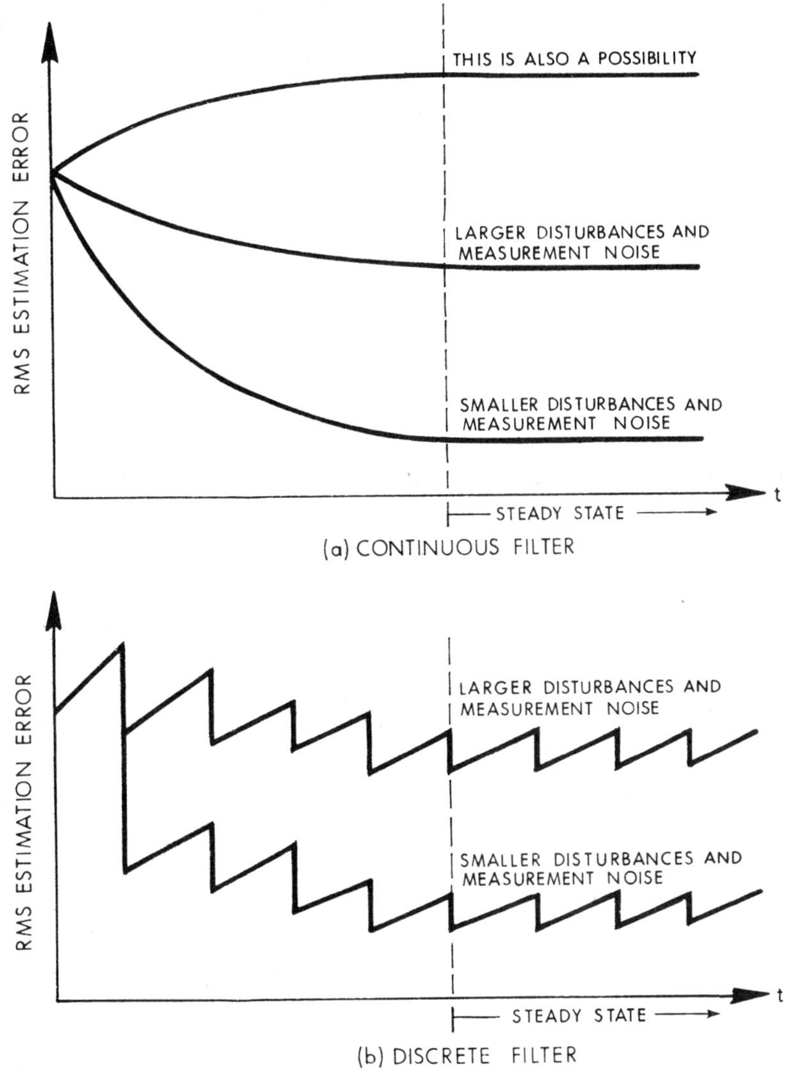

Figure 4.4-1 Behavior of the RMS Error in the Kalman Filter Estimate of a Particular State Variable

Optimal Prediction – Optimal prediction can be thought of, quite simply, in terms of optimal filtering *in the absence of measurements.* This, in turn, is equivalent to optimal filtering with arbitrarily large measurement errors (thus $R^{-1} \rightarrow 0$ and hence $K \rightarrow 0$). Therefore, if measurements are unavailable beyond some time, t_0, the optimal prediction of $\underline{x}(t)$ for $t \geq t_0$ given $\hat{\underline{x}}(t_0)$ must be obtained from [Eqs. (4.3-10) and (4.3-12)]:

$$\hat{\underline{x}}(t) = \Phi(t,t_0)\hat{\underline{x}}(t_0) \qquad \text{discrete} \tag{4.4-4}$$

$$\dot{\hat{\underline{x}}}(t) = F(t)\hat{\underline{x}}(t) \qquad \text{continuous} \tag{4.4-5}$$

The corresponding equations for uncertainty in the optimal predictions, given $P(t_0)$, are Eq. (4.3-3) for a single time stage in the discrete case and Eq. (4.3-9) in the continuous case.

Note that both the discrete and continuous Kalman filters contain an exact *model* of the system in their formulation (i.e., the F or Φ matrices). This provides the mechanism by which past information is extrapolated into the future for the purpose of prediction.

System Model Contains Deterministic Inputs – When the system under observation is excited by a deterministic time-varying input, $\underline{u}$, whether due to a control being intentionally applied or a deterministic disturbance which occurs, these known inputs must be accounted for by the optimal estimator. It is easy to see that the modification shown in Table 4.4-1 must be made in order for the estimators to remain unbiased. By subtracting $\underline{x}$ from $\hat{\underline{x}}$ in both discrete and continuous cases, it is observed that the resultant equations for $\tilde{\underline{x}}$ are precisely those obtained before. Hence, the procedures for computing P, K, etc., remain unchanged.

TABLE 4.4-1 MODIFICATION TO ACCOUNT FOR KNOWN INPUTS ($\underline{u}_{k-1}$ OR $\underline{u}(t)$)

System Model	Estimator
Discrete	
$\underline{x}_k = \Phi_{k-1}\underline{x}_{k-1} + \underline{w}_{k-1} + \Lambda_{k-1}\underline{u}_{k-1}$	$\hat{\underline{x}}_k(+) = \Phi_{k-1}\hat{\underline{x}}_{k-1}(+) + \Lambda_{k-1}\underline{u}_{k-1}$
$\underline{z}_k = H_k\underline{x}_k + \underline{v}_k$	$+ K_k[\underline{z}_k - H_k\Phi_{k-1}\hat{\underline{x}}_{k-1}(+)]$
Continuous	
$\dot{\underline{x}}(t) = F(t)\underline{x}(t) + G(t)\underline{w}(t) + L(t)\underline{u}(t)$	$\dot{\hat{\underline{x}}}(t) = F(t)\hat{\underline{x}}(t) + L(t)\underline{u}(t)$
$\underline{z}(t) = H(t)\underline{x}(t) + \underline{v}(t)$	$+ K(t)[\underline{z}(t) - H(t)\hat{\underline{x}}(t)]$

Stochastic Controllability – In the absence of measurements and with perfect *a priori* information, the *continuous* system matrix Riccati equation is

$$\dot{P} = FP + PF^T + GQG^T, \qquad P(0) = 0 \tag{4.4-6}$$

for which the solution is*

$$P(t) = \int_0^t \Phi(t,\tau)\, G(\tau)\, Q(\tau)\, G^T(\tau)\, \Phi^T(t,\tau)\, d\tau \tag{4.4-7}$$

where $\Phi(t,\tau)$ is the transition matrix corresponding to F. If the integral is positive definite for some $t > 0$, then $P(t) > 0$ – i.e., the process noise excites all the states in the system. The system is said to be *uniformly completely controllable* when the integral is positive definite and bounded for some $t > 0$. The property of stochastic controllability is important for establishing stability of the filter equations and for obtaining a unique steady-state value of P. When the system is stationary, and if Q is positive definite, this criterion of complete controllability can be expressed algebraically; the result is exactly that discussed in Section 3.5.

In the case of *discrete* systems, the condition for complete controllability is expressed as

$$\beta_1 I \leqslant \sum_{i=k-N}^{k-1} \Phi(k, i+1) Q_i \Phi^T(k, i+1) \leqslant \beta_2 I \tag{4.4-8}$$

for some value of $N > 0$, where $\beta_1 > 0$ and $\beta_2 > 0$.

Stochastic Observability – In the absence of process noise and *a priori* information, the *continuous* system matrix Riccati equation is given by

$$\dot{P} = FP + PF^T - PH^T R^{-1} HP, \qquad P(0) \to \infty \tag{4.4-9}$$

This can be rewritten as

$$\dot{P}^{-1} = -P^{-1} F - F^T P^{-1} + H^T R^{-1} H, \qquad P^{-1}(0) = 0 \tag{4.4-10}$$

using the matrix identity, $\dot{P}^{-1} = -P^{-1} \dot{P} P^{-1}$. The solution to this *linear* equation in P^{-1} is

$$P^{-1}(t) = \int_0^t \Phi^T(\tau, t) H^T(\tau) R^{-1}(\tau) H(\tau) \Phi(\tau, t)\, d\tau \tag{4.4-11}$$

---

*This is easily verified by substitution into Eq. (4.4-6), using Leibniz' rule

$$\frac{d}{dt} \int_0^t A(t,\tau)\, d\tau = A(t,t) + \int_0^t \frac{\partial}{\partial t} A(t,\tau)\, d\tau$$

where $\Phi(t,\tau)$ is the transition matrix corresponding to F. If the integral is positive definite for some $t > 0$, then $P^{-1}(t) > 0$, and it follows that $0 < P(t) < \infty$ – i.e., through processing measurements it is possible to acquire information (decrease the estimation error variance) about states that are initially completely unknown (Ref. 6). The system is said to be *uniformly completely observable* when the integral is positive definite and bounded for some $t > 0$. When the linear system is stationary, this criterion of complete observability can be expressed algebraically; the result is exactly that discussed in Section 3.5 (Ref. 8).

In the case of *discrete* systems, the condition for uniform complete observability is

$$\alpha_1 I \leqslant \sum_{i=k-N}^{k} \Phi^T(i, k)\, H_i^T R_i^{-1} H_i\, \Phi(i, k) \leqslant \alpha_2 I \tag{4.4-12}$$

for some value of $N > 0$, where $\alpha_1 > 0$ and $\alpha_2 > 0$.

Stability – One consideration of both practical and theoretical interest is the stability of the Kalman filter. Stability refers to the behavior of state estimates when measurements are suppressed. For example, in the continuous case, the "unforced" filter equation takes the form

$$\dot{\hat{\underline{x}}}(t) = [F(t) - K(t)\, H(t)]\, \hat{\underline{x}}(t) \tag{4.4-13}$$

It is desirable that the solution of Eq. (4.4-13) be *asymptotically stable*; – i.e., loosely speaking, $\hat{\underline{x}}(t) \to \underline{0}$ as $t \to \infty$, for any initial condition $\hat{\underline{x}}(0)$. This will insure that any unwanted component of $\hat{\underline{x}}$ caused by disturbances driving Eq. (4.4-13) – such as computational errors arising from finite word length in a digital computer – are bounded.

Optimality of the Kalman filter does not guarantee its stability. However, one key result exists which assures both stability (more precisely, uniform asymptotic stability) of the filter and uniqueness of the behavior of P(t) for large t, independently of P(0). It requires stochastic uniform complete observability, stochastic uniform complete controllability, bounded Q and R (from above and below), and bounded F (from above). References 1 and 5 provide details of this theorem and other related mathematical facts. It is important to note that complete observability and controllability requirements are quite restrictive and, in many cases of practical significance, these conditions are not fulfilled; but Kalman filters, designed in the normal way, operate satisfactorily. This is attributed to the fact that the solution to Eq. (4.4-13) frequently tends toward zero over a finite time interval of interest, even though it may not be asymptotically stable in the strict sense of the definition. Perhaps, from a *practical* viewpoint, the key issues pertaining to various forms of instability are those associated with modeling errors and implementation considerations. These are discussed in Chapters 7 and 8.

4.5 CORRELATED MEASUREMENT ERRORS

Measurements may contain errors whose correlation times are significant. Let us assume that, through the use of a shaping filter, these measurement errors are described as the output of a first-order vector differential equation forced by white noise. One might argue that the technique of state vector augmentation could then be used to recast this problem into a form where the solution has already been obtained; but it is readily demonstrated (Refs. 9-12) that this is not the case in continuous-time systems, and undesirable in the case of discrete-time systems.

STATE VECTOR AUGMENTATION

Consider the continuous system and measurement described by

$$\dot{\underline{x}} = F\underline{x} + G\underline{w}, \qquad \underline{w} \sim N(\underline{0}, Q) \tag{4.5-1}$$

$$\underline{z} = H\underline{x} + \underline{v} \tag{4.5-2}$$

where

$$\dot{\underline{v}} = E\underline{v} + \underline{w}_1, \qquad \underline{w}_1 \sim N(\underline{0}, Q_1) \tag{4.5-3}$$

The augmented state vector $\underline{x}'^T = [\underline{x} \;\; \underline{v}]^T$ satisfies the differential equation

$$\dot{\underline{x}}' = \begin{bmatrix} \dot{\underline{x}} \\ \dot{\underline{v}} \end{bmatrix} = \left[\begin{array}{c|c} F & O \\ \hline O & E \end{array}\right] \begin{bmatrix} \underline{x} \\ \underline{v} \end{bmatrix} + \left[\begin{array}{c|c} G & O \\ \hline O & I \end{array}\right] \begin{bmatrix} \underline{w} \\ \underline{w}_1 \end{bmatrix} \tag{4.5-4}$$

and the measurement equation becomes

$$\underline{z} = [H \,|\, I]\,\underline{x}' \tag{4.5-5}$$

In this reformulated problem, the *equivalent* measurement noise is zero. Correspondingly, the *equivalent* R matrix is singular and thus the Kalman gain matrix, $K = PH^TR^{-1}$, required for the optimal state estimator, does not exist.

CONTINUOUS TIME, R SINGULAR

There is another approach to this problem which avoids both the difficulty of singular R and the undesirability of working with a higher order (i.e., augmented) system. From Eq. (4.5-3), where we see that $\dot{\underline{v}} - E\underline{v}$ is a white noise process, we are led to consider the *derived* measurement $\underline{z}_1$,

$$\begin{aligned}\underline{z}_1 &= \dot{\underline{z}} - E\underline{z}\\ &= \dot{H}\underline{x} + H\dot{\underline{x}} - EH\underline{x} - E\underline{v} + \dot{\underline{v}}\\ &= (\dot{H} + HF - EH)\underline{x} + (HG\underline{w} + \underline{w}_1)\\ &= H_1\underline{x} + \underline{v}_1\end{aligned} \tag{4.5-6}$$

where the definitions of H_1 and $\underline{v}_1$ are apparent. For this derived measurement, the noise $\underline{v}_1$ is indeed white; it is also correlated with the process noise $\underline{w}$. The corresponding quantities R_1 and C_1 are directly obtained as $(E[\underline{w}_1(t)\underline{w}^T(\tau)] = 0)$

$$R_1 = HGQG^TH^T + Q_1 \tag{4.5-7}$$

$$C_1 = QG^TH^T \tag{4.5-8}$$

and thus the Kalman gain matrix for this *equivalent* problem, from Table 4.3-1, is

$$\begin{aligned}K_1 &= [PH_1{}^T + GC_1]R_1{}^{-1}\\ &= [P(\dot{H} + HF - EH)^T + GQG^TH^T][HGQG^TH^T + Q_1]^{-1}\end{aligned} \tag{4.5-9}$$

The equations for $\hat{\underline{x}}(t)$ and $P(t)$ are

$$\dot{\hat{\underline{x}}} = F\hat{\underline{x}} + K_1(\dot{\underline{z}} - E\underline{z} - H_1\hat{\underline{x}}) \tag{4.5-10}$$

$$\dot{P} = FP + PF^T + GQG^T - K_1R_1K_1{}^T \tag{4.5-11}$$

There remain two aspects of this solution which warrant further discussion; namely, the need for *differentiation* in the derived measurement and appropriate *initial values* of $\hat{\underline{x}}$,(t) and P(t). Assuming $K_1(t)$ exists and $\dot{K}_1(t)$ is piecewise continuous, use of the identity

$$\frac{d}{dt}(K_1\underline{z}) = \dot{K}_1\underline{z} + K_1\dot{\underline{z}} \tag{4.5-12}$$

enables rewriting Eq. (4.5-10) in the form

$$\frac{d}{dt}(\hat{\underline{x}} - K_1\underline{z}) = F\hat{\underline{x}} - \dot{K}_1\underline{z} - K_1(E\underline{z} + H_1\hat{\underline{x}}) \tag{4.5-13}$$

In this manner, the need to differentiate the measurements is avoided. Figure 4.5-1 is a block diagram of the corresponding optimal estimator. Regarding the initial estimate and error covariance matrix, note that the instant after measurement data are available ($t = 0^+$), the following *discontinuities* occur,

$$\hat{\underline{x}}(0^+) = \hat{\underline{x}}(0) + P(0)H^T(0)[H(0)P(0)H^T(0) + R(0)]^{-1}[\underline{z}(0) - H(0)\hat{\underline{x}}(0)] \quad (4.5\text{-}14)$$

and

$$P(0^+) = P(0) - P(0)H^T(0)[H(0)P(0)H^T(0) + R(0)]^{-1}H(0)P(0) \quad (4.5\text{-}15)$$

where $E[\underline{v}(0)\underline{v}^T(0)] = R(0)$. The initial condition on the filter in Fig. 4.5-1 is dependent upon the initial measurement and cannot be determined *a priori*, viz:

$$\text{initial condition} = \hat{\underline{x}}(0^+) - K_1(0)\underline{z}(0) \quad (4.5\text{-}16)$$

For additional details, including the treatment of general situations, see Refs. 9 and 11.

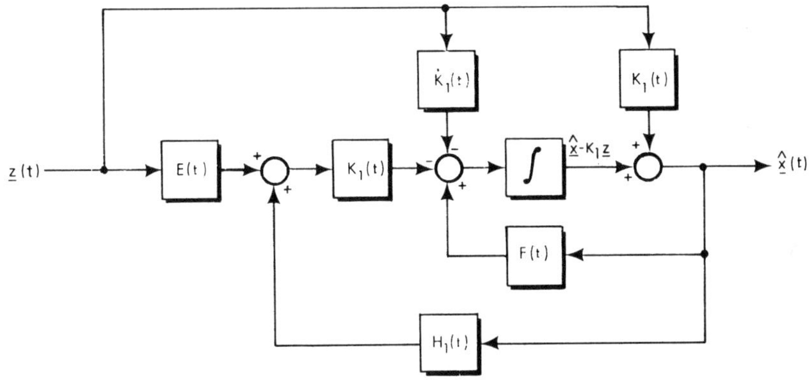

Figure 4.5-1 Correlated Noise Optimal Filter, Continuous Time

DISCRETE TIME, R_k SINGULAR

State vector augmentation can be employed in the discrete-time problem with correlated noise to arrive at an equivalent measurement-noise-free problem, as in Eqs. (4.5-4) and (4.5-5). Let us suppose that the equivalent problem is described by

$$\underline{x}_k = \Phi_{k-1}\underline{x}_{k-1} + \underline{w}_{k-1} \quad (4.5\text{-}17)$$

$$\underline{z}_k = H_k\underline{x}_k \quad (4.5\text{-}18)$$

Although the measurement noise is zero (hence, $R_k = 0$ and R_k^{-1} does not exist), apparently there is no difficulty because the discrete Kalman gain matrix does not explicitly involve R_k^{-1}. However, consider the expression for updating the error covariance matrix (Table 4.2-1),

$$P_k(+) = P_k(-) - P_k(-)H_k^T [H_k P_k(-) H_k^T]^{-1} H_k P_k(-) \quad (4.5\text{-}19)$$

It is easy to see from this expression that

$$H_k P_k(+) H_k^T = 0 \quad (4.5\text{-}20)$$

$P_k(+)$ must be singular, of course, since certain linear combinations of the state vector elements are known exactly. Error covariance extrapolation is performed according to (Table 4.2-1)

$$P_{k+1}(-) = \Phi_k P_k(+)\, \Phi_k^T + Q_k \quad (4.5\text{-}21)$$

Therefore it follows that if $\Phi_k \sim I$ (i.e., as it would be, in a continuous system which is sampled at points close in time relative to the system time constants) and if Q_k is small, the covariance update may become *ill-conditioned* [i.e., $P_{k+1}(-) \approx P_k(+)$]. To alleviate this problem, *measurement differencing* may be employed, which is analogous to measurement differentiation previously discussed. Details of this procedure are available in Ref. 12.

4.6 SOLUTION OF THE RICCATI EQUATION

The optimal filtering covariance equations can only be solved analytically for simple problems. Various numerical techniques are available for more complicated problems. Numerical integration is the direct method of approach for time-varying systems (i.e., when the matrix Riccati equation has time-varying coefficients). For an nth order system, there are $n(n+1)/2$ variables involved [the distinct elements of the $n \times n$ symmetric matrix $P(t)$].

Systems for which the matrix Riccati equation has constant (or piecewise-constant) coefficients, or can be approximated as such, can be treated as follows: For the nth order nonlinear equation

$$\dot{P} = FP + PF^T + GQG^T - PH^T R^{-1} HP, \qquad P(t_0) \text{ given} \quad (4.6\text{-}1)$$

the transformations (Ref. 5)

$$\underline{\lambda} = P\underline{y} \quad (4.6\text{-}2)$$

and

$$\dot{\underline{y}} = -F^T\underline{y} + H^T R^{-1} HP\underline{y} \quad (4.6\text{-}3)$$

result in a system of linear differential equations. Using the above, we compute

$$\begin{aligned}\dot{\underline{\lambda}} &= \dot{P}\underline{y} + P\dot{\underline{y}} \\ &= FP\underline{y} + PF^T\underline{y} + GQG^T\underline{y} - PH^T R^{-1} HP\underline{y} - PF^T\underline{y} + PH^T R^{-1} HP\underline{y} \quad (4.6\text{-}4)\end{aligned}$$

Therefore

$$\dot{\underline{\lambda}} = F\underline{\lambda} + GQG^T\underline{y} \tag{4.6-5}$$

and the equivalent $2n^{th}$ order *linear* system equations are:

$$\begin{bmatrix} \dot{\underline{y}} \\ \dot{\underline{\lambda}} \end{bmatrix} = \underbrace{\begin{bmatrix} -F^T & H^TR^{-1}H \\ GQG^T & F \end{bmatrix}}_{M} \begin{bmatrix} \underline{y} \\ \underline{\lambda} \end{bmatrix} \tag{4.6-6}$$

By denoting the transition matrix for this linear system as Φ (namely, $\Phi(t_0+\tau,t_0) = \Phi(\tau) = e^{M\tau}$), we can write

$$\begin{bmatrix} \underline{y}(t_0+\tau) \\ \underline{\lambda}(t_0+\tau) \end{bmatrix} = \underbrace{\left[\begin{array}{c|c} \Phi_{yy}(\tau) & \Phi_{y\lambda}(\tau) \\ \hline \Phi_{\lambda y}(\tau) & \Phi_{\lambda\lambda}(\tau) \end{array}\right]}_{\Phi(\tau)} \begin{bmatrix} \underline{y}(t_0) \\ \underline{\lambda}(t_0) \end{bmatrix} \tag{4.6-7}$$

where $\Phi(\tau)$ is shown partitioned into square $n \times n$ matrices. Writing out the expressions for $\underline{y}(t_0+\tau)$ and $\underline{\lambda}(t_0+\tau)$ and employing Eq. (4.6-2) yields

$$P(t_0+\tau) = [\Phi_{\lambda y}(\tau) + \Phi_{\lambda\lambda}(\tau)P(t_0)]\ [\Phi_{yy}(\tau) + \Phi_{y\lambda}(\tau)P(t_0)]^{-1} \tag{4.6-8}$$

Thus, the nth order nonlinear matrix Riccati equation has been converted to an equivalent $2n^{th}$ order linear matrix equation and solved. Once $\Phi(\tau)$ is computed, P may be generated as a function of time by repeated use of Eq. (4.6-8). The solution to the Riccati equation is obtained without any truncation errors, and is subject only to roundoff errors in computing Φ. Note the importance of periodically replacing P with its symmetric part, $(P + P^T)/2$, to avoid errors due to asymmetry. This method is often faster than direct numerical integration. Although this technique is equally applicable to general time-varying systems, $M = M(t)$; thus, Φ is a function of both t_0 and τ in these cases. The added difficulty in computing $\Phi(t_0+\tau,t_0)$ is usually not justifiable when compared to other numerical methods of solution.

Solution of the linear variance equation

$$\dot{P} = FP + PF^T + GQG^T \quad , \quad P(t_0) \text{ given} \tag{4.6-9}$$

is easily obtained by noting that $H^TR^{-1}H = 0$ in this case and thus $\Phi_{yy}(\tau) = (e^{-F\tau})^T$, $\Phi_{\lambda\lambda}(\tau) = e^{F\tau}$, and $\Phi_{y\lambda}(\tau) = 0$. The desired result, obtained from Eq. (4.6-8), is

$$P(t_0+\tau) = \Phi_{\lambda y}(\tau)\,\Phi_{\lambda\lambda}{}^T(\tau) + \Phi_{\lambda\lambda}(\tau)P(t_0)\,\Phi_{\lambda\lambda}{}^T(\tau) \tag{4.6-10}$$

and when all the eigenvalues of F have negative real parts, the unique steady state solution of the linear variance equation is

$$P(\infty) = \lim_{\tau \to \infty} \Phi_{\lambda y}(\tau)\Phi_{\lambda\lambda}{}^{T}(\tau) \tag{4.6-11}$$

Example 4.6-1

For a certain class of integrating gyroscopes. all units have a constant but *a priori* unknown drift rate, ϵ, once thermally stabilized. The gyroscopes are instrumented to stabilize a single-axis test table. Continuous indications of table angle, θ, which is a direct measure of the integrated gyroscope drift rate, are available. The table angle readout has an error, e, which is well described by an exponential autocorrelation function model of standard deviation σ ($\widehat{\text{sec}}$) and short correlation time T (sec). Design an efficient real-time drift rate test data processor.

The equation of motion of the test table, neglecting servo errors, is

$$\theta = \epsilon t + \theta_0 \tag{4.6-12}$$

and the measurement is described by

$$z = \theta + e \tag{4.6-13}$$

with

$$\dot{e} = -\frac{1}{T}\, e + w \tag{4.6-14}$$

where w is a white noise of zero mean and spectral density $q = 2\sigma^2/T$ ($\widehat{\text{sec}}^2$/sec).

There are two ways to formulate the data processor. One is to augment the two differential equations implied by Eq. (4.6-12) with Eq. (4.6-14), resulting in a third-order system (three-row state vector), and to proceed as usual. Another is to *approximate* the relatively high frequency correlated noise by a white noise process and thus delete Eq. (4.6-14), resulting in a second-order system. In this case the spectral density of the noise (now *directly* representing e) is $r = 2\sigma^2 T$ ($\widehat{\text{sec}}^2$ sec). We proceed using the latter approach.

The differential equations corresponding to Eq. (4.6-12) are:

$$\dot{\theta} = \epsilon$$

$$\dot{\epsilon} = 0$$

or

$$\begin{bmatrix} \dot{\theta} \\ \dot{\epsilon} \end{bmatrix} = \underbrace{\begin{bmatrix} 0 & 1 \\ 0 & 0 \end{bmatrix}}_{F} \underbrace{\begin{bmatrix} \theta \\ \epsilon \end{bmatrix}}_{\underline{x}} \tag{4.6-15}$$

and the measurement is expressed as

$$z = [1 \;\; 0] \begin{bmatrix} \theta \\ \epsilon \end{bmatrix} + e \tag{4.6-16}$$

(see Fig. 4.6-1). Four methods of attack now suggest themselves.

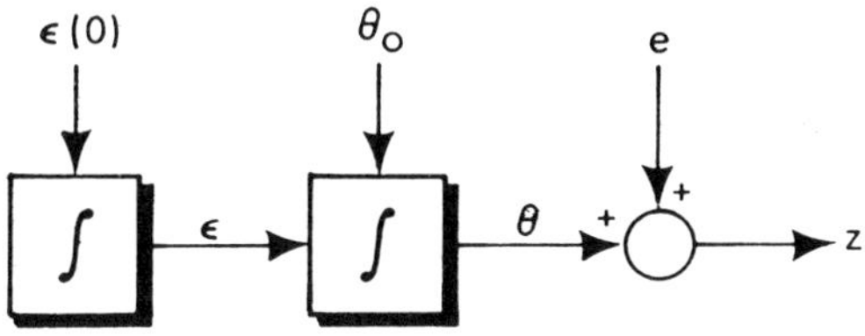

Figure 4.6-1 Model of the System and Measurement

Method 1 – Having identified the system and measurement equations, the Riccati equation

$$\dot{P} = FP + PF^T - PH^TR^{-1}HP$$

must be solved for P(t). Note the absence of the term GQG^T in this case. With F and H as given in Eqs. (4.6-15) and (4.6-16), the Riccati equation becomes

$$\begin{bmatrix} \dot{p}_{11} & \dot{p}_{12} \\ \dot{p}_{12} & \dot{p}_{22} \end{bmatrix} = \begin{bmatrix} p_{12} & p_{22} \\ 0 & 0 \end{bmatrix} + \begin{bmatrix} p_{12} & 0 \\ p_{22} & 0 \end{bmatrix} - \frac{1}{r} \begin{bmatrix} p_{11}^2 & p_{11}p_{12} \\ p_{11}p_{12} & p_{12}^2 \end{bmatrix}$$

or

$$\dot{p}_{11} = 2p_{12} - p_{11}^2/r$$

$$\dot{p}_{12} = p_{22} - p_{11}p_{12}/r$$

$$\dot{p}_{22} = -p_{12}^2/r$$

which is a set of coupled nonlinear differential equations. While this set of equations can be solved directly (albeit, with considerable difficulty), there are better methods of approach.

Method 2 – Following Eq. (4.6-6), form the matrix M to obtain

$$M = \left[\begin{array}{cc|cc} 0 & 0 & 1/r & 0 \\ -1 & 0 & 1 & 0 \\ \hline 0 & 0 & 0 & 1 \\ 0 & 0 & 0 & 0 \end{array}\right]$$

The transition matrix corresponding to M is computed next. In this case the matrix exponential series is finite, truncating after four terms, with the result

$$\Phi(\tau) = \left[\begin{array}{cc|cc} 1 & 0 & \tau/r & \tau^2/2r \\ -\tau & 1 & -\tau^2/2r & -\tau^3/6r \\ \hline 0 & 0 & 1 & \tau \\ 0 & 0 & 0 & 1 \end{array}\right]$$

The partitioned matrices Φ_{yy}, $\Phi_{y\lambda}$, $\Phi_{\lambda y}$ and $\Phi_{\lambda\lambda}$ are readily identified. With $t_0 = 0$ and $\tau =$ t, and assuming P(0) is diagonal,

$$P(0) = \begin{bmatrix} p_{11}(0) & 0 \\ 0 & p_{22}(0) \end{bmatrix} \tag{4.6-17}$$

Eq. (4.6-8) yields

$$P(t) = \frac{1}{\Delta(t)} \begin{bmatrix} p_{11}(0)+t^2p_{22}(0)+t^3p_{11}(0)p_{22}(0)/3r & tp_{22}(0)+t^2p_{11}(0)p_{22}(0)/2r \\ tp_{22}(0)+t^2p_{11}(0)p_{22}(0)/2r & p_{22}(0)+tp_{11}(0)p_{22}(0)/r \end{bmatrix} \tag{4.6-18a}$$

where

$$\Delta(t) = 1 + tp_{11}(0)/r + t^3p_{22}(0)/3r + t^4p_{11}(0)p_{22}(0)/12r^2 \tag{4.6-18b}$$

This is the complete answer for P(t). K(t) can now be computed and the optimal filter has thus been determined. In the case where there is no *a priori* information, $p_{11}(0) = p_{22}(0) \rightarrow \infty$ and the limit of the above result yields

$$P(t) = r\begin{bmatrix} 4/t & 6/t^2 \\ 6/t^2 & 12/t^3 \end{bmatrix} \tag{4.6-19}$$

The Kalman gain is thus given by

$$\begin{aligned} K(t) &= P(t)H^TR^{-1} \\ &= r\begin{bmatrix} 4/t & 6/t^2 \\ 6/t^2 & 12/t^3 \end{bmatrix}\begin{bmatrix} 1 \\ 0 \end{bmatrix}\frac{1}{r} \\ &= \begin{bmatrix} 4/t \\ 6/t^2 \end{bmatrix} \end{aligned}$$

and the optimal filter is as illustrated in Fig. 4.6-2.

In the limit as $t \rightarrow \infty$, our theory says that $K(t) \rightarrow 0$ corresponding to the covariance matrix limit $P(t) \rightarrow 0$. This steady state would not be one of ideal knowledge if there were any other error sources driving Eq. (4.6-15). Although we have not modeled any such errors, it is certain that they will indeed exist and hence influence the drift rate test. Therefore, the theoretical prediction $P(t) \rightarrow 0$ should be viewed as unrealistically optimistic, and should in practice be adjusted to account for other expected noise terms.

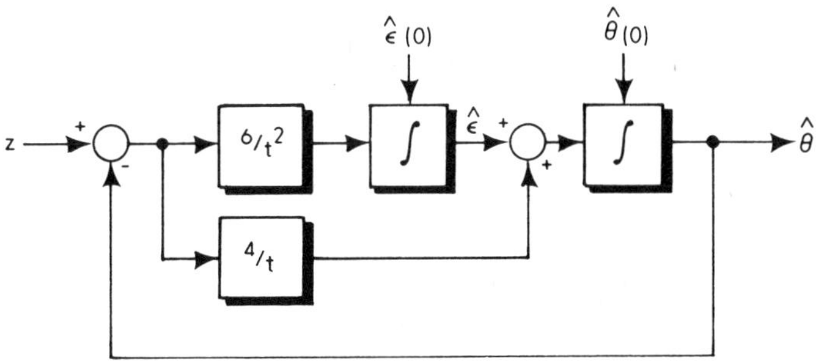

Figure 4.6-2 Optimal Filter Configuration

Method 3 – Notice that there is no noise term driving the state vector differential equation. In this case the *nonlinear* Riccati equation in P is identical to the following *linear* differential equation in P^{-1}

$$\dot{P}^{-1} = -P^{-1}F - F^TP^{-1} + H^TR^{-1}H$$

Equivalence is readily shown by premultiplying and postmultiplying both sides of this equation by P and noting the relationship $\dot{P} = -P\dot{P}^{-1}P$. Denoting the elements of P^{-1} by a_{ij}, we write

$$\begin{bmatrix} \dot{a}_{11} & \dot{a}_{12} \\ \dot{a}_{12} & \dot{a}_{22} \end{bmatrix} = -\begin{bmatrix} a_{11} & a_{12} \\ a_{12} & a_{22} \end{bmatrix}\begin{bmatrix} 0 & 1 \\ 0 & 0 \end{bmatrix} - \begin{bmatrix} 0 & 0 \\ 1 & 0 \end{bmatrix}\begin{bmatrix} a_{11} & a_{12} \\ a_{12} & a_{22} \end{bmatrix} + \begin{bmatrix} 1/r & 0 \\ 0 & 0 \end{bmatrix}$$

$$= \begin{bmatrix} 1/r & -a_{11} \\ -a_{11} & -2a_{12} \end{bmatrix}$$

Equivalently,

$$\dot{a}_{11} = 1/r$$

$$\dot{a}_{12} = -a_{11}$$

$$\dot{a}_{22} = -2a_{12}$$

The solutions are,

$$a_{11} = \frac{t}{r} + a_{11}(0)$$

$$a_{12} = -\frac{t^2}{2r} - a_{11}(0)t + a_{12}(0)$$

$$a_{22} = \frac{t^3}{3r} + a_{11}(0)t^2 - 2a_{12}(0)t + a_{22}(0)$$

In the case where P(0) is given by Eq. (4.6-17), the corresponding elements of $P^{-1}(0)$ are $a_{11}(0) = 1/p_{11}(0)$, $a_{22}(0) = 1/p_{22}(0)$ and $a_{12}(0) = 0$. Thus,

$$P^{-1}(t) = \begin{bmatrix} t/r + 1/p_{11}(0) & -t^2/2r - t/p_{11}(0) \\ -t^2/2r - t/p_{11}(0) & t^3/3r + t^2/p_{11}(0) + 1/p_{22}(0) \end{bmatrix}$$

which, upon inversion, yields Eq. (4.6-18).

Method 4 – It has been mentioned that the choice of state variables is not unique. The present example serves well for exploration of this point. Suppose we use, as an alternative system state vector,

$$\underline{x}_a = \begin{bmatrix} \theta_0 \\ \epsilon \end{bmatrix}$$

Briefly, it follows that

$$H = [1 \quad t] \quad , \quad F = 0$$

and consequently that

$$\dot{P}_a^{-1} = \frac{1}{r}\begin{bmatrix} 1 & t \\ t & t^2 \end{bmatrix}$$

Integrating and inverting, assuming no *a priori* information (i.e., $P^{-1}(0) = 0$), yields

$$P_a = r\begin{bmatrix} 4/t & -6/t^2 \\ -6/t^2 & 12/t^3 \end{bmatrix}$$

This result *can* be reconciled with Eq. (4.6-19). The procedure is left as an exercise for the reader.

4.7 STATISTICAL STEADY STATE–THE WIENER FILTER

In the case where system and measurement dynamics are linear, constant coefficient equations (F, G, H are not functions of time) and the driving noise statistics are stationary (Q, R are not functions of time), the filtering process may reach a "steady state" wherein P is constant. *Complete observability* has been shown (Ref. 5) to be a sufficient condition for the existence of a steady-state solution. *Complete controllability* will assure that the steady-state solution is *unique*. Thus, for $\dot{P} = 0$, we have

$$FP_\infty + P_\infty F^T + GQG^T - P_\infty H^T R^{-1} HP_\infty = 0 \qquad (4.7\text{-}1)$$

where P_∞ denotes the steady-state value of P. In this steady state, the rate at which uncertainty builds (GQG^T) is just balanced by (a) the rate at which new information enters the system ($\underset{\infty}{P}H^T R^{-1} HP_\infty$), and (b) the system dissipation due to damping (expressed in F) (Ref. 6).

The corresponding steady-state optimal filter is given by

$$\dot{\hat{\underline{x}}}(t) = F\hat{\underline{x}}(t) + K_\infty [\underline{z}(t) - H\hat{\underline{x}}(t)] \tag{4.7-2}$$

where K_∞ is constant ($K_\infty = P_\infty H^T R^{-1}$). This equation may be rewritten

$$\dot{\hat{\underline{x}}}(t) - (F - K_\infty H)\,\hat{\underline{x}}(t) = K_\infty \underline{z}(t) \tag{4.7-3}$$

Laplace transforming both sides and neglecting initial conditions yields

$$(sI - F + K_\infty H)\,\hat{\underline{x}}(s) = K_\infty \underline{z}(s) \tag{4.7-4}$$

where s is the Laplace transform variable. Thus,

$$\hat{\underline{x}}(s) = [(sI - F + K_\infty H)^{-1} K_\infty]\ \underline{z}(s) \tag{4.7-5}$$

The quantity in brackets (representing the transfer function), which operates on $\underline{z}(s)$ to produce $\hat{\underline{x}}(s)$, is the *Wiener optimal filter*. For example, in the scalar case, Eq. (4.7-5) may be written

$$\frac{\hat{x}(s)}{z(s)} = \frac{k_\infty}{s + (k_\infty h - f)} \tag{4.7-6}$$

which is the optimum filter in conventional transfer function form.

Underlying Wiener filter design (Ref. 13), is the so-called *Wiener-Hopf* (integral) *equation*, its solution through spectral factorization and the *practical* problem of synthesizing the theoretically optimal filter from its impulse response. The contribution of Kalman and Bucy was recognition of the fact that the integral equation could be converted into a nonlinear differential equation, whose solution contains all the necessary information for design of the optimal filter. The problem of spectral factorization in the Wiener filter is analogous to the requirement for solving n(n+1)/2 coupled nonlinear algebraic equations in the Kalman filter [Eq. (4.7-1)].

REFERENCES

1. Kalman, R.E., "New Methods in Wiener Filtering," Chapter 9, *Proc. of the First Symposium on Engineering Applications of Random Function Theory and Probability*, John Wiley and Sons, Inc., New York, 1963.
2. Kalman, R.E., "A New Approach to Linear Filtering and Prediction Problems," *Journal of Basic Engineering* (ASME), Vol. 82D, March 1960, pp. 35-45.
3. Smith, G.L., "The Scientific Inferential Relationships Between Statistical Estimation, Decision Theory, and Modern Filter Theory," *Proc. JACC*, Rensselaer Polytechnic Inst., June 1965, pp. 350-359.

4. Laning, J.H., Jr. and Battin, R.H., *Random Processes in Automatic Control*, McGraw-Hill Book Company, Inc., New York, 1956, pp. 269-275.

5. Kalman, R.E. and Bucy, R., "New Results in Linear Filtering and Prediction," *Journal of Basic Engineering* (ASME), Vol. 83D, 1961, pp. 95-108.

6. Bryson, A.E., Jr., and Ho, Y.C., *Applied Optimal Control*, Blaisdell Publishing Company, Waltham, Mass., 1969, pp. 366-368.

7. Sorenson, H.W. and Stubberud, A.R., "Linear Estimation Theory," Chapter 1 of *Theory and Applications of Kalman Filtering*, C.T. Leondes, Editor, NATO AGARD-ograph 139, February 1970.

8. Kalman, R.E., Ho, Y.C., and Narendra, K.S., "Controllability of Linear Dynamical Systems," *Contributions to Differential Equations*, Vol. I, McMillan, New York, 1961.

9. Bryson, A.E., Jr. and Johansen, D.E., "Linear Filtering for Time-Varying Systems Using Measurements Containing Colored Noise," *IEEE Trans. on Automatic Control*, Vol. AC-10, No. 1, January 1965, pp. 4-10.

10. Deyst, J.J., "A Derivation of the Optimum Continuous Linear Estimator for Systems with Correlated Measurement Noise," *AIAA Journal*, Vol. 7, No. 11, September 1969, pp. 2116-2119.

11. Stear, E.B. and Stubberud, A.R., "Optimal Filtering for Gauss-Markov Noise," *Int. J. Control*, Vol. 8, No. 2, 1968, pp. 123-130.

12. Bryson, A.E., Jr. and Henrikson, L.J., "Estimation Using Sampled Data Containing Sequentially Correlated Noise," *J. Spacecraft and Rockets*, Vol. 5, No. 6, June 1968, pp. 662-665.

13. Wiener, N., *The Extrapolation, Interpolation and Smoothing of Stationary Time Series*, John Wiley & Sons, Inc., New York, 1949.

14. Kasper, J.F., Jr., "A Skywave Correction Adjustment Procedure for Improved OMEGA Accuracy," *Proceedings of the ION National Marine Meeting*, New London, Connecticut, October 1970.

15. Sutherland, A.A., Jr. and Gelb, A., "Application of the Kalman Filter to Aided Inertial Systems," Naval Weapons Center, China Lake, California, NWC TP 4652, August 1968.

16. Lee, R.C.K., *Optimal Estimation, Identification, and Control*, Research Monograph No. 28, MIT Press, Cambridge, Mass. 1964.

17. Potter, J.E. and Fraser, D.C., "A Formula for Updating the Determinant of the Covariance Matrix," *AIAA Journal*, Vol. 5, No. 7, July 1967, pp. 1352-1354.

PROBLEMS

Problem 4-1

Repeat Example 1.0-1 by treating the two measurements (a) sequentially and (b) simultaneously, within the framework of Kalman filtering. Assume no *a priori* information.

Problem 4-2

Use the matrix inversion lemma to easily solve Problems 1-1 and 1-3 in a minimum number of steps.

Problem 4-3

Repeat Problem 1-4 by reformulating it as a Kalman filter problem and considering (a) simultaneous and (b) sequential measurement processing. Assume no *a priori* information about x_0.

Problem 4-4

The weighted-least-squares estimate, $\hat{\underline{x}}(-)$, corresponding to an original linear measurement set $\underline{z}_0$, measurement matrix H_0 and weighting matrix R_0 is given by [Eq. (4.0-5)]

$$\hat{\underline{x}}(-) = \left(H_0^T R_0^{-1} H_0\right)^{-1} H_0^T R_0^{-1} \underline{z}_0$$

Suppose an additional measurement set, $\underline{z}$, becomes available. Defining the following matrices for the complete measurement set,

$$H_1 = \begin{bmatrix} H_0 \\ H \end{bmatrix}, \quad \underline{z}_1 = \begin{bmatrix} \underline{z}_0 \\ \underline{z} \end{bmatrix}, \quad R_1 = \left[\begin{array}{c|c} R_0 & 0 \\ \hline 0 & R \end{array}\right]$$

the new estimate, $\hat{\underline{x}}(+)$, can be found as

$$\hat{\underline{x}}(+) = \left(H_1^T R_1^{-1} H_1\right)^{-1} H_1^T R_1^{-1} \underline{z}_1$$

Using the definitions for H_1, R_1, $\underline{z}$, above, show that $\hat{\underline{x}}(+)$ can be manipulated into the form ($P^{-1}(-) = H_0^T R_0^{-1} H_0$):

$$\hat{\underline{x}}(+) = \hat{\underline{x}}(-) + P(+) H^T R^{-1} [\underline{z} - H\hat{\underline{x}}(-)]$$

$$P^{-1}(+) = P^{-1}(-) + H^T R^{-1} H$$

Problem 4-5

Choose $\hat{\underline{x}}$ to minimize the particular scalar loss function

$$J = [\hat{\underline{x}} - \underline{x}(-)]^T P^{-1}(-) [\hat{\underline{x}} - \underline{x}(-)] + (\underline{z} - H\hat{\underline{x}})^T R^{-1} (\underline{z} - H\hat{\underline{x}})$$

and directly obtain the recursive, weighted-least-squares estimator,

$$\hat{\underline{x}}(+) = \hat{\underline{x}}(-) + P(+) H^T R^{-1} [\underline{z} - H\hat{\underline{x}}(-)]$$

$$P^{-1}(+) = P^{-1}(-) + H^T R^{-1} H$$

Problem 4-6

For the particular linear vector measurement equation, $\underline{z} = H\underline{x} + \underline{v}$, where $\underline{v} \sim N(\underline{0}, R)$ is independent of $\underline{x}$, the conditional probability $p(\underline{z}|\underline{x})$ can be written as

$$p(\underline{z}|\underline{x}) = \frac{p(\underline{x},\underline{z})}{p(\underline{x})} = \frac{p(\underline{x},\underline{v})}{p(\underline{x})} = p(\underline{v})$$

Demonstrate that the estimate, $\hat{\underline{x}}$, which maximizes $p(\underline{z}|\underline{x})$ (i.e., the maximum likelihood estimate) is found by minimizing the expression $(\underline{z} - H\hat{\underline{x}})^T R^{-1} (\underline{z} - H\hat{\underline{x}})$. Show that

$$\hat{\underline{x}} = \left(H^T R^{-1} H\right)^{-1} H^T R^{-1} \underline{z}$$

State the recursive form of this estimate.

Problem 4-7

Consider a gaussian vector $\underline{x}$, where $\underline{x} \sim N(\hat{\underline{x}}(-), P(-))$, and a linear measurement, $\underline{z} = H\underline{x}+\underline{v}$, where the gaussian measurement noise $\underline{v}$ is independent of $\underline{x}$ and $\underline{v} \sim N(\underline{0},R)$.

a) Show that $\underline{z} \sim N(H\hat{\underline{x}}(-), HP(-)H^T+R)$

b) Show that the *a posteriori* density function of $\underline{x}$ is given by (Ref. 16)

$$p(\underline{x}|\underline{z}) = \frac{p(\underline{x})p(\underline{v})}{p(\underline{z})}$$

c) By direct substitution into this relationship, obtain

$$p(\underline{x}|\underline{z}) = c \exp\left\{-\frac{1}{2}\left[[\underline{x}-\hat{\underline{x}}(-)]^T P^{-1}(-)[\underline{x}-\hat{\underline{x}}(-)] + (\underline{z}-H\underline{x})^T R^{-1}(\underline{z}-H\underline{x}) - [\underline{z}-H\hat{\underline{x}}(-)]^T [HP(-)H^T+R]^{-1}[\underline{z}-H\hat{\underline{x}}(-)]\right]\right\}$$

where c is a constant.

d) Complete the square in the expression in braces to obtain the form

$$p(\underline{x}|\underline{z}) = c \exp\left\{-\frac{1}{2}\left[[\underline{x}-\hat{\underline{x}}(+)] P^{-1}(+)[\underline{x}-\hat{\underline{x}}(+)]\right]\right\}$$

and thus, identify as the Bayesian maximum likelihood estimate (maximum *a posteriori* estimate)

$$\hat{\underline{x}}(+) = \hat{\underline{x}}(-) + P(+)H^TR^{-1}[\underline{z}-H\hat{\underline{x}}(-)]$$

$$P^{-1}(+) = P^{-1}(-) + H^TR^{-1}H$$

Problem 4-8

For the system and measurement equations,

$$\dot{\underline{x}} = F\underline{x} + G\underline{w}, \qquad \underline{w} \sim N(\underline{0},Q)$$

$$\underline{z} = H\underline{x} + \underline{v}, \qquad \underline{v} \sim N(\underline{0},R)$$

consider a linear filter described by

$$\dot{\hat{\underline{x}}} = K'\hat{\underline{x}} + K\underline{z}$$

where K' and K are to be chosen to optimize the estimate, $\hat{\underline{x}}$. First, by requiring that the estimate be *unbiased*, show that $K' = F - KH$, and thus obtain

$$\dot{\hat{\underline{x}}} = F\hat{\underline{x}} + K(\underline{z} - H\hat{\underline{x}})$$

Next, show that the covariance equation of the estimation *error* is

$$\dot{P} = (F - KH)P + P(F - KH)^T + GQG^T + KRK^T$$

Finally, choose K to yield a maximum rate of decrease of error by minimizing the scalar cost function, $J=\text{trace}[\dot{P}]$, and find the result:

$$K = PH^TR^{-1}$$

Problem 4-9

a) Can a Kalman filter separate two biases in the absence of *a priori* information? Given *a priori* information?

b) Can a Kalman filter separate two sinusoids of frequency ω_0 given no *a priori* information? Given *a priori* information about amplitude and/or phase?

c) Can a Kalman filter separate two markov processes with correlation time τ given no *a priori* information? Given *a priori* information?

Problem 4-10

A random variable, x, may take on any values in the range $-\infty$ to ∞. Based on a sample of k values, x_i, $i=1,2,\ldots,k$, we wish to compute the sample mean, $\hat{m}_k$, and sample variance, $\hat{\sigma}_k^2$, as estimates of the population mean, m, and variance, σ^2. Show that *unbiased* estimators for these quantities are:

$$\hat{m}_k = \frac{1}{k}\sum_{i=1}^{k} x_i$$

$$\hat{\sigma}_k^2 = \frac{1}{k-1}\sum_{i=1}^{k}(x_i-\hat{m}_k)^2$$

and recast these expressions in *recursive* form.

Problem 4-11

A simple dynamical system and measurement are given by

$$\dot{x} = ax + w\,, \qquad w \sim N(0, q)$$

$$z = bx + v\,, \qquad v \sim N(0, r)$$

Show that the optimal filter error variance is given by

$$p(t) = \frac{(ap_0 + q)\sinh \beta t + \beta p_0 \cosh \beta t}{\left(\frac{b^2}{r} p_0 - a\right)\sinh \beta t + \beta \cosh \beta t}$$

where

$$\beta = a\sqrt{1 + \frac{b^2 q}{a^2 r}}$$

Demonstrate that the steady-state value of p(t) is given by

$$p_\infty = \frac{ar}{b^2}\left(1 + \sqrt{1 + \frac{b^2 q}{a^2 r}}\right)$$

independent of p_0. Draw a block diagram of the optimal filter and discuss its steady-state behavior.

Problem 4-12

A second-order system and scalar measurement are illustrated in Fig. 4-1, where $w \sim N(0,q)$. Draw a block diagram of the Kalman optimal filter for this system, and show that the steady-state Kalman gain matrix is

$$K_\infty = \begin{bmatrix} \sqrt{\frac{q}{r}} \\ \frac{\beta}{\alpha}\left(-1 + \sqrt{1 + \frac{2\alpha}{\beta^2}\sqrt{\frac{q}{r}}}\right) \end{bmatrix}$$

Repeat the calculation for the case where simultaneous measurements of x_1 and x_2 are made. Assume uncorrelated measurement errors.

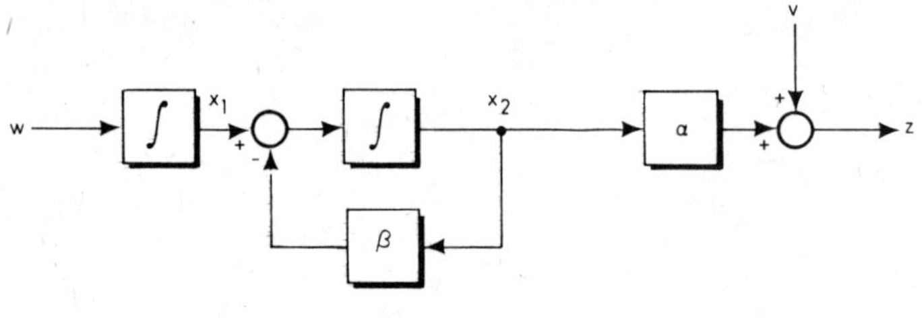

Figure 4-1 Example Second-Order System

Problem 4-13

Show that the optimal filter for detecting a sine wave in white noise, based upon the measurement

$$z(t) = x_1(t)\cos(t-T) + x_2(t)\sin(t-T) + v(t)$$

where $v \sim N(0,r)$, is as shown in Fig. 4-2, where (Ref. 1)

$$k_{11}(t) = \frac{2(t-t_0) + \sin 2(t-t_0)}{4r\,[(t-t_0)^2 - \sin^2(t-t_0)]}$$

$$k_{12}(t) = \frac{\sin^2(t-t_0)}{2r\,[(t-t_0)^2 - \sin^2(t-t_0)]}$$

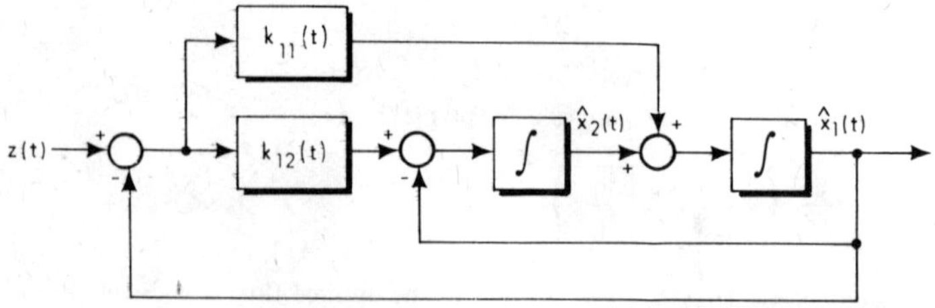

Figure 4-2 Optimal Sine Wave Estimator

Problem 4-14

The output of an integrator driven by white noise w [where $w \sim N(0,q)$] is sampled every Δ seconds (where $\Delta = t_{k+1} - t_k$ = constant), in the presence of a measurement noise, v_k [where $v_k \sim N(0, r_0)$]. Assume that there is no *a priori* information. Calculate $p_k(+)$ and $p_{k+1}(-)$ for k = 0,1,2, etc. and thus, demonstrate that, for *all* k,

$$p_k(+) = r_0 \quad \text{and} \quad p_{k+1}(-) = r_0 + q\Delta \quad \text{for} \quad q\Delta \gg r_0$$

and

$$p_k(+) = \frac{r_0}{k+1} = p_{k+1}(-) \qquad \text{for } r_0 \gg q\Delta$$

Sketch and physically interpret error variance curves for each of these cases.

Problem 4-15

Reformulate Example 4.2-4 using the alternate state vector $\underline{x}_a = [\delta p(0)\ \delta v(0)\ \delta a(0)]^T$. Make appropriate arguments about initial conditions and employ the matrix inversion lemma to arrive at the following result:

$$P_a(T^+) = \frac{1}{\Delta_a(T)} \left[\begin{array}{c|c|c} \dfrac{\sigma_p^2}{p_{22}(0)p_{33}(0)} + \dfrac{T^2}{p_{33}(0)} + \dfrac{T^4}{4p_{22}(0)} & -\dfrac{T}{p_{33}(0)} & -\dfrac{T^2}{2p_{22}(0)} \\ \hline & \dfrac{\sigma_p^2}{p_{11}(0)p_{33}(0)} + \dfrac{1}{p_{33}(0)} + \dfrac{T^4}{4p_{11}(0)} & -\dfrac{T^3}{2p_{11}(0)} \\ \hline & & \dfrac{\sigma_p^2}{p_{11}(0)p_{22}(0)} + \dfrac{1}{p_{22}(0)} + \dfrac{T^2}{p_{11}(0)} \end{array} \right]$$

where

$$\Delta_a(T) = \frac{1}{p_{11}(0)p_{22}(0)p_{33}(0)} \left[\sigma_p^2 + p_{11}(0) + p_{22}(0)T^2 + p_{33}(0)\frac{T^4}{4} \right]$$

and $p_{11}(0) = E[\delta p^2(0)]$, $p_{22}(0) = E[\delta v^2(0)]$, $p_{33}(0) = E[\delta a^2(0)]$. Reconcile this result with the *exact* version of Eq. (4.2-25).

Problem 4-16

By measuring the line of sight to a star, a spacecraft stellar navigation system can measure two of the three angles comprising $\underline{\theta}$, the navigation coordinate frame misalignment. For $\underline{\theta} = [\theta_1\ \theta_2\ \theta_3]^T$, $P(0) = \sigma^2 I$, demonstrate the value of a "single-star fix" (i.e., a measurement of θ_1 and θ_2). Then, assume another measurement on a different star (i.e., a measurement of θ_1 and θ_3), and thus, demonstrate the value of a "two-star fix." Assume that each component of $\underline{\theta}$ is observed with an uncorrelated measurement error, viz. (i=1, 2 or 3)

$$z_i = \theta_i + v_i, \qquad v_i \sim N(0, \sigma_i^2)$$

Specifically, show that ($\sigma_i^2 \ll \sigma^2$ for i = 1, 2, 3)

$$\text{trace } P \begin{pmatrix} \text{single} \\ \text{star} \\ \text{fix} \end{pmatrix} \approx \sigma^2$$

and

$$\text{trace } P \begin{pmatrix}\text{two}\\ \text{star}\\ \text{fix}\end{pmatrix} \approx \frac{\sigma_1^2}{2} + \sigma_2^2 + \sigma_3^2$$

Problem 4-17

A *polynomial tracking filter* is designed to optimally estimate the state of the system described by $\ddot{x}(t) = 0$, given scalar measurements $z = x + v$, $v \sim N(0,r)$. Assuming no *a priori* information and measurements spaced τ time units apart, show that $(k = 1, 2, \ldots)$

$$P_{k+1}(+) = \frac{2r}{(k+1)(k+2)} \begin{bmatrix} 2k+1 & \dfrac{3}{\tau} \\ \dfrac{3}{\tau} & \dfrac{6}{k\tau^2} \end{bmatrix}$$

and

$$K_{k+1} = \begin{bmatrix} \dfrac{2(2k+1)}{(k+1)(k+2)} \\ \dfrac{6}{\tau(k+1)(k+2)} \end{bmatrix}$$

(Hint: It may prove useful to employ the relationship $P(0)^{-1} = \lim_{\epsilon \to 0} \epsilon I$, and to solve by induction.)

Problem 4-18

A system and measurement are given as,

$$\dot{\underline{x}} = F\underline{x} + G\underline{w}, \qquad \underline{w} \sim N(\underline{0},Q)$$

$$\underline{z} = H\underline{x} + \underline{v}\ , \qquad \underline{v} \sim N(\underline{0},R)$$

Show that the *optimal differentiator* associated with a particular output of the system,

$$\underline{y} = M\underline{x}$$

is

$$\hat{\dot{\underline{y}}} = (\dot{M} + MF)\,\hat{\underline{x}}$$

Why is it incorrect to compute $\hat{\dot{\underline{y}}}$ by forming $\hat{\underline{y}} = M\hat{\underline{x}}$ and then differentiating the result?

Problem 4-19

Manipulate the discrete covariance matrix update equation into the form

$$H_k P_k(+) = R_k \left[H_k P_k(-) H_k^T + R_k\right]^{-1} H_k P_k(-)$$

and thus show that, when H_k is square and nonsingular,

$$|P_k(+)| = \frac{|P_k(-)|\,|R_k|}{|H_k P_k(-) H_k^T + R_k|}$$

This formula for updating the determinant of the covariance matrix has been shown to be valid *independent* of whether H_k is square (Ref. 17).

Problem 4-20

Observations of a constant parameter x are made through a digital instrument with quantization levels of width q. A reasonable approach for "small" q is to model the quantizer as a noise source whose distribution is uniform over $(-q/2, q/2)$, and which is uncorrelated with x – i.e.,

$$z = x + v, \qquad v \text{ is uniform over } \left(-\frac{q}{2}, \frac{q}{2}\right)$$

Find the optimal linear estimator for x, given that

$$E[x] = m, \; E[x^2] = \sigma^2$$

Problem 4-21

Observations z of the constant parameter x are corrupted by multiplicative noise – i.e., a scale factor error, η,

$$z = (1 + \eta)\, x$$

where

$$E[x^2] = \sigma_x{}^2, \; E[\eta^2] = \sigma_\eta{}^2$$

$$E[x] = E[\eta] = E[\eta x] = 0$$

(a) Find the optimal linear estimate of x based on a measurement $z(\hat{x} = kz)$. (b) What is the mean square error in the estimate?

Problem 4-22

Design an optimal linear filter to separate a noise n(t) from a signal s(t) when the spectral densities for the signal and noise are given by:

$$\Phi_{ss}(\omega) = \frac{1}{\omega^2 + 1}, \; \Phi_{nn}(\omega) = \frac{2\omega^2}{\omega^4 + 1}$$

(Hint: this problem in Wiener filtering can be solved as the steady-state portion of a Kalman filtering problem).

Problem 4-23

Consider a satellite in space which is spinning at a constant, but unknown, rate. The angular position is measured every T seconds, viz.:

$$z_k = \theta_k + v_k, \qquad k = 1, 2, 3 \ldots$$

$$E\left[v_k{}^2\right] = (5 \text{ deg})^2$$

where θ_k is the angular position at $t = kT$, and v_k is the measurement error. The uncertainties in initial conditions are described by

$$E\left[\theta_0^2\right] = (20\text{ deg})^2,\quad E\left[\dot{\theta}_0^2\right] = (20\text{ deg/sec})^2$$

$$E[\theta_0] = E[\dot{\theta}_0] = E[\theta_0\dot{\theta}_0] = 0$$

Write the system state equations and then the linear filter equations to give an optimal estimate of the position and velocity after each observation.

Problem 4-24

An RC filter with time constant τ is excited by white noise, and the output is measured every T seconds. The output at the sample times obeys the equation

$$x_k = e^{-\frac{T}{\tau}} \cdot x_{k-1} + w_{k-1}, \qquad k = 1, 2, \ldots$$

where

$$E[x_0] = 1,\ E\left[x_0^2\right] = 2$$

$$E[w_k] = 0,\ E[w_j w_k] = \begin{cases} 2 & j = k \\ 0 & j \neq k \end{cases}$$

$$T = \tau = 0.1 \text{ sec}$$

The measurements are described by

$$z_k = x_k + v_k, \qquad k = 1, 2, \ldots$$

where v_k is a white sequence and has the following probability density function:

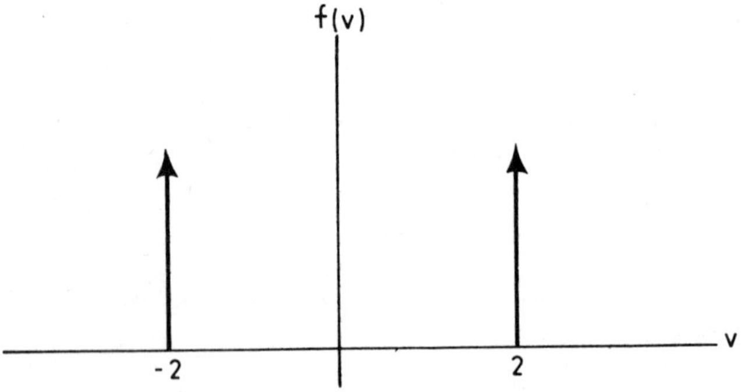

Figure 4-3

Find the best linear estimate (a) of x_1 based on z_1, and (b) x_2 based on z_1 and z_2; when $z_1 = 1.5$ and $z_2 = 3.0$.

Problem 4-25

Design a data processor to determine the position of a ship at sea. Examine one dimensional motion (e.g., North-South) under the following assumptions: the ship's velocity relative to the water is constant, but unknown with mean m_s and variance σ_s^2; the current has a constant mean value m_c. The random component of current can be modeled as integrated white noise of spectral density q_c and initial variance σ_c^2; position measurements are made every T hours by radio or sextant methods. The errors in these readings are independent of each other and have mean zero and variance σ_R^2 (miles)^2; the initial position estimate (T hours before the first measurement) is x_0, with uncertainty σ_0^2.

a) Set up the state equations for this system. Show clearly what differential equations are needed.

b) Set up the filter equations to give a *continuous* estimate of position and the error in this estimate. Be sure to specify the contents of all matrices and vectors, and all initial conditions.

Problem 4-26

The differential equation for the altitude hold mode of an airplane autopilot is given by

$$\ddot{h}(t) + 0.006\dot{h}(t) + 0.003h(t) = 0.3[\dot{h}_c(t) + 0.01h_c(t)]$$

where h represents altitude and h_c is commanded altitude. The altitude command h_c is modelled as a constant h_{c_0} plus gaussian white noise $\delta h_c(t)$ in the command channel

$$h_c(t) = h_{c_0} + \delta h_c(t)$$

The constant h_{c_0} is a normal random variable with statistics

$$\text{mean} = 10{,}000 \text{ ft}$$
$$\text{variance} = 250{,}000 \text{ ft}^2$$

Noise in the command channel has the following statistics

$$\delta h_c \sim N(0, 400 \text{ ft}^2 \text{ sec})$$

and δh_c is independent of all other variables.

Discrete measurements of altitude are available every 10 seconds and we wish to process them to obtain the minimum variance estimate of altitude. The altitude measurements contain random errors

$$z(t_k) = h(t_k) + v_h(t_k)$$

where $z(t_k)$ is measured altitude and $v_h(t_k)$ is a white sequence of measurement errors,

$$v_h \sim N(0, 100 \text{ ft}^2)$$

Determine the difference equations defining the minimum variance estimator of h(t). Write these equations out in scalar form.

Problem 4-27

Consider the circuit in Fig. 4-4. It has been constructed and sealed into the proverbial black box. Capacitor C_1 has a very low voltage rating and it is desired to monitor the voltage across C_1 to determine when it exceeds the capacitor limit.

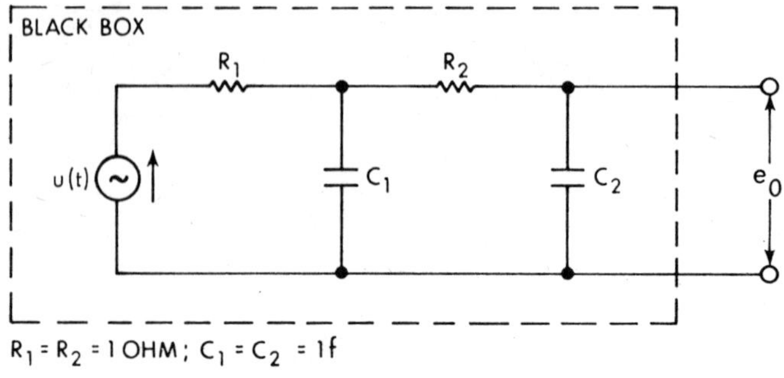

Figure 4-4

The only measurement that can be made on this system is the output voltage, e_0. However, thanks to an exceedingly good voltmeter, perfect measurements can be made of this voltage at discrete times. In order to estimate the voltage across C_1, assume that u(t) can be described as

$$u \sim N(0, 2\ \text{volt}^2\ \text{sec})$$

Determine an expression for the optimal estimate of the voltage across C_1. Assume that the system starts up with no charge in the capacitors. Plot the variance of the error in the estimate as a function of time, taking measurements every half second for two seconds.

Problem 4-28

The motion of a unit mass, in an inverse square law force field, is governed by a pair of second-order equations in the radius r and the angle θ.

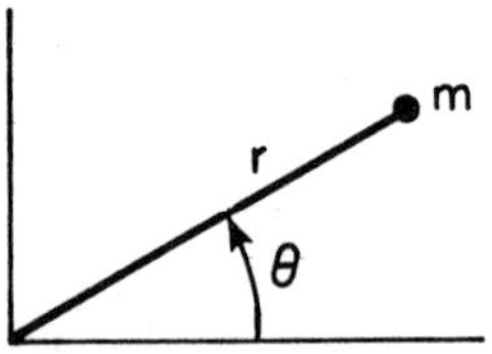

Figure 4-5

If we assume that the unit mass has the capability of thrusting in the radial direction with a thrust u_1 and in the tangential direction with a thrust u_2, then we have

$$\ddot{r}(t) = r(t)\,\dot{\theta}(t)^2 - \frac{G_0}{r^2(t)} + u_1(t)$$

$$\ddot{\theta}(t) = -2\,\dot{\theta}(t)\,\frac{\dot{r}(t)}{r(t)} + \left[\frac{1}{r(t)}\right] u_2(t)$$

If $u_1(t) = u_2(t) = 0$, these equations admit the solution

$$r(t) = R \qquad (R \text{ constant})$$

$$\theta(t) = \omega t \qquad (\omega \text{ constant})$$

where $R^3\,\omega^2 = G_0$ – i.e., circular orbits are possible. Let x_1, x_2, x_3, and x_4 be given by the relationships

$$x_1 = r - R, \qquad x_2 = \dot{r}, \qquad x_3 = R\,(\theta - \omega t), \qquad x_4 = R\,(\dot{\theta} - \omega)$$

and show that the linearized equations of motion about the nominally circular solution are

$$\begin{bmatrix} \dot{x}_1(t) \\ \dot{x}_2(t) \\ \dot{x}_3(t) \\ \dot{x}_4(t) \end{bmatrix} = \begin{bmatrix} 0 & 1 & 0 & 0 \\ 3\omega^2 & 0 & 0 & 2\omega \\ 0 & 0 & 0 & 1 \\ 0 & -2\omega & 0 & 0 \end{bmatrix} \begin{bmatrix} x_1(t) \\ x_2(t) \\ x_3(t) \\ x_4(t) \end{bmatrix} + \begin{bmatrix} 0 & 0 \\ 1 & 0 \\ 0 & 0 \\ 0 & 1 \end{bmatrix} \begin{bmatrix} u_1(t) \\ u_2(t) \end{bmatrix}$$

Note that there is no process noise included in the above state equations.

It is desired to measure these small orbital deviations from observations on the ground. Two proposals are presented. (a) In an effort to keep the measurement stations rather simple and inexpensive, only angle (x_3) measurements will be made. However, the designer realizes the very likely possibility of measurement errors and includes an optimal filter in his proposal for estimating the states. The measurement may be represented as

$$z(t) = x_3(t) + v_3(t), \qquad v_3 \sim N(0, q_3)$$

(b) The second design proposes to use measurements of range (x_1). In this case

$$z(t) = x_1(t) + v_1(t), \qquad v_1(t) \sim N(0, q_1)$$

It is your task to determine which of these proposals is superior.

Problem 4-29

Consider the scalar moving average time-series model,

$$z_k = r_k + r_{k-1}$$

where $\{r_k\}$ is a unit-variance, white gaussian sequence. Show that the optimal one-step predictor for this model is (assume $P_0 = 1$)

$$\hat{z}_{k+1}(-) = \frac{k+1}{k+2}\,[z_k - \hat{z}_k(-)]$$

(Hint: use the state-space formulation of Section 3.4)

5. OPTIMAL LINEAR SMOOTHING

Smoothing is a non-real-time data processing scheme that uses all measurements between 0 and T to estimate the state of a system at a certain time t, where $0 \leqslant t \leqslant T$. The smoothed estimate of $\underline{x}(t)$ based on all the measurements between 0 and T is denoted by $\hat{\underline{x}}(t|T)$. An optimal smoother can be thought of as a suitable combination of two optimal filters. One of the filters, called a "forward filter," operates on all the data before time t and produces the estimate $\hat{\underline{x}}(t)$; the other filter, called a "backward filter," operates on all the data after time t and produces the estimate $\hat{\underline{x}}_b(t)$. Together these two filters utilize *all* the available information; see Fig. 5.0-1. The two estimates they provide have

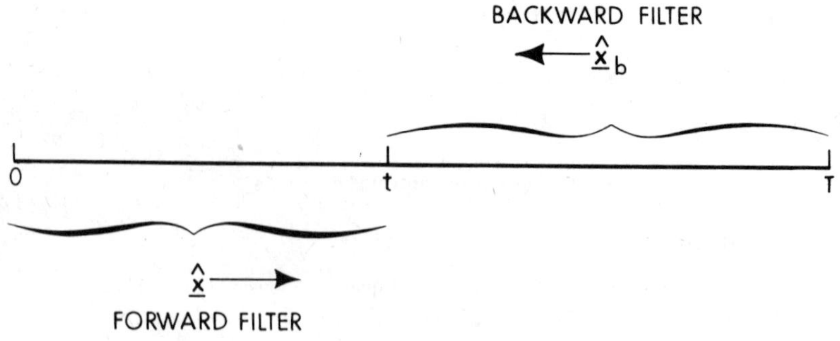

Figure 5.0-1 Relationship of Forward and Backward Filters

uncorrelated errors, since process and measurement noises are assumed white. This suggests that the optimal combination of $\hat{\underline{x}}(t)$ and $\hat{\underline{x}}_b(t)$ will, indeed, yield the optimal smoother; proof of this assertion can be found in Ref. 1.

Three types of smoothing are of interest. In *fixed-interval smoothing*, the initial and final times 0 and T are fixed and the estimate $\hat{\underline{x}}(t|T)$ is sought, where t varies from 0 to T. In *fixed-point smoothing*, t is fixed and $\hat{\underline{x}}(t|T)$ is sought as T increases. In *fixed-lag smoothing*, $\hat{\underline{x}}(T-\Delta|T)$ is sought as T increases, with Δ held fixed.

In this chapter the two-filter form of optimal smoother is used as a point of departure. Fixed-interval, fixed-point and fixed-lag smoothers are derived for the continuous-time case, with corresponding results presented for the discrete-time case, and several examples are discussed.

5.1 FORM OF THE OPTIMAL SMOOTHER

Following the lead of the previous chapter, we seek the optimal smoother in the form

$$\hat{\underline{x}}(t|T) = A\hat{\underline{x}}(t) + A'\hat{\underline{x}}_b(t) \tag{5.1-1}$$

where A and A′ are weighting matrices to be determined. Replacing each of the estimates in this expression by the corresponding true value plus an estimation error, we obtain

$$\tilde{\underline{x}}(t|T) = [A + A' - I]\ \underline{x}(t) + A\tilde{\underline{x}}(t) + A'\tilde{\underline{x}}_b(t) \tag{5.1-2}$$

For unbiased filtering errors, $\tilde{\underline{x}}(t)$ and $\tilde{\underline{x}}_b(t)$, we wish to obtain an unbiased smoothing error, $\tilde{\underline{x}}(t|T)$; thus, we set the expression in brackets to zero. This yields

$$A' = I - A \tag{5.1-3}$$

and, consequently,

$$\tilde{\underline{x}}(t|T) = A\tilde{\underline{x}}(t) + (I - A)\,\tilde{\underline{x}}_b(t) \tag{5.1-4}$$

Computing the smoother error covariance, we find

$$\begin{aligned} P(t|T) &= E[\tilde{\underline{x}}(t|T)\,\tilde{\underline{x}}^T(t|T)] \\ &= AP(t)\,A^T + (I - A)\,P_b(t)(I - A)^T \end{aligned} \tag{5.1-5}$$

where product terms involving $\tilde{\underline{x}}(t)$ and $\tilde{\underline{x}}_b(t)$ do not appear. P(t|T) denotes the smoother error covariance matrix, while P(t) and $P_b(t)$ denote forward and backward optimal filter error covariance matrices, respectively.

OPTIMIZATION OF THE SMOOTHER

Once again, following the previous chapter, we choose that value of A which minimizes the trace of P(t|T). Forming this quantity, differentiating with respect to A and setting the result to zero, we find

$$0 = 2AP + 2(I - A)P_b(-I) \tag{5.1-6}$$

or

$$A = P_b(P + P_b)^{-1} \tag{5.1-7}$$

and, correspondingly

$$I - A = P(P + P_b)^{-1} \tag{5.1-8}$$

Inserting these results into Eq. (5.1-5), we obtain

$$P(t|T) = P_b(P + P_b)^{-1} P(P + P_b)^{-1} P_b + P(P + P_b)^{-1} P_b(P + P_b)^{-1} P \tag{5.1-9}$$

By systematically combining factors in each of the two right-side terms of this equation, we arrive at a far more compact result. The algebraic steps are sketched below,

$$\begin{aligned} P(t|T) &= P_b(P+P_b)^{-1} P \left(I + P_b^{-1} P\right)^{-1} + P(P+P_b)^{-1} P_b \left(P^{-1} P_b + I\right)^{-1} \\ &= P_b(P+P_b)^{-1} \left(P^{-1} + P_b^{-1}\right)^{-1} + P(P+P_b)^{-1} (P^{-1} + P_b^{-1}) \\ &= \left(P^{-1} + P_b^{-1}\right)^{-1} \end{aligned} \tag{5.1-10}$$

or

$$P^{-1}(t|T) = P^{-1}(t) + P_b^{-1}(t) \tag{5.1-11}$$

From Eq. (5.1-11), $P(t|T) \leqslant P(t)$, which means that the smoothed estimate of $\underline{x}(t)$ is always better than or equal to its filtered estimate. This is shown graphically in Fig. 5.1-1. Performing similar manipulations on Eq. (5.1-1), we find

$$\begin{aligned} \hat{\underline{x}}(t|T) &= A\hat{\underline{x}}(t) + (I - A)\, \hat{\underline{x}}_b(t) \\ &= P_b(P + P_b)^{-1}\, \hat{\underline{x}}(t) + P(P + P_b)^{-1}\, \hat{\underline{x}}_b(t) \\ &= \left(P^{-1} + P_b^{-1}\right)^{-1} P^{-1} \hat{\underline{x}}(t) + \left(P^{-1} + P_b^{-1}\right)^{-1} P_b^{-1} \hat{\underline{x}}_b(t) \\ &= P(t|T) \left[P^{-1}(t)\, \hat{\underline{x}}(t) + P_b^{-1}(t)\, \hat{\underline{x}}_b(t)\right] \end{aligned} \tag{5.1-12}$$

Equations (5.1-11) and (5.1-12) are the results of interest.

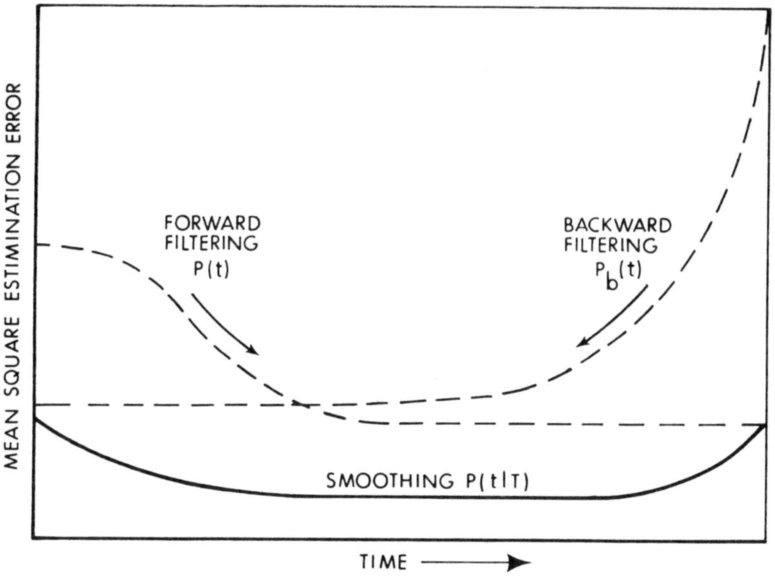

Figure 5.1-1 Advantage of Performing Optimal Smoothing

REINTERPRETATION OF PREVIOUS RESULTS

It is interesting to note that we could have arrived at these expressions by interpretation of optimal filter relationships. In the subsequent analogy, estimates $\hat{\underline{x}}_b$ from the backward filter will be thought of as providing "measurements" with which to update the forward filter. In the corresponding "measurement equation," H = I, as the total state vector is estimated by the backward filter. Clearly, the "measurement error" covariance matrix is then represented by P_b. From Eq (4.2-19), in which $P_k^{-1}(-)$ and $P_k^{-1}(+)$ are now interpreted as $P^{-1}(t)$ and $P^{-1}(t|T)$, respectively, we obtain

$$P^{-1}(t|T) = P^{-1}(t) + P_b^{-1}(t) \tag{5.1-13}$$

Equations (4.2-16b) and (4.2-20) provide the relationships

$$K_k = P_k(+)\,H_k^T R_k^{-1} \rightarrow P(t|T)\,P_b^{-1}(t)$$

$$(I - K_k H_k) = P_k(+)\,P_k^{-1}(-) \rightarrow P(t|T)\,P^{-1}(t)$$

which, when inserted in Eq. (4.2-5) yield the result

$$\hat{\underline{x}}(t|T) = P(t|T)\left[P^{-1}(t)\,\hat{\underline{x}}(t) + P_b^{-1}(t)\,\hat{\underline{x}}_b(t)\right], \tag{5.1-14}$$

where $\hat{\underline{x}}_k(-)$ and $\hat{\underline{x}}_k(+)$ have been interpreted as $\hat{\underline{x}}(t)$ and $\hat{\underline{x}}(t|T)$, respectively. Thus, we arrive at the same expressions for the optimal smoother and its error covariance matrix as obtained previously.

5.2 OPTIMAL FIXED-INTERVAL SMOOTHER

The forward-backward filter approach provides a particularly simple mechanism for arriving at a set of optimal smoother equations. Other formulations are also possible, one of which is also presented in this section. For the moment, we restrict our attention to time-invariant systems.

FORWARD-BACKWARD FILTER FORMULATION OF THE OPTIMAL SMOOTHER

For a system and measurement given by

$$\dot{\underline{x}} = F\underline{x} + G\underline{w}, \qquad \underline{w} \sim N(\underline{0}, Q)$$

$$\underline{z} = H\underline{x} + \underline{v}, \qquad \underline{v} \sim N(\underline{0}, R) \tag{5.2-1}$$

the equations defining the *forward* filter are, as usual,

$$\dot{\hat{\underline{x}}} = F\hat{\underline{x}} + PH^T R^{-1} [\underline{z} - H\hat{\underline{x}}], \qquad \hat{\underline{x}}(0) = \hat{\underline{x}}_0 \tag{5.2-2}$$

$$\dot{P} = FP + PF^T + GQG^T - PH^T R^{-1} HP, \qquad P(0) = P_0 \tag{5.2-3}$$

The equations defining the *backward* filter are quite similar. Since this filter runs backward in time, it is convenient to set $\tau = T - t$. Writing Eq. (5.2-1) in terms of τ, gives*

$$\frac{d}{d\tau}\underline{x} = -\frac{d}{dt}\underline{x}$$

$$= -F\underline{x} - G\underline{w} \tag{5.2-4}$$

$$\underline{z}(\tau) = H\underline{x} + \underline{v} \tag{5.2-5}$$

for $0 \leqslant \tau \leqslant T$. By analogy with the forward filter, the equations for the backward filter can be written changing F to –F and G to –G. This results in

$$\frac{d}{d\tau}\hat{\underline{x}}_b = -F\hat{\underline{x}}_b + P_b H^T R^{-1} [\underline{z} - H\hat{\underline{x}}_b] \tag{5.2-6}$$

$$\frac{d}{d\tau} P_b = -FP_b - P_b F^T + GQG^T - P_b H^T R^{-1} HP_b \tag{5.2-7}$$

At time t = T, the smoothed estimate must be the same as the forward filter estimate. Therefore, $\hat{\underline{x}}(T|T) = \hat{\underline{x}}(T)$ and $P(T|T) = P(T)$. The latter result, in combination with Eq. (5.1-11), yields the boundary condition on P_b^{-1},

*In this chapter, a dot denotes differentiation with respect to (forward) time t. Differentiation with respect to backward time is denoted by $d/d\tau$.

$$P_b^{-1}(t = T) = 0 \quad \text{or} \quad P_b^{-1}(\tau = 0) = 0 \tag{5.2-8}$$

but the boundary condition on $\hat{\underline{x}}_b(T)$ is yet unknown. One way of avoiding this problem is to transform Eq. (5.2-6) by defining the new variable

$$\underline{s}(t) = P_b^{-1}(t)\,\hat{\underline{x}}_b(t) \tag{5.2-9}$$

where, since $\hat{\underline{x}}_b(T)$ is finite, it follows that

$$\underline{s}(t = T) = \underline{0} \quad \text{or} \quad \underline{s}(\tau = 0) = \underline{0} \tag{5.2-10}$$

Computational considerations regarding the equations above lead us to their reformulation in terms of P_b^{-1}. Using the relationship

$$\frac{d}{d\tau} P_b^{-1} = -P_b^{-1}\left(\frac{d}{d\tau} P_b\right) P_b^{-1} \tag{5.2-11}$$

Eq. (5.2-7) can be written as

$$\frac{d}{d\tau} P_b^{-1} = P_b^{-1}F + F^T P_b^{-1} - P_b^{-1} G Q G^T P_b^{-1} + H^T R^{-1} H \tag{5.2-12}$$

for which Eq. (5.2-8) is the appropriate boundary condition. Differentiating Eq. (5.2-9) with respect to τ and employing Eqs. (5.2-6) and (5.2-12) and manipulating, yields

$$\frac{d}{d\tau} \underline{s} = (F^T - P_b^{-1} G Q G^T)\,\underline{s} + H^T R^{-1} \underline{z} \tag{5.2-13}$$

for which Eq. (5.2-10) is the appropriate boundary condition. Equations (5.1-11, 12) and (5.2-2, 3, 12, 13) define the optimal smoother. See Table 5.2-1, in which alternate expressions for $\hat{\underline{x}}(t|T)$ and $P(t|T)$, which obviate the need for unnecessary matrix inversions, are also presented (Ref. 1). These can be verified by algebraic manipulation. The results presented in Table 5.2-1 are for the general, time-varying case.

ANOTHER FORM OF THE EQUATIONS

Several other forms of the smoothing equations may also be derived. One is the Rauch-Tung-Striebel form (Ref. 3), which we utilize in the sequel. This form, which does not involve backward filtering *per se*, can be obtained by differentiating Eqs. (5.1-11) and (5.1-12) and using Eq. (5.2-12). It is given by Eqs. (5.2-2) and (5.2-3) and*

*From this point on, all discussion pertains to the general time-varying case unless stated otherwise. However, for notational convenience, explicit dependence of F, G, H, Q, R upon t may not be shown.

TABLE 5.2-1 SUMMARY OF CONTINUOUS, FIXED-INTERVAL OPTIMAL LINEAR SMOOTHER EQUATIONS, TWO-FILTER FORM

System Model	$\dot{\underline{x}}(t) = F(t)\underline{x}(t) + G(t)\underline{w}(t), \qquad \underline{w}(t) \sim N[\underline{0}, Q(t)]$
Measurement Model	$\underline{z}(t) = H(t)\underline{x}(t) + \underline{v}(t), \qquad \underline{v}(t) \sim N[\underline{0}, R(t)]$
Initial Conditions	$E[\underline{x}(0)] = \hat{\underline{x}}_0, \quad E[(\underline{x}(0) - \hat{\underline{x}}_0)(\underline{x}(0) - \hat{\underline{x}}_0)^T] = P_0$
Other Assumptions	$E[\underline{w}(t_1)\underline{v}^T(t_2)] = 0$ for all t_1, t_2; $R^{-1}(t)$ exists
Forward Filter	$\dot{\hat{\underline{x}}}(t) = F(t)\hat{\underline{x}}(t) + P(t)H^T(t)R^{-1}(t)[\underline{z}(t) - H(t)\hat{\underline{x}}(t)], \qquad \hat{\underline{x}}(0) = \hat{\underline{x}}_0$
Error Covariance Propagation	$\dot{P}(t) = F(t)P(t) + P(t)F^T(t) + G(t)Q(t)G^T(t)$ $-P(t)H^T(t)R^{-1}(t)H(t)P(t), \qquad P(0) = P_0$
Backward Filter ($\tau = T-t$)	$\frac{d}{d\tau}\underline{s}(T-\tau) = [F^T(T-\tau) - P_b^{-1}(T-\tau)G(T-\tau)Q(T-\tau)G^T(T-\tau)]\underline{s}(T-\tau)$ $+ H^T(T-\tau)R^{-1}(T-\tau)\underline{z}(T-\tau), \qquad \underline{s}(0) = \underline{0}$
Error Covariance Propagation ($\tau = T-t$)	$\frac{d}{d\tau}P_b^{-1}(T-\tau) = P_b^{-1}(T-\tau)F(T-\tau) + F^T(T-\tau)P_b^{-1}(T-\tau)$ $-P_b^{-1}(T-\tau)G(T-\tau)Q(T-\tau)G^T(T-\tau)P_b^{-1}(T-\tau)$ $+ H^T(T-\tau)R^{-1}(T-\tau)H(T-\tau), \qquad P_b^{-1}(0) = 0$
Optimal Smoother	$\hat{\underline{x}}(t\vert T) = P(t\vert T)[P^{-1}(t)\hat{\underline{x}}(t) + \underline{s}(t)]$ $= [I + P(t)P_b^{-1}(t)]\hat{\underline{x}}(t) + P(t\vert T)\underline{s}(t)$
Error Covariance Propagation	$P(t\vert T) = [P^{-1}(t) + P_b^{-1}(t)]^{-1}$ $= P(t) - P(t)P_b^{-1}(t)[I + P(t)P_b^{-1}(t)]^{-1}P(t)$

$$\dot{\hat{\underline{x}}}(t|T) = F\hat{\underline{x}}(t|T) + GQG^TP^{-1}(t)[\hat{\underline{x}}(t|T) - \hat{\underline{x}}(t)] \tag{5.2-14}$$

$$\dot{P}(t|T) = [F + GQG^TP^{-1}(t)]\,P(t|T) + P(t|T)[F + GQG^TP^{-1}(t)]^T - GQG^T \tag{5.2-15}$$

Equations (5.2-14) and (5.2-15) are integrated *backwards* from $t = T$ to $t = 0$, with starting conditions given by $\hat{\underline{x}}(T|T) = \hat{\underline{x}}(T)$ and $P(T|T) = P(T)$. Figure 5.2-1 is a block diagram of the optimal smoother. Note that the operation which produces the smoothed state estimate does *not* involve the processing of actual measurement data. It does utilize the *complete* filtering solution, however, so that problem must be solved first. Thus, fixed-interval smoothing cannot be done real-time, on-line. It must be done after all the measurement data are collected. Note also that $P(t|T)$ is a continuous time function even where $P(t)$ may be discontinuous, as can be seen from Eq. (5.2-15).

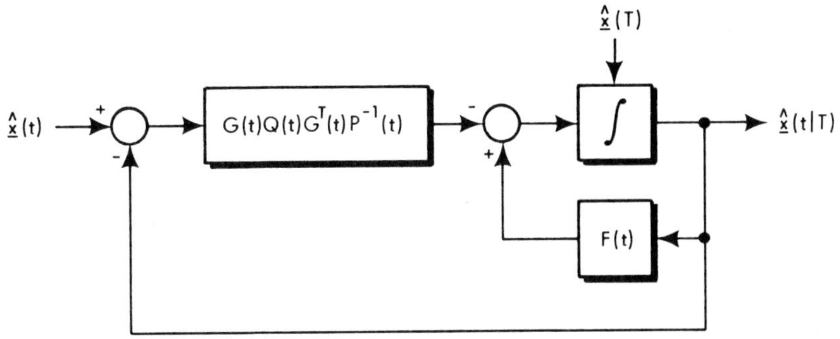

Figure 5.2-1 Diagram of Rauch-Tung-Striebel Fixed-Interval Continuous Optimal Smoother ($t \leq T$)

All smoothing algorithms depend, in some way, on the forward filtering solution. Therefore, accurate filtering is prerequisite to accurate smoothing. Since fixed-interval smoothing is done off-line, after the data record has been obtained, computation speed is not usually an important factor. However, since it is often necessary to process long data records involving many measurements, computation error due to computer roundoff is an important factor. Hence, it is desirable to have recursion formulas that are relatively insensitive to computer roundoff errors. These are discussed in Chapter 8 in connection with optimal filtering; the extension to optimal smoothing is straightforward (Ref. 1).

The continuous-time and corresponding discrete-time (Ref. 3) fixed-interval Rauch-Tung-Striebel optimal smoother equations are summarized in Table 5.2-2. In the discrete-time case the intermediate time variable is k, with final time denoted by N. $P_{k|N}$ corresponds to $P(t|T)$, $\hat{\underline{x}}_{k|N}$ corresponds to $\hat{\underline{x}}(t|T)$ and the single subscripted quantities $\hat{\underline{x}}_k$ and P_k refer to the discrete optimal filter solution. Another, equivalent fixed-interval smoother is given in Ref. 4 and the case of correlated measurement noise is treated in Ref. 5.

SMOOTHABILITY

A state is said to be *smoothable* if an optimal smoother provides a state estimate superior to that obtained when the final optimal filter estimate is extrapolated backwards in time. In Ref. 2, it is shown that *only those states which are controllable by the noise driving the system state vector are smoothable.* Thus, constant states are not smoothable, whereas randomly time-varying states are smoothable. This smoothability condition is explored below.

Consider the case where there are no system disturbances. From Eq. (5.2-14), we find ($Q = 0$)

$$\dot{\hat{\underline{x}}}(t|T) = F\hat{\underline{x}}(t|T) \tag{5.2-16}$$

TABLE 5.2-2 SUMMARY OF RAUCH-TUNG-STRIEBEL FIXED-INTERVAL OPTIMAL SMOOTHER EQUATIONS

Continuous-Time

(See Table 4.3-1 for the required continuous optimal filter terms)

Smoothed State Estimate	$\dot{\hat{\underline{x}}}(t\|T) = F(t)\hat{\underline{x}}(t\|T) + G(t)Q(t)G^T(t)P^{-1}(t)[\hat{\underline{x}}(t\|T) - \hat{\underline{x}}(t)]$ where T is fixed, $t \leqslant T$, and $\hat{\underline{x}}(T\|T) = \hat{\underline{x}}(T)$.
Error Covariance Matrix Propagation	$\dot{P}(t\|T) = [F(t) + G(t)Q(t)G^T(t)P^{-1}(t)]P(t\|T)$ $+ P(t\|T)[F(t) + G(t)Q(t)G^T(t)P^{-1}(t)]^T - G(t)Q(t)G^T(t)$ where $P(T\|T) = P(T)$

Discrete-Time

(See Table 4.2-1 for the required discrete optimal filter terms)

Smoothed State Estimate	$\hat{\underline{x}}_{k\|N} = \hat{\underline{x}}_k(+) + A_k[\hat{\underline{x}}_{k+1\|N} - \hat{\underline{x}}_{k+1}(-)]$ where $A_k = P_k(+)\Phi_k^T P_{k+1}^{-1}(-)$, $\hat{\underline{x}}_{N\|N} = \hat{\underline{x}}_N(+)$ for $k = N - 1$.
Error Covariance Matrix Propagation	$P_{k\|N} = P_k(+) + A_k[P_{k+1\|N} - P_{k+1}(-)]\ A_k^T$ where $P_{N\|N} = P_N(+)$ for $k = N - 1$

The solution is $(\hat{\underline{x}}(T|T) = \hat{\underline{x}}(T))$

$$\hat{\underline{x}}(t|T) = \Phi(t,T)\hat{\underline{x}}(T) \tag{5.2-17}$$

Thus, the optimal fixed-interval smoother estimate, when $Q = 0$, is the final optimal filter estimate extrapolated backwards in time. The corresponding smoothed state error covariance matrix behavior is governed by

$$\dot{P}(t|T) = FP(t|T) + P(t|T)F^T \tag{5.2-18}$$

for which the solution is $[P(T|T) = P(T)]$

$$P(t|T) = \Phi(t,T)\,P(T)\,\Phi^T(t,T) \tag{5.2-19}$$

If, in addition, $F = 0$, it follows that $\Phi(t,T) = I$ and hence, that

$$\hat{\underline{x}}(t|T) = \hat{\underline{x}}(T) \tag{5.2-20}$$

and

$$P(t|T) = P(T) \tag{5.2-21}$$

for all $t \leqslant T$. That is, *smoothing offers no improvement over filtering when $F = Q = 0$.* This corresponds to the case in which a constant vector is being estimated with no process noise present. Identical results clearly apply to the m constant states of an nth order system ($n > m$); hence, the validity of the smoothability condition.

Example 5.2-1

This spacecraft tracking problem was treated before in Example 4.3-2. The underlying equations are:

$$\dot{x} = w, \qquad w \sim N(0,q)$$

$$z = x + v, \qquad v \sim N(0,r)$$

and the steady-state optimal filter solution was shown to be $p(t) = \alpha$, where $\alpha = \sqrt{rq}$. Examine the steady-state, fixed-interval optimal smoother both in terms of (1) forward-backward optimal filters, and (2) Rauch-Tung-Striebel form.

Part 1 – The forward filter Riccati equation is (f=0, g=h=1)

$$\dot{p} = q - p^2/r$$

which, in the steady state ($\dot{p}=0$), yields $p = \sqrt{rq} = \alpha$. The backward filter Riccati equation is from Eq. (5.2-7)

$$\frac{d}{d\tau} p_b = q - {p_b}^2/r$$

which has the steady state $p_b = \sqrt{rq} = \alpha$. Thus, we find, for the smoothed covariance

$$p(t|T) = \left(p^{-1}(t) + p_b^{-1}(t)\right)^{-1}$$

$$= \frac{\alpha}{2}$$

which is half the optimal filter covariance. Consequently,

$$\hat{x}(t|T) = p(t|T)\ [\hat{x}(t)/p(t) + \hat{x}_b(t)/p_b(t)]$$

$$= \frac{1}{2}\ [\hat{x}(t) + \hat{x}_b(t)] \tag{5.2-22}$$

The smoothed estimate of x is the *average* of forward plus backward estimates, in steady state.

Part 2 – In Rauch-Tung-Striebel form, the steady-state smoothed covariance matrix differential equation (Table 5.2-2, T fixed, $t \leqslant T$) is

$$\dot{p}(t|T) = \frac{2q}{\alpha} p(t|T) - q$$

for which the solution is ($q/\alpha = \beta$, $p(T|T) = \alpha$)

$$p(t|T) = \frac{\alpha}{2}(1 + e^{-2\beta(T-t)}) \quad , \quad t \leq T$$

This result is plotted in Fig. 5.2-2. For T–t sufficiently large (i.e., $T-t > 2/\beta$), the backward sweep is in steady state. In this case, we obtain $p(t|T) = \alpha/2$, as before. The corresponding differential equation for the smoothed state estimate, from Table 5.2-2, is

$$\dot{\hat{x}}(t|T) = \beta[\hat{x}(t|T) - \hat{x}(t)]$$

This can be shown to be identical to Eq. (5.2-22) by differentiation of the latter with respect to time and manipulation of the resulting equation.

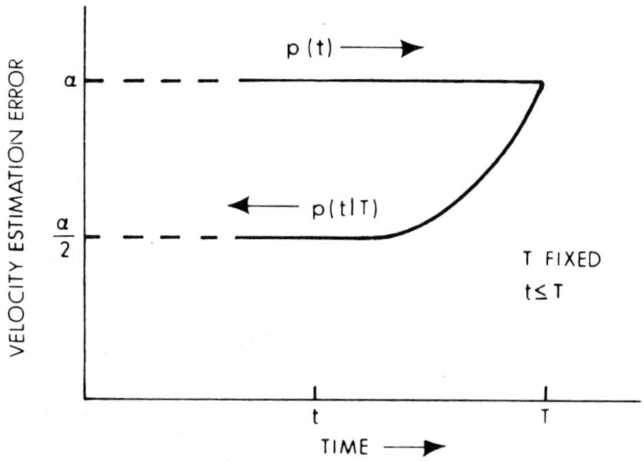

Figure 5.2-2 Optimal Filter and Smoother Covariances for Example 5.2-1

Example 5.2-2

This example describes an investigation of the applicability of fixed-interval optimal smoothing to gyroscope testing. In the test considered, a gyroscope is mounted on a servoed turntable and samples of the table angle, which is a measure of the integrated gyroscope drift rate, ϵ, are recorded. The gyroscope drift rate is assumed to be a linear combination of a random bias b, a random walk, a random ramp (slope m), and a first-order markov process; this is shown in block diagram form in Fig. 5.2-3. The available measurements are the samples, θ_k, corrupted by a noise sequence, v_k, and measurement data are to be batch processed after the test.

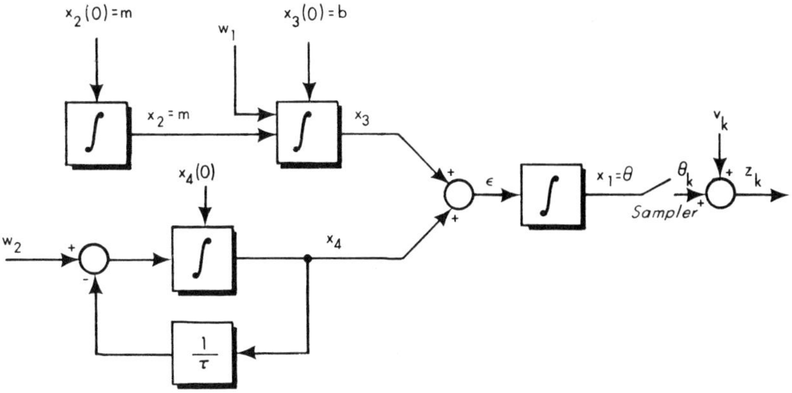

Figure 5.2-3 Block Diagram of Gyroscope Drift Rate Model and Test Measurement

Consider the aspects of observability and smoothability. The equations governing this system are

$$\begin{bmatrix} \dot{x}_1 \\ \dot{x}_2 \\ \dot{x}_3 \\ \dot{x}_4 \end{bmatrix} = \underbrace{\begin{bmatrix} 0 & 0 & 1 & 1 \\ 0 & 0 & 0 & 0 \\ 0 & 1 & 0 & 0 \\ 0 & 0 & 0 & -1/\tau \end{bmatrix}}_{F} \underbrace{\begin{bmatrix} x_1 \\ x_2 \\ x_3 \\ x_4 \end{bmatrix}}_{\underline{x}} + \begin{bmatrix} 0 \\ 0 \\ w_1 \\ w_2 \end{bmatrix}$$

and

$$z_k = \underbrace{[1\ \ 0\ \ 0\ \ 0]}_{H} \begin{bmatrix} x_1 \\ x_2 \\ x_3 \\ x_4 \end{bmatrix} + v_k$$

The test for observability involves determination of the rank of the matrix Ξ (Section 3.5), where in this case

$$\Xi = \left[H^T \,\vdots\, F^T H^T \,\vdots\, (F^T)^2 H^T \,\vdots\, (F^T)^3 H^T \right]$$

$$= \begin{bmatrix} 1 & 0 & 0 & 0 \\ 0 & 0 & 1 & 0 \\ 0 & 1 & 0 & 0 \\ 0 & 1 & -1/\tau & 1/\tau^2 \end{bmatrix}$$

It is easily shown that the rank of Ξ is four, equal to the dimension of the system; hence, the system is completely observable. Performing the test for observability, as illustrated above, can lend insight into the problem at hand and can help avoid attempting impossible tasks. For example, it is tempting to add another integrator (state variable) to the system shown in Fig. 5.2-3, in order to separate the bias and random walk components, with the hope of separately identifying them through filtering and smoothing. When such a five-dimensional system is formulated, and the appropriate F(5 X 5) and H(1 X 5) are considered, the resulting Ξ (5 X 5) matrix has rank four. This five-dimensional system is not completely observable because two state variables have the same dynamic relationship to the measured quantity; it is impossible to distinguish between the bias drift and the initial condition of the random walk component. Thus, these two components should be combined at one integrator, as in Fig. 5.2-3, where no such distinction is made. In this example, the randomly varying states x_1, x_3, and x_4 are smoothable; the constant state, x_2, is not.

Figure 5.2-4 shows normalized estimation errors for the drift rate ϵ over a 20-hour period, based upon drift rate samples taken once per hour. Results are shown for the complete system and for a simpler three-dimensional system which does not include the markov process component of gyroscope drift rate. In the latter case, state x_3 is the entire drift rate, which is the sum of a ramp, a bias, and a random walk. The filtered estimate of x_3 is much improved over that in the four-dimensional situation, and reaches equilibrium after approximately four hours. The estimate at each point is evidently based primarily on the current measurement and the four previous measurements. The fixed-interval smoothed estimate reduces the rms error by almost half, being largely based on the current measurement, the preceding four measurements, and the subsequent four measurements.

A numerical example based on simulated real data is presented in order to graphically illustrate the difference between a set of filtered and smoothed estimates of gyroscope drift measurements. The example corresponds to the three-dimensional no-markov-process case. A set of "real" data is generated by simulating this three-dimensional linear system and using a random number generator to produce initial conditions and two white noise

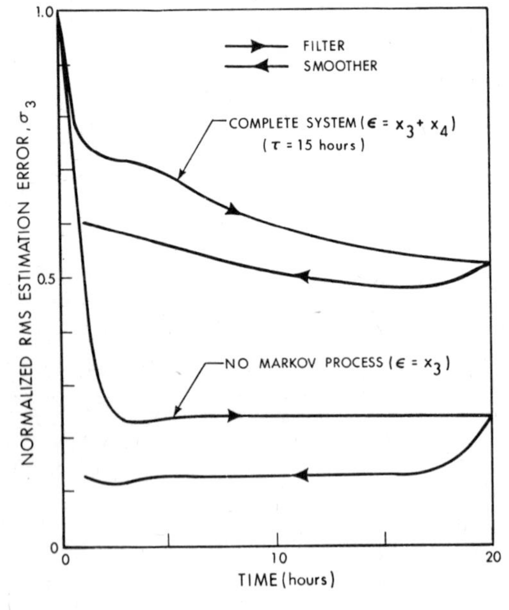

Figure 5.2-4 Error in Filtered and Smoothed Estimates With and Without the Markov Process (Ref. 7)

sequences, one representing the random walk input and one representing the sequence of measurement errors. Figure 5.2-5 compares a sample of a "real" time history of state x_3 (which in this case is ϵ, the entire drift rate) with filtered and smoothed estimates. The solid line connects the "real" hourly values of x_3. The dashed line indicates the real bias drift b (the initial value) and the ramp slope m (the slope of the dashed line). The departure from the dashed line is due to the random walk component. Each filtered estimate of x_3 is based on the partial set of measurements (z_1 to z_k). Each smoothed estimate of x_3 is based on the full set of measurements (z_1 to z_N).

The relative advantage of the smoother is most apparent in the first two hours, where the filter has only one or two data points to work with. The smoother is also more accurate, generally, throughout the 20-hour time span due to its ability to "look ahead" at the subsequent data points. The example also shows that the filter tends to lag the real data whenever it takes a major "swing" up or down. The smoother does not exhibit this type of behavior.

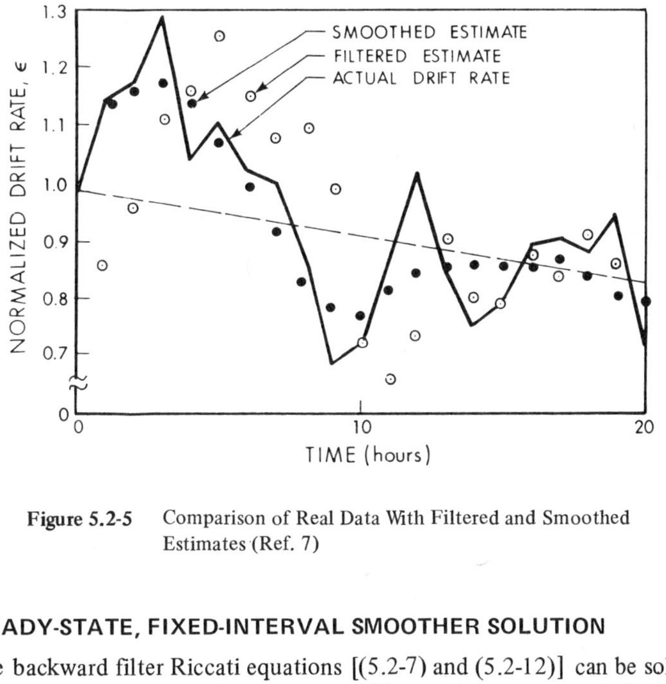

Figure 5.2-5 Comparison of Real Data With Filtered and Smoothed Estimates (Ref. 7)

A STEADY-STATE, FIXED-INTERVAL SMOOTHER SOLUTION

The backward filter Riccati equations [(5.2-7) and (5.2-12)] can be solved by transforming the n $\times$ n nonlinear matrix differential equations to 2n $\times$ 2n linear matrix differential equations, precisely as was done in Section 4.6. Therefore, this approach warrants no further discussion here. The linear smoother covariance equation given in Eq. (5.2-15) can also be treated in a manner similar to that used in Section 4.6; this is briefly treated below. Defining the transformations

$$\underline{\lambda} = P(t|T)\, \underline{y} \tag{5.2-23a}$$

and

$$\dot{\underline{y}} = -[F + GQG^T P^{-1}(t)]^T \underline{y} \tag{5.2-23b}$$

we find

$$\begin{bmatrix} \dot{\underline{y}} \\ \dot{\underline{\lambda}} \end{bmatrix} = \begin{bmatrix} -[F + GQG^T P^{-1}(t)]^T & 0 \\ -GQG^T & F + GQG^T P^{-1}(t) \end{bmatrix} \begin{bmatrix} \underline{y} \\ \underline{\lambda} \end{bmatrix} \tag{5.2-24}$$

Since the boundary condition for the smoothing process is specified at $t = T$, we let

$$\tau = T - t \tag{5.2-25}$$

and use τ as the independent variable in Eq. (5.2-24). Thus, we obtain

$$\begin{bmatrix} \frac{d}{d\tau} \underline{y} \\ \frac{d}{d\tau} \underline{\lambda} \end{bmatrix} = \begin{bmatrix} [F + GQG^T P^{-1}(t)]^T & 0 \\ GQG^T & -[F + GQG^T P^{-1}(t)] \end{bmatrix} \begin{bmatrix} \underline{y} \\ \underline{\lambda} \end{bmatrix} \tag{5.2-26}$$

An expression similar to Eq. (4.6-8) may, in principle, be written for the recursive solution to the smoothing covariance equation. As before, this formulation of the solution is only of practical value when all elements of the square matrix in Eq. (5.2-26) are constant. However, if the system is observable and time-invariant, P(t) tends to a stationary limit, and the constant value of P(t), denoted P_∞, can be employed in Eq. (5.2-26). If the time interval under consideration is sufficiently large and if P_∞ is employed in Eq. (5.2-26), the latter would also have a stationary solution. In many cases of practical interest it is precisely the stationary filtering and smoothing solutions which are sought; in these cases the solution to Eq. (5.2-26) is obtained in the form of Eq. (4.6-8) by *iterating* until changes in the diagonal elements of $P(t|T)$ are sufficiently small.

5.3 OPTIMAL FIXED-POINT SMOOTHER

When the smoothing solution is sought only for a specific time of interest, it is more efficient to reformulate the smoother equations to perform the task than to accomplish it through use of the fixed-interval smoother. To do this, we first write Eq. (5.2-14) in the form

$$\dot{\hat{\underline{x}}}(t|T) = [F + GQG^T P^{-1}(t)]\, \hat{\underline{x}}(t|T) - GQG^T P^{-1}(t)\, \hat{\underline{x}}(t) \tag{5.3-1}$$

for which the solution is $[\hat{\underline{x}}(T|T) = \hat{\underline{x}}(T)]$

$$\underline{\hat{x}}(t|T) = \Phi_s(t,T)\,\underline{\hat{x}}(T) - \int_T^t \Phi_s(t,\tau)GQG^T P^{-1}(\tau)\underline{\hat{x}}(\tau)\,d\tau \tag{5.3-2}$$

where $\Phi_s(t,\tau)$ is the transition matrix corresponding to $F + GQG^TP^{-1}(t)$, that is,

$$\dot{\Phi}_s(t,T) = [F + GQG^TP^{-1}(t)]\ \Phi_s(t,T), \qquad \Phi_s(t,t) = I \tag{5.3-3}$$

Equation (5.3-2) is readily verified as the solution to Eq. (5.3-1) by differentiation with respect to t and use of Leibniz' rule.*

Now consider t fixed, and let T increase. Differentiating Eq. (5.3-2), making use of Leibniz' rule, we obtain

$$\begin{aligned}\frac{d\underline{\hat{x}}(t|T)}{dT} &= \frac{d\Phi_s(t,T)}{dT}\underline{\hat{x}}(T) + \Phi_s(t,T)\frac{d\underline{\hat{x}}(T)}{dT} - [\underline{0} - \Phi_s(t,T)GQG^TP^{-1}(T)\underline{\hat{x}}(T) + \underline{0}]\\ &= -\Phi_s(t,T)[F + GQG^TP^{-1}(T)]\,\underline{\hat{x}}(T) + \Phi_s(t,T)\big\{F\underline{\hat{x}}(T) + K(T)[\underline{z}(T)\\ &\quad - H(T)\underline{\hat{x}}(T)]\big\} + \Phi_s(t,T)GQG^TP^{-1}(T)\underline{\hat{x}}(T)\\ &= \Phi_s(t,T)K(T)[\underline{z}(T) - H(T)\underline{\hat{x}}(T)]\end{aligned} \tag{5.3-4}$$

where we have used the known optimal filter differential equation for $\underline{\hat{x}}(T)$, and the relationship $(T \geqslant t)$

$$\frac{d\Phi_s(t,T)}{dT} = -\Phi_s(t,T)[F(T) + G(T)Q(T)G^T(T)P^{-1}(T)]\quad,\quad \Phi_s(t,t) = I \tag{5.3-5}$$

The latter is easily shown by differentiating the expression $\Phi_s(t,T)\Phi_s(T,t) = I$ with respect to t and using Eq. (5.3-3).

To establish the fixed-point, optimal smoother covariance matrix differential equation, write the solution to Eq. (5.2-15) as $[P(T|T) = P(T)]$

$$P(t|T) = \Phi_s(t,T)P(T)\Phi_s^T(t,T) - \int_T^t \Phi_s(t,\tau)G(\tau)Q(\tau)G^T(\tau)\Phi_s^T(t,\tau)\,d\tau \tag{5.3-6}$$

---

* $$\frac{d}{dt}\int_{a(t)}^{b(t)} \underline{f}(t,\tau)\,d\tau = \underline{f}[t,b(t)]\frac{db(t)}{dt} - \underline{f}[t,a(t)]\frac{da(t)}{dt} + \int_{a(t)}^{b(t)} \frac{\partial}{\partial t}\underline{f}(t,\tau)\,d\tau$$

This, too, may be verified directly by differentiation. As before, we now consider t fixed and differentiate with respect to T to obtain

$$\frac{dP(t|T)}{dT} = \frac{d\Phi_s(t,T)}{dT} P(T)\Phi_s^T(t,T) + \Phi_s(t,T)\frac{dP(T)}{dT}\Phi_s^T(t,T) + \Phi_s(t,T)P(T)\frac{d\Phi_s^T(t,T)}{dT} - (0 - \Phi_s(t,T)GQG^T\Phi_s^T(t,T) + 0) \tag{5.3-7}$$

which ultimately simplifies to

$$\frac{dP(t|T)}{dT} = -\Phi_s(t,T)P(T)H^T(T)R^{-1}(T)H(T)P(T)\Phi_s^T(t,T) \tag{5.3-8}$$

In applications, the fixed-point optimal smoother is perhaps most often used to estimate the initial state of a dynamic system or process – i.e., orbit injection conditions for a spacecraft. The continuous-time and corresponding discrete-time (Ref. 6) optimal fixed-point smoother equations are summarized in Table 5.3-1.

TABLE 5.3-1 SUMMARY OF FIXED-POINT OPTIMAL SMOOTHER EQUATIONS

Continuous-Time

(See Table 4.3-1 for the required continuous optimal filter terms)

Smoothed State Estimate	$\frac{d\hat{\underline{x}}(t\|T)}{dT} = \Phi_s(t,T)K(T)[\underline{z}(T) - H(T)\hat{\underline{x}}(T)]$ $\frac{d\Phi_s(t,T)}{dT} = -\Phi_s^T(t,T)[F(T) + G(T)Q(T)G^T(T)P^{-1}(T)]$ where t is fixed, $T \geqslant t$, $\hat{\underline{x}}(t\|t) = \hat{\underline{x}}(t)$ and $\Phi_s(t,t) = I$
Error Covariance Matrix Propagation	$\frac{dP(t\|T)}{dT} = -\Phi_s(t,T)P(T)H^T(T)R^{-1}(T)H(T)P(T)\Phi_s^T(t,T)$ where $P(t\|t) = P(t)$

Discrete-Time

(See Table 4.2-1 for the required discrete optimal filter terms)

Smoothed State Estimate	$\hat{\underline{x}}_{k\|N} = \hat{\underline{x}}_{k\|N-1} + B_N[\hat{\underline{x}}_N(+) - \hat{\underline{x}}_N(-)]$ where $B_N = \prod_{i=k}^{N-1} A_i, \quad A_i = P_i(+)\Phi_i^T P_{i+1}^{-1}(-)$ $\hat{\underline{x}}_{k\|k} = \hat{\underline{x}}_k;\ N = k+1,\ k+2, \ldots$
Error Covariance Matrix Propagation	$P_{k\|N} = P_{k\|N-1} + B_N[P_k(+) - P_k(-)]B_N^T$ where $P_{k\|k} = P_k(+)$

Example 5.3-1

Determine the steady-state behavior of the fixed-point optimal smoother for the system considered in Example 5.2-1. The transition matrix required for the fixed-point optimal smoother is governed (Table 5.3-1, t fixed, $T \geqslant t$) by

$$\frac{d\varphi_S(t,T)}{dT} = -\beta\varphi_S(t,T), \qquad \varphi_S(t,t) = I$$

Thus,

$$\varphi_S(t,T) = e^{-\beta(T-t)} \tag{5.3-9}$$

Using this result in the equation for fixed-point optimal smoother covariance propagation (Table 5.3-1), we find

$$\frac{dp(t|T)}{dT} = -\alpha\beta e^{-2\beta(T-t)}, \qquad p(t|t) = \alpha$$

for which the solution,

$$p(t|T) = \frac{\alpha}{2}\left(1 + e^{-2\beta(T-t)}\right) \tag{5.3-10}$$

is obtained directly by integration. This equation is identical to that plotted in Fig. 5.2-2, but the interpretation differs in that now t is fixed and T is increasing. When examining a point sufficiently far in the past (i.e., $T - t \geqslant 2/\beta$), the fixed-point optimal smoother error is in the steady state described by $p(t|T) = \alpha/2$, $T > t + 2/\beta$. The differential equation for the fixed-point optimal smoother state estimate (Table 5.3-1) is

$$\frac{d\hat{x}(t|T)}{dT} = \beta e^{-\beta(T-t)}[z(T) - \hat{x}(T)], \qquad \hat{x}(t|t) = \hat{x}(t)$$

for which the solution is computed forward in time from t until the present, T.

5.4 OPTIMAL FIXED-LAG SMOOTHER

In cases where a running smoothing solution that lags the most recent measurement by a constant time delay, Δ, is sought, the fixed-lag smoother is used. The derivation closely follows that of Section 5.3. From Eq. (5.3-2), with $t = T - \Delta$, we get

$$\underline{\hat{x}}(T-\Delta|T) = \Phi_L(T-\Delta, T)\underline{\hat{x}}(T) - \int_T^{T-\Delta} \Phi_L(T-\Delta,\tau)GQG^TP^{-1}(\tau)\underline{\hat{x}}(\tau)d\tau \tag{5.4-1}$$

Differentiation with respect to T and combining terms, yields

$$\frac{d\hat{\underline{x}}(T-\Delta|T)}{dT} = F(T-\Delta) + G(T-\Delta)Q(T-\Delta)G^T(T-\Delta)P^{-1}(T-\Delta)]\,\hat{\underline{x}}(T-\Delta|T)$$

$$-G(T-\Delta)Q(T-\Delta)G^T(T-\Delta)P^{-1}(T-\Delta)\hat{\underline{x}}(T-\Delta)$$

$$+\,\Phi_L(T-\Delta,T)K(T)[\underline{z}(T) - H(T)\hat{\underline{x}}(T)] \qquad (5.4\text{-}2)$$

where the relationship $[(\Phi_L(0,\Delta) = \Phi_s(0,\Delta)]$

$$\frac{d\Phi_L(T-\Delta,T)}{dT} = [F(T-\Delta)+G(T-\Delta)Q(T-\Delta)G^T(T-\Delta)P^{-1}(T-\Delta)]\,\Phi_L(T-\Delta,T)$$

$$-\,\Phi_L(T-\Delta,T)[F(T) + G(T)Q(T)G^T(T)P^{-1}(T)] \qquad (5.4\text{-}3)$$

has been used ($T \geqslant \Delta$). This is easily verified by differentiation of the expression

$$\Phi_L(T-\Delta, T) = \Phi_s(T-\Delta, t)\Phi_s(t,T)$$

with respect to T. The initial condition for $\hat{\underline{x}}(T-\Delta|T)$ is the fixed-point solution, $\hat{\underline{x}}(0|\Delta)$. When measurements are received they are processed by a fixed-point algorithm until $T = \Delta$, at which point the fixed-lag algorithm is initialized and subsequently takes over. The filtering solution, of course, is carried throughout. The corresponding solution for the fixed-lag smoother covariance matrix is

$$\frac{dP(T-\Delta|T)}{dT} = [F(T-\Delta)+G(T-\Delta)Q(T-\Delta)G^T(T-\Delta)P^{-1}(T-\Delta)]\,P(T-\Delta|T)$$

$$+\,P(T-\Delta|T)[F(T-\Delta)+G(T-\Delta)Q(T-\Delta)G^T(T-\Delta)P^{-1}(T-\Delta)]^T$$

$$-\,\Phi_L(T-\Delta,T)P(T)H^T(T)R^{-1}(T)H(T)P(T)\Phi_L{}^T(T-\Delta,T)$$

$$-\,G(T-\Delta)Q(T-\Delta)G^T(T-\Delta) \qquad (5.4\text{-}4)$$

for which the initial condition is $P(0|\Delta)$, the optimal fixed-point smoothing error covariance matrix evaluated at $T = \Delta$.

In practice, the fixed-lag optimal smoother is used as a "refined," albeit delayed, optimal filter. Applications in communications and telemetry, among others, are suggested. The continuous-time and corresponding discrete-time (Ref. 6) fixed-lag smoother equations are summarized in Table 5.4-1.

Example 5.4-1

Determine the steady-state behavior of the fixed-lag optimal smoother for the system considered in Example 5.2-1. The transition matrix required for the fixed-lag optimal smoother (Table 5.4-1, $T - t = \Delta$ is fixed, $T \geqslant \Delta$) is

TABLE 5.4-1 SUMMARY OF FIXED-LAG OPTIMAL SMOOTHER EQUATIONS

Continuous-Time

(See Table 4.3-1 for the required continuous optimal filter terms)

Smoothed State Estimate	$\frac{d\hat{\underline{x}}(T-\Delta\|T)}{dT} = [F(T-\Delta)+G(T-\Delta)Q(T-\Delta)G^T(T-\Delta)P^{-1}(T-\Delta)]\hat{\underline{x}}(T-\Delta\|T)$ $-G(T-\Delta)Q(T-\Delta)G^T(T-\Delta)P^{-1}(T-\Delta)\hat{\underline{x}}(T-\Delta)$ $+\Phi_L(T-\Delta,T)K(T)[\underline{z}(T) - H(T)\hat{\underline{x}}(T)]$ where $T \geqslant \Delta$, Δ fixed, and $\hat{\underline{x}}(0\|\Delta)$ is the initial condition obtained from the optimal fixed-point smoother, and $\frac{d\Phi_L(T-\Delta,T)}{dT} = [F(T-\Delta)+G(T-\Delta)Q(T-\Delta)G^T(T-\Delta)P^{-1}(T-\Delta)]\,\Phi_L(T-\Delta,T)$ $-\Phi_L(T-\Delta,T)[F(T)+G(T)Q(T)G^T(T)P^{-1}(T)]$ where $\Phi_L(0,\Delta) = \Phi_s(0,\Delta)$
Error Covariance Matrix Propagation	$\frac{dP(T-\Delta\|T)}{dT} = [F(T-\Delta)+G(T-\Delta)Q(T-\Delta)G^T(T-\Delta)P^{-1}(T-\Delta)]P(T-\Delta\|T)$ $+P(T-\Delta\|T)[F(T-\Delta)+G(T-\Delta)Q(T-\Delta)G^T(T-\Delta)P^{-1}(T-\Delta)]^T$ $-\Phi_L(T-\Delta,T)P(T)H^T(T)R^{-1}(T)H(T)P(T)\Phi_L^T(T-\Delta,T)$ $-G(T-\Delta)Q(T-\Delta)G^T(T-\Delta)$ where $P(0\|\Delta)$ is the initial condition obtained from the optimal fixed-point smoother.

Discrete-Time

(See Table 4.2-1 for the required discrete optimal filter terms)

Smoothed State Estimate	$\hat{\underline{x}}_{k+1\|k+1+N} = \Phi_k\hat{\underline{x}}_{k\|k+N} + Q_k\Phi_k^T P_k^{-1}(+)[\underline{x}_{k\|k+N} - \hat{\underline{x}}_k(+)]$ $+B_{k+1+N}K_{k+1+N}[\underline{z}_{k+1+N} - H_{k+1+N}\Phi_{k+N}\hat{\underline{x}}_{k+N}(+)]$ where $B_{k+1+N} = \prod_{i=k+1}^{k+N} A_i$, $\quad A_i = P_i(+)\Phi_i^T P_{i+1}^{-1}(-)$ k=0, 1, 2, . . . and $\hat{\underline{x}}(0\|N)$ is the initial condition.
Error Covariance Matrix Propagation	$P_{k+1\|k+1+N} = P_{k+1}(-) - B_{k+1+N}K_{k+1+N}H_{k+1+N}P_{k+1+N}(-)B_{k+1+N}^T$ $-A_k^{-1}[P_k(+) - P_{k\|k+N}]\,(A_k^T)^{-1}$ where the initial condition is $P(0\|N)$.

$$\frac{d}{dT}\varphi_L(T-\Delta,T) = 0$$

under the assumption that $p(T - \Delta) = p(T) = \alpha$. The solution, a constant, is given by the initial condition obtained from fixed-point smoothing [Eq. (5.3-9)],

$$\varphi_L(T-\Delta, T) = e^{-\beta\Delta}$$

The fixed-lag smoother covariance matrix, therefore, behaves according to (Table 5.4-1)

$$\frac{dp(T-\Delta|T)}{dT} = 2\beta p(T-\Delta|T) - q(1 + e^{-2\beta\Delta})$$

Employing the initial condition $p(0|\Delta)$ obtained from the fixed-point solution, Eq. (5.3-10), we find

$$p(T-\Delta|T) = \frac{\alpha}{2}(1 + e^{-2\beta\Delta})$$

Thus, for Δ sufficiently large (i.e., $\Delta > 2/\beta$), the delayed estimate of the state has the accuracy of the steady-state smoother. This, of course, is the reason for its utility. The corresponding delayed state estimate (Table 5.4-1) is given by

$$\frac{d\hat{x}(T-\Delta|T)}{dT} = \beta\left(\hat{x}(T-\Delta|T) - \hat{x}(T-\Delta) + e^{-\beta\Delta}[z(T) - \hat{x}(T)]\right)$$

subject to the initial condition, $\hat{x}(0|\Delta)$, which is obtained from the optimal fixed-point smoother.

REFERENCES

1. Fraser, D.C., and Potter, J.E., "The Optimum Linear Smoother as a Combination of Two Optimum Linear Filters," *IEEE Trans. on Automatic Control*, Vol. 7, No. 8, August 1969, pp. 387-390.

2. Fraser, D.C., "A New Technique for the Optimal Smoothing of Data," Ph.D. Thesis, Massachusetts Institute of Technology, January 1967.

3. Rauch, H.E., Tung, F., and Striebel, C.T., "Maximum Likelihood Estimates of Linear Dynamic Systems," *AIAA Journal*, Vol. 3, No. 8, August 1965, pp. 1445-1450.

4. Bryson, A.E., and Frazier, M., "Smoothing for Linear and Nonlinear Dynamic Systems," Technical Report, ASD-TDR-63-119, Wright Patterson Air Force Base, Ohio, September 1962, pp. 353-364.

5. Mehra, R.K., and Bryson, A.E., "Linear Smoothing Using Measurements Containing Correlated Noise with an Application to Inertial Navigation," *IEEE Trans. on Automatic Control*, Vol. AC-13, No. 5, October 1968, pp. 496-503.

6. Meditch, J.S., *Stochastic Optimal Linear Estimation and Control*, McGraw-Hill Book Co., Inc., New York, 1969.

7. Nash, R.A., Jr., Kasper, J.F., Jr., Crawford, B.S., and Levine, S.A., "Application of Optimal Smoothing to the Testing and Evaluation of Inertial Navigation Systems and Components," *IEEE Trans. on Automatic Control*, Vol. AC-16, No. 6, December 1971, pp. 806-816.

8. Nash, R.A., Jr., "The Estimation and Control of Terrestrial Inertial Navigation System Errors Due to Vertical Deflections," *IEEE Trans. on Automatic Control*, Vol. AC-13, No. 4, August 1968, pp. 329-338.

PROBLEMS

Problem 5-1

Choose $\underline{x}$ to minimize the scalar loss function

$$J = (\underline{x} - \hat{\underline{x}})^T P^{-1} (\underline{x} - \hat{\underline{x}}) + (\underline{x} - \hat{\underline{x}}_b)^T P_b^{-1} (\underline{x} - \hat{\underline{x}}_b)$$

and directly obtain the forward-backward filter form of the optimal smoother,

$$\hat{\underline{x}}(t|T) = P(t|T)[P^{-1}(t)\hat{\underline{x}}(t) + P_b^{-1}(t)\hat{\underline{x}}_b(t)]$$

$$P(t|T) = [P^{-1}(t) + P_b^{-1}(t)]^{-1}$$

Problem 5-2

Derive the Rauch-Tung-Striebel smoother equations [Eqs. (5.2-14) and (5.2-15)] by following the steps outlined in the text.

Problem 5-3

For the variable $\underline{\lambda}(t)$, defined by

$$\hat{\underline{x}}(t|T) = \hat{\underline{x}}(t) - P(t)\,\underline{\lambda}(t)$$

show that

$$\frac{d\underline{\lambda}(t)}{dt} = -[F - P(t)\,H^T R^{-1} H]^T \underline{\lambda}(t) + H^T R^{-1}[\underline{z}(t) - H\hat{\underline{x}}(t)]$$

where $\underline{\lambda}(T) = \underline{0}$. For $\Lambda(t) = E[\underline{\lambda}(t)\,\underline{\lambda}^T(t)]$, show that

$$P(t|T) = P(t) - P(t)\,\Lambda(t)\,P(t)$$

where

$$\dot{\Lambda}(t) = -[F - P(t)H^T R^{-1} H]^T \Lambda(t) - \Lambda(t)[F - P(t)H^T R^{-1} H] + H^T R^{-1} H$$

and $\Lambda(T) = 0$. These are the Bryson-Frazier smoother equations (Ref. 4).

Problem 5-4

A scalar system and measurement are described by

$$\dot{x} = ax + w, \qquad w \sim N(0,q)$$

$$z = bx + v, \qquad v \sim N(0,r)$$

(a) Show that the forward filter steady-state error covariance, obtained from the Riccati equation by setting $\dot{p}(t) = 0$, is

$$p_\infty = \frac{ar}{b^2}\left(1 + \sqrt{1 + \frac{b^2 q}{a^2 r}}\right)$$

(b) Next, obtain the steady-state fixed-interval smoother error covariance [denoted $p_\infty(t|T)$] by setting $\dot{p}(t|T) = 0$, as

$$p_\infty(t|T) = \frac{p_\infty}{2(1 + \frac{a}{q} p_\infty)}$$

(c) Show that this result can be written in the alternate form ($\gamma^2 = b^2 q/a^2 r \geqslant 0$)

$$\frac{p_\infty(t|T)}{p_\infty} = \frac{1}{2}\left(\frac{\gamma^2}{1 + \gamma^2 + \sqrt{1 + \gamma^2}}\right)$$

In this form it is apparent that the smoother error variance is *always* less than half the filter error variance.

Problem 5-5

It is a fact that smoothed covariances are *not* necessarily *symmetric* about the midpoint of the sampling interval (see, for example, Ref. 8). A simple illustration of this point can be made with the second-order system of Fig. 5-1, where $w \sim N(0, 2\alpha\sigma^2)$. Assume that the system is in steady state, and that at t=0 a *perfect* measurement of x_1 is made. Show that, while $E[\tilde{x}_1^2(t)]$ *is* symmetric with respect to t=0, $E[\tilde{x}_2^2(t)]$ is *not* symmetric with respect to t=0, viz:

$$E[\tilde{x}_1^2(t)] = \sigma^2(1 - e^{-2\alpha|t|})$$

$$E[\tilde{x}_2^2(t)] = \frac{\sigma^2}{\beta(\alpha+\beta)}\left(1 - \frac{\beta}{\alpha+\beta} e^{2\alpha t}\right), \quad t \leqslant 0$$

$$= \frac{\sigma^2}{\beta(\alpha+\beta)}\left(1 - \frac{\beta[(\alpha+\beta)e^{-\alpha t} - 2\beta e^{-\beta t}]^2}{(\alpha+\beta)(\alpha-\beta)^2}\right), \quad t \geqslant 0$$

(*Hint:* It may prove simplest to *seek* estimates in the form $\hat{x}_1(t) = k_1(t)\, x_1(0)$, $\hat{x}_2(t) = k_2(t) x_2(0)$, and to choose $k_1(t)$ and $k_2(t)$ to minimize $E[\tilde{x}_1^2(t)]$ and $E[\tilde{x}_2^2(t)]$, respectively.)

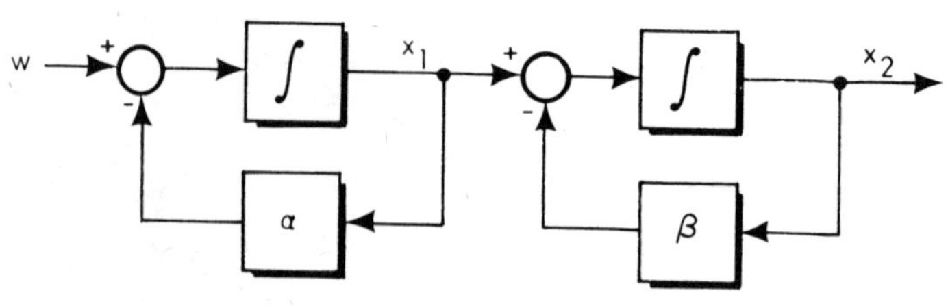

Figure 5-1

Problem 5-6

The output of an integrator driven by white noise w [where $w \sim N(0,q)$] is sampled every Δ seconds (where $\Delta = t_{k+1} - t_k$ = constant) in the presence of a measurement noise v_k

[where $v_k \sim N(0,r_0)$]. Assume that three measurements are made, corresponding to k = 0, 1, 2. Further assume that there is no *a priori* information.

a) Show that a *fixed-interval* optimal smoother yields ($\gamma = q\,\Delta/r_0$)

$$p_{1|2} = r_0\left(\frac{1+\gamma}{3+\gamma}\right)$$

$$p_{0|2} = r_0\,\frac{1+3\gamma+\gamma^2}{(1+\gamma)(3+\gamma)}$$

b) Check these results by formulating the optimal smoother in terms of a continuous optimal forward filter updated at the measurement times by a continuous optimal backward filter, and sketch the filtering and smoothing covariances.

c) Show that a *fixed-point* optimal smoother for the initial condition yields

$$p_{0|1} = r_0\left(\frac{1+\gamma}{2+\gamma}\right)$$

$$p_{0|2} = r_0\,\frac{1+3\gamma+\gamma^2}{(1+\gamma)(3+\gamma)}$$

Problem 5-7

Design an optimal linear smoother to separate a noise n(t) from a signal s(t) when the spectral densities for the signal and noise are given by:

$$\Phi_{ss}(\omega) = \frac{1}{\omega^2+1}, \quad \Phi_{nn}(\omega) = \frac{2\omega^2}{\omega^4+1}$$

(*Hint:* This problem in Wiener smoothing can be solved as the steady-state portion of an optimal linear smoothing problem.)

6. NONLINEAR ESTIMATION

This chapter extends the discussion of optimal estimation for linear systems to the more general case described by the nonlinear stochastic differential equation

$$\underline{\dot{x}}(t) = \underline{f}(\underline{x}(t),t) + \underline{w}(t) \tag{6.0-1}$$

The vector $\underline{f}$ is a nonlinear function of the state and $\underline{w}(t)$ is zero mean gaussian noise having spectral density matrix Q(t). We shall investigate the problem of estimating $\underline{x}(t)$ from sampled nonlinear measurements of the form

$$\underline{z}_k = \underline{h}_k(\underline{x}(t_k)) + \underline{v}_k, \qquad k = 1,2, \ldots \tag{6.0-2}$$

where $\underline{h}_k$ depends upon both the index k and the state at each sampling time, and $\{\underline{v}_k\}$ is a white random sequence of zero mean gaussian random variables with associated covariance matrices $\{R_k\}$. This constitutes a class of estimation problems for nonlinear systems having continuous dynamics and discrete-time measurements.

For several reasons, the problem of filtering and smoothing for nonlinear systems is considerably more difficult and admits a wider variety of solutions than does the linear estimation problem. First of all, in the linear gaussian case the optimal estimate of $\underline{x}(t)$ for most reasonable Bayesian optimization criteria is the conditional mean defined in Eq. (4.0-9). Furthermore, the gaussian property implies that the conditional mean can be computed from a unique linear operation on the measurement data – e.g., the Kalman filter algorithm. Consequently, there is little theoretical justification for using a different data processing technique, unless a nonBayesian optimization criterion is preferred. By contrast, in the nonlinear problem $\underline{x}(t)$ is generally not gaussian; hence, many Bayesian criteria lead to estimates that are different from the conditional mean. In addition, optimal estimation algorithms for nonlinear systems often cannot be expressed in closed form, requiring methods for approximating optimal nonlinear filters.

One further complication associated with general nonlinear estimation problems arises in the structure of the system nonlinearities. Theoretical treatments of this subject often deal with a more general version of Eq. (6.0-1), namely

$$\dot{\underline{x}}(t) = \underline{f}(\underline{x}(t),t) + G(\underline{x}(t),t)\underline{w}(t) \tag{6.0-3}$$

where $G(\underline{x}(t),t)$ is a nonlinear matrix function of $\underline{x}(t)$, and $\underline{w}(t)$ is again (formally) a vector white noise process. In this case a theory for estimating $\underline{x}(t)$ cannot be developed within the traditional framework of mean square stochastic calculus because the right side of Eq. (6.0-3) is not integrable in the mean square sense, owing to the statistical properties of the term $G(\underline{x}(t),t)\underline{w}(t)$. This difficulty is overcome by formulating the nonlinear filtering problem within the context of Ito calculus (Refs. 1 and 2) which provides consistent mathematical rules for integrating Eq. (6.0-3). However, a theoretical discussion of the latter topic is beyond the scope of this book.

The main goal of this chapter is to provide insight into principles of nonlinear estimation theory which will be useful for most practical problems. In this spirit, we circumvent the mathematical issues raised by Eq. (6.0-3) using the following argument. Most *physical* nonlinear systems can be represented by a differential equation of the form

$$\dot{\underline{x}}_1(t) = \underline{f}_1(\underline{x}_1(t),\underline{x}_2(t),t) \tag{6.0-4}$$

where $\underline{x}_2(t)$ is a *bandlimited* (nonwhite) random forcing function having *bounded* rms value – i.e., there is no such thing as white noise in nature. We shall *model* $\underline{x}_2(t)$ as a gaussian random process generated by the linear system

$$\dot{\underline{x}}_2(t) = F_2(t)\underline{x}_2(t) + \underline{w}_2(t) \tag{6.0-5}$$

where $\underline{w}_2(t)$ is gaussian white noise. Combining Eqs. (6.0-4) and (6.0-5) and defining

$$\underline{x}(t) \triangleq \begin{bmatrix} \underline{x}_1(t) \\ --- \\ \underline{x}_2(t) \end{bmatrix}$$

we obtain the augmented equations of motion

$$\underline{\dot{x}}(t) = \begin{bmatrix} \underline{\dot{x}}_1(t) \\ \hline \underline{\dot{x}}_2(t) \end{bmatrix} = \begin{bmatrix} \underline{f}_1(\underline{x}_1(t),\underline{x}_2(t),t) \\ \hline F_2(t)\,\underline{x}_2(t) \end{bmatrix} + \begin{bmatrix} \underline{0} \\ \hline \underline{w}_2(t) \end{bmatrix} \tag{6.0-6}$$

having the same form as Eq. (6.0-1). Because the white noise term in Eq. (6.0-6) is independent of $\underline{x}(t)$, the manipulations associated with mean square stochastic calculus can be applied. A more detailed discussion of this point is provided in Ref. 1.

Within the framework of the model in Eqs. (6.0-1) and (6.0-2), this chapter considers some estimation criteria that lead to practical techniques for estimating the state of a nonlinear system. Sections 6.1 and 6.2 discuss filtering and smoothing algorithms for *minimum variance estimators* – i.e., those which calculate the conditional mean of $\underline{x}(t)$. In Section 6.1 emphasis is placed on Taylor series approximation methods for computing the estimate. Section 6.2 describes the use of statistical linearization approximations. Section 6.3 briefly treats the topic of *nonlinear least-squares estimation*, a technique that avoids the need to specify statistical models for the noise processes, $\underline{w}(t)$ and $\underline{v}_k$, in Eqs. (6.0-1) and (6.0-2). Finally, Section 6.4 discusses a practical analytical technique for *analyzing* nonlinear stochastic systems, based upon statistical linearization arguments.

6.1 NONLINEAR MINIMUM VARIANCE ESTIMATION

THE EXTENDED KALMAN FILTER

Given the equations of motion and measurement data in Eqs. (6.0-1) and (6.0-2), we seek algorithms for calculating the minimum variance estimate of $\underline{x}(t)$ as a function of time and the accumulated measurement data. Recall from Chapter 4 that the minimum variance estimate is always the conditional mean of the state vector, regardless of its probability density function. Now suppose that the measurement at time t_{k-1} has just been processed and the corresponding value $\hat{\underline{x}}(t_{k-1})$ of the conditional mean is known. Between times t_{k-1} and t_k, no measurements are taken and the state propagates according to Eq. (6.0-1). By formally integrating the latter, we obtain

$$\underline{x}(t) = \underline{x}(t_{k-1}) + \int_{t_{k-1}}^{t} \underline{f}(\underline{x}(\tau),\tau)\,d\tau + \int_{t_{k-1}}^{t} \underline{w}(\tau)\,d\tau \tag{6.1-1}$$

Taking the expectation of both sides of Eq. (6.1-1) conditioned on all the measurements taken up until time t_{k-1}, interchanging the order of expectation and integration, and differentiating produces

$$\frac{d}{dt}\,E[\underline{x}(t)] = E[\underline{f}(\underline{x}(t),t)]\,, \qquad t_{k-1} \leqslant t < t_k \tag{6.1-2}$$

with the initial condition

$$E[\underline{x}(t_{k-1})] = \hat{\underline{x}}(t_{k-1})$$

Therefore, on the interval $t_{k-1} \leqslant t < t_k$, the conditional mean of $\underline{x}(t)$ is the solution to Eq. (6.1-2), which can be written more compactly as

$$\dot{\hat{\underline{x}}}(t) = \hat{\underline{f}}(\underline{x}(t),t), \qquad t_{k-1} \leqslant t < t_k \tag{6.1-3}$$

where the caret (ˆ) denotes the expectation operation. Similarly, a differential equation for the estimation error covariance matrix

$$P(t) \triangleq E\left[[\hat{\underline{x}}(t) - \underline{x}(t)]\,[\hat{\underline{x}}(t) - \underline{x}(t)]^T\right] \tag{6.1-4}$$

is derived by substituting for $\underline{x}(t)$ in Eq. (6.1-4) from Eq. (6.1-1), interchanging the order of expectation and integration, and differentiating. The result is

$$\dot{P}(t) = \widehat{\underline{x}\,\underline{f}^T} - \hat{\underline{x}}\,\hat{\underline{f}}^T + \widehat{\underline{f}\,\underline{x}^T} - \hat{\underline{f}}\,\hat{\underline{x}}^T + Q(t), \qquad t_{k-1} \leqslant t < t_k \tag{6.1-5}$$

where the dependence of $\underline{x}$ upon t, and $\underline{f}$ upon $\underline{x}$ and t, is suppressed for notational convenience.

Equations (6.1-3) and (6.1-5) are generalizations of the propagation equations for the linear estimation problem. If $\hat{\underline{x}}(t)$ and P(t) can be calculated, they will provide both an estimate of the state vector between measurement times and a measure of the estimation accuracy. Now observe that the differential equations for $\hat{\underline{x}}(t)$ and P(t) depend upon the entire probability density function* $p(\underline{x},t)$ for $\underline{x}(t)$. Recall that for *linear systems* $\underline{f}(\underline{x}(t), t) = F(t)\,\underline{x}(t)$ so that Eq. (6.1-3) reduces to

$$\begin{aligned}\dot{\hat{\underline{x}}}(t) &= \widehat{F(t)\,\underline{x}(t)}\\ &= F(t)\,\hat{\underline{x}}(t)\\ &= \underline{f}(\hat{\underline{x}}, t)\end{aligned}$$

That is, $\dot{\hat{\underline{x}}}(t)$ depends only upon F(t) and $\hat{\underline{x}}(t)$; similarly, by substituting the linear form for $\underline{f}$ into Eq. (6.1-5) it follows that $\dot{P}(t)$ depends only upon F(t), P(t), and Q(t).** Therefore, the differential equations for the estimate and its covariance matrix can be readily integrated. However, in the more general *nonlinear* case

$$\hat{\underline{f}}(\underline{x},t) = \int_{-\infty}^{\infty} \cdots \int_{-\infty}^{\infty} \underline{f}(\underline{x},t)\, p(\underline{x},t)\, dx_1 \ldots dx_n \neq \underline{f}(\hat{\underline{x}},t)$$

---

* The symbol $p(\underline{x},t)$ is used to denote the probability density function in this chapter to distinguish it from the dynamic nonlinearities, $\underline{f}(\underline{x}, t)$.

**See Problem 6-2.

Thus, in order to compute $\hat{\underline{f}}(\underline{x},t)$, $p(\underline{x},t)$ must be known.

To obtain practical estimation algorithms, methods of computing the mean and covariance matrix which do not depend upon knowing $p(\underline{x},t)$ are needed. A method often used to achieve this goal is to expand $\underline{f}$ in a Taylor series about a known vector $\bar{\underline{x}}(t)$ that is close to $\underline{x}(t)$. In particular, if $\underline{f}$ is expanded about the current estimate (conditional mean) of the state vector, then $\bar{\underline{x}} = \hat{\underline{x}}$ and

$$\underline{f}(\underline{x},t) = \underline{f}(\hat{\underline{x}},t) + \left.\frac{\partial \underline{f}}{\partial \underline{x}}\right|_{\underline{x}=\hat{\underline{x}}} (\underline{x}-\hat{\underline{x}}) + \ldots \tag{6.1-6}$$

where it is assumed the required partial derivatives exist. Taking the expectation of both sides of Eq. (6.1-6) produces

$$\hat{\underline{f}}(\underline{x},t) = \underline{f}(\hat{\underline{x}},t) + \underline{0} + \ldots$$

The first-order approximation to $\hat{\underline{f}}(\underline{x}(t), t)$ is obtained by dropping all but the first term of the power series for $\hat{\underline{f}}$ and substituting the result into Eq. (6.1-3); this produces*

$$\dot{\hat{\underline{x}}}(t) = \underline{f}(\hat{\underline{x}}(t)), \qquad t_{k-1} \leqslant t < t_k \tag{6.1-7}$$

Similarly, an approximate differential equation for the estimation error covariance matrix is obtained by substituting the first two terms of the expansion for $\underline{f}$ into Eq. (6.1-5), carrying out the indicated expectation operations, and combining terms; the result is

$$\dot{P}(t) = F(\hat{\underline{x}}(t), t)\, P(t) + P(t)\, F^T(\hat{\underline{x}}(t), t) + Q(t), \qquad t_{k-1} \leqslant t < t_k \tag{6.1-8}$$

where $F(\hat{\underline{x}}(t), t)$ is the matrix whose ij^{th} element is given by

$$f_{ij}(\hat{\underline{x}}(t),t) \triangleq \left.\frac{\partial f_i(\underline{x}(t),t)}{\partial x_j(t)}\right|_{\underline{x}(t)=\hat{\underline{x}}(t)}$$

Equations (6.1-7) and (6.1-8) are approximate expressions for propagating the conditional mean of the state and its associated covariance matrix. Being linearized about $\hat{\underline{x}}(t)$, they have a structure similar to the Kalman filter propagation equations for linear systems. Consequently they are referred to as the *extended Kalman filter* propagation equations. Higher-order, more exact approximations to the optimal nonlinear filter can be achieved using more terms in the Taylor series expansion for the nonlinearities, and by deriving recursive relations for the higher moments of $\underline{x}$. Methods of this type are discussed

*Up to this point $\hat{\underline{x}}$ has denoted the exact conditional mean; henceforth, $\hat{\underline{x}}$ denotes any estimate of the state that is an approximation to the conditional mean.

subsequently. Other techniques that depend upon finding functional approximations to the conditional probability density function of $\underline{x}$, and which are not treated here, are discussed in Refs. 19 and 20.

To obtain a complete filtering algorithm, update equations which account for measurement data are needed. To develop update equations, assume that the estimate of $\underline{x}(t)$ and its associated covariance matrix have been propagated using Eqs. (6.1-7) and (6.1-8), and denote the solutions at time t_k by $\hat{\underline{x}}_k(-)$ and $P_k(-)$. When the measurement $\underline{z}_k$ is taken, an improved estimate of the state is sought. Motivated by the linear estimation problem discussed in Chapter 4, we require that the updated estimate be a *linear function* of the measurement – i.e.,

$$\hat{\underline{x}}_k(+) = \underline{a}_k + K_k \underline{z}_k \tag{6.1-9}$$

where the vector $\underline{a}_k$ and the "gain" matrix K_k are to be determined. Proceeding with arguments similar to those used in Section 4.2, we define the estimation errors just before and just after update, $\tilde{\underline{x}}_k(-)$ and $\tilde{\underline{x}}_k(+)$, respectively, by

$$\begin{aligned} \tilde{\underline{x}}_k(+) &\triangleq \hat{\underline{x}}_k(+) - \underline{x}_k \\ \tilde{\underline{x}}_k(-) &\triangleq \hat{\underline{x}}_k(-) - \underline{x}_k \end{aligned} \tag{6.1-10}$$

Then Eqs. (6.1-9) and (6.1-10) are combined with Eq. (6.0-2) to produce the following expression for the estimation error:

$$\tilde{\underline{x}}_k(+) = \underline{a}_k + K_k \underline{h}_k(\underline{x}_k) + K_k \underline{v}_k + \tilde{\underline{x}}_k(-) - \hat{\underline{x}}_k(-) \tag{6.1-11}$$

One condition required is that the estimate be unbiased – i.e., $E[\tilde{\underline{x}}_k(+)] = \underline{0}$. This is consistent with the fact that the desired estimate is an approximation to the conditional mean. Applying the latter requirement to Eq. (6.1-11) and recognizing that

$$E[\tilde{\underline{x}}_k(-)] = E[\underline{v}_k] = \underline{0}$$

we obtain

$$\underline{a}_k + K_k \hat{\underline{h}}_k(\underline{x}_k) - \hat{\underline{x}}_k(-) = \underline{0} \tag{6.1-12}$$

Solving Eq. (6.1-12) for $\underline{a}_k$ and substituting the result into Eq. (6.1-9) yields

$$\hat{\underline{x}}_k(+) = \hat{\underline{x}}_k(-) + K_k [\underline{z}_k - \hat{\underline{h}}_k(\underline{x}_k)] \tag{6.1-13}$$

In addition, Eqs. (6.1-11) and (6.1-12) can be combined to express the estimation error in the form

$$\tilde{\underline{x}}_k(+) = \tilde{\underline{x}}_k(-) + K_k [\underline{h}_k(\underline{x}_k) - \hat{\underline{h}}_k(\underline{x}_k)] + K_k \underline{v}_k \tag{6.1-14}$$

To determine the optimal gain matrix K_k, the same procedure used for the linear estimation problem in Section 4.2 is employed. First an expression is obtained for the estimation error covariance matrix $P_k(+)$ in terms of K_k; then K_k is chosen to minimize an appropriate function of $P_k(+)$. Applying the definition

$$P_k(+) = E[\underline{\tilde{x}}_k(+)\,\underline{\tilde{x}}_k(+)^T]$$

to Eq. (6.1-14), recognizing that $\underline{v}_k$ is uncorrelated with $\underline{\tilde{x}}_k(-)$ and $\underline{x}_k$, using the relations

$$P_k(-) = E[\underline{\tilde{x}}_k(-)\,\underline{\tilde{x}}_k(-)^T]$$

$$R_k = E[\underline{v}_k \underline{v}_k^T]$$

and assuming that $P_k(+)$ is independent of $\underline{z}_k$, we obtain

$$\begin{aligned} P_k(+) = P_k(-) &+ K_k\, E\left[[\underline{h}_k(\underline{x}_k) - \hat{\underline{h}}_k(\underline{x}_k)]\,[\underline{h}_k(\underline{x}_k) - \hat{\underline{h}}_k(\underline{x}_k)]^T\right] K_k^T \\ &+ E\left[\underline{\tilde{x}}_k(-)\,[\underline{h}_k(\underline{x}_k) - \hat{\underline{h}}_k(\underline{x}_k)]^T\right] K_k^T \\ &+ K_k\, E\left[[\underline{h}_k(\underline{x}_k) - \hat{\underline{h}}_k(\underline{x}_k)]\,\underline{\tilde{x}}_k(-)^T\right] + K_k R_k K_k^T \end{aligned} \tag{6.1-15}$$

The estimate being sought – the approximate conditional mean of $\underline{x}(t)$ – is a minimum variance estimate; that is, it minimizes the class of functions

$$J_k = E\left[\underline{\tilde{x}}_k(+)^T\, S\underline{\tilde{x}}_k(+)\right]$$

for any positive semidefinite matrix S. Consequently, we can choose S = I, and write

$$J_k = E\left[\underline{\tilde{x}}_k(+)^T\, \underline{\tilde{x}}_k(+)\right] = \text{trace}\,[P_k(+)] \tag{6.1-16}$$

Taking the trace of both sides of Eq. (6.1-15), substituting the result into (6.1-16) and solving the equation

$$\frac{\partial J_k}{\partial K_k} = 0$$

for K_k yields the desired optimal gain matrix,

$$\begin{aligned} K_k = -E&\left[\underline{\tilde{x}}_k(-)\,[\underline{h}_k(\underline{x}_k) - \hat{\underline{h}}_k(\underline{x}_k)]^T\right] \\ &\times\left\{E\left[[\underline{h}_k(\underline{x}_k) - \hat{\underline{h}}_k(\underline{x}_k)]\,[\underline{h}_k(\underline{x}_k) - \hat{\underline{h}}_k(\underline{x}_k)]^T\right] + R_k\right\}^{-1} \end{aligned} \tag{6.1-17}$$

Substituting Eq. (6.1-17) into Eq. (6.1-15) produces, after some manipulation,

$$P_k(+) = P_k(-) + K_k\, E\left[[\underline{h}_k(\underline{x}_k) - \hat{\underline{h}}_k(\underline{x}_k)]\, \tilde{\underline{x}}_k(-)^T\right] \qquad (6.1\text{-}18)$$

Equations (6.1-13), (6.1-17), and (6.1-18) together provide updating algorithms for the estimate when a measurement is taken. However, they are impractical to implement in this form because they depend upon the probability density function for $\underline{x}(t)$ to calculate $\hat{\underline{h}}_k$. To simplify the computation, expand $\underline{h}_k(\underline{x}_k)$ in a power series about $\hat{\underline{x}}_k(-)$ as follows:

$$\underline{h}_k(\underline{x}_k) = \underline{h}_k(\hat{\underline{x}}_k(-)) + H_k(\hat{\underline{x}}_k(-))(\underline{x}_k - \hat{\underline{x}}_k(-)) + \ldots \qquad (6.1\text{-}19)$$

where

$$H_k(\hat{\underline{x}}_k(-)) = \left.\frac{\partial \underline{h}_k(\underline{x})}{\partial \underline{x}}\right|_{\underline{x}=\hat{\underline{x}}_k(-)} \qquad (6.1\text{-}20)$$

Truncating the above series after the first two terms, substituting the resulting approximation for $\underline{h}_k(\underline{x}_k)$ into Eqs. (6.1-13), (6.1-17), and (6.1-18), and carrying out the indicated expectation operations produces the *extended Kalman filter* measurement update equations

$$\hat{\underline{x}}_k(+) = \hat{\underline{x}}_k(-) + K_k[\underline{z}_k - \underline{h}_k(\hat{\underline{x}}_k(-))]$$

$$K_k = P_k(-)\, H_k^T(\hat{\underline{x}}_k(-))\left[H_k(\hat{\underline{x}}_k(-))\, P_k(-)\, H_k^T(\hat{\underline{x}}_k(-)) + R_k\right]^{-1}$$

$$P_k(+) = [I - K_k H_k(\hat{\underline{x}}_k(-))]\, P_k(-) \qquad (6.1\text{-}21)$$

Equations (6.1-7), (6.1-8), and (6.1-21) constitute the extended Kalman filtering algorithm for nonlinear systems with discrete measurements. A summary of the mathematical model and the filter equations is given in Table 6.1-1; the extension of these results to the case of continuous measurements is given in Table 6.1-2. A comparison of these results with the conventional Kalman filter discussed in Section 4.2 indicates an important difference; the gains K_k in Eq. (6.1-21) are actually random variables depending upon the estimate $\hat{\underline{x}}(t)$ through the matrices $F(\hat{\underline{x}}(t), t)$ and $H_k(\hat{\underline{x}}_k(-))$. This results from the fact that we have chosen to linearize $\underline{f}$ and $\underline{h}_k$ about the current estimate of $\underline{x}(t)$. Hence, the sequence $\{K_k\}$ must be computed in real time; it cannot be precomputed before the measurements are collected and stored in a computer memory. Furthermore, the sequence of (approximate) estimation error covariance matrices $\{P_k\}$ is also random, depending upon the time-history of $\hat{\underline{x}}(t)$ – i.e., the estimation accuracy achieved is trajectory dependent. The reader can verify that when the system dynamics and measurements are linear, the extended Kalman filter reduces to the conventional Kalman filter.

TABLE 6.1-1 SUMMARY OF CONTINUOUS-DISCRETE EXTENDED KALMAN FILTER

System Model	$\dot{\underline{x}}(t) = \underline{f}(\underline{x}(t),t) + \underline{w}(t)\ ; \quad \underline{w}(t) \sim N(\underline{0}, Q(t))$
Measurement Model	$\underline{z}_k = \underline{h}_k(\underline{x}(t_k)) + \underline{v}_k\ ; \quad k = 1, 2, \ldots ; \quad \underline{v}_k \sim N(\underline{0}, R_k)$
Initial Conditions	$\underline{x}(0) \sim N(\hat{\underline{x}}_0, P_0)$
Other Assumptions	$E\left[\underline{w}(t)\,\underline{v}_k^T\right] = 0$ for all k and all t
State Estimate Propagation	$\dot{\hat{\underline{x}}}(t) = \underline{f}(\hat{\underline{x}}(t),t)$
Error Covariance Propagation	$\dot{P}(t) = F(\hat{\underline{x}}(t),t)\,P(t) + P(t)\,F^T(\hat{\underline{x}}(t),t) + Q(t)$
State Estimate Update	$\hat{\underline{x}}_k(+) = \hat{\underline{x}}_k(-) + K_k[\underline{z}_k - \underline{h}_k(\hat{\underline{x}}_k(-))]$
Error Covariance Update	$P_k(+) = [I - K_k H_k(\hat{\underline{x}}_k(-))]\ P_k(-)$
Gain Matrix	$K_k = P_k(-)\,H_k^T(\hat{\underline{x}}_k(-))\left[H_k(\hat{\underline{x}}_k(-))\,P_k(-)\,H_k^T(\hat{\underline{x}}_k(-)) + R_k\right]^{-1}$
Definitions	$F(\hat{\underline{x}}(t),t) = \left.\dfrac{\partial \underline{f}(\underline{x}(t),t)}{\partial \underline{x}(t)}\right\vert_{\underline{x}(t)=\hat{\underline{x}}(t)}$ $H_k(\hat{\underline{x}}(-)) = \left.\dfrac{\partial \underline{h}_k(\underline{x}(t_k))}{\partial \underline{x}(t_k)}\right\vert_{\underline{x}(t_k)=\hat{\underline{x}}(-)}$

TABLE 6.1-2 SUMMARY OF CONTINUOUS EXTENDED KALMAN FILTER

System Model	$\dot{\underline{x}}(t) = \underline{f}(\underline{x}(t),t) + \underline{w}(t)\ ; \ \underline{w}(t) \sim N(\underline{0}, Q(t))$
Measurement Model	$\underline{z}(t) = \underline{h}(\underline{x}(t),t) + \underline{v}(t)\ ; \ \underline{v}(t) \sim N(\underline{0}, R(t))$
Initial Conditions	$\underline{x}(0) \sim N(\hat{\underline{x}}_0, P_0)$
Other Assumptions	$E\left[\underline{w}(t)\,\underline{v}^T(\tau)\right] = 0$ for all t and all τ
State Estimate Equation	$\dot{\hat{\underline{x}}}(t) = \underline{f}(\hat{\underline{x}}(t),t) + K(t)\,[\underline{z}(t) - \underline{h}(\hat{\underline{x}}(t),t)]$
Error Covariance Equation	$\dot{P}(t) = F(\hat{\underline{x}}(t),t)\,P(t) + P(t)\,F^T(\hat{\underline{x}}(t),t) + Q(t)$ $- P(t)\,H^T(\hat{\underline{x}}(t),t)\,R^{-1}(t)\,H(\hat{\underline{x}}(t),t)\,P(t)$
Gain Equation	$K(t) = P(t)\,H^T(\hat{\underline{x}}(t),t)\,R^{-1}(t)$
Definitions	$F(\hat{\underline{x}}(t),t) = \left.\dfrac{\partial \underline{f}(\underline{x}(t),t)}{\partial \underline{x}(t)}\right\vert_{\underline{x}(t)=\hat{\underline{x}}(t)}$ $H(\hat{\underline{x}}(t),t) = \left.\dfrac{\partial \underline{h}(\underline{x}(t),t)}{\partial \underline{x}(t)}\right\vert_{\underline{x}(t)=\hat{\underline{x}}(t)}$

It should be noted that a *linearized Kalman filtering algorithm*, for which the filter gains *can* be precalculated, is developed if the nonlinear functions $\underline{f}$ and $\underline{h}_k$ in Eqs. (6.0-1) and (6.0-2) are linearized about a vector $\underline{\bar{x}}(t)$ which is specified prior to processing the measurement data. If the stages in the extended Kalman filter derivation are repeated, with $\underline{\bar{x}}$ substituted for $\underline{\hat{x}}$ throughout, the filtering algorithm given in Table 6.1-3 is obtained. Generally speaking, this procedure yields less filtering accuracy than the extended Kalman filter because $\underline{\bar{x}}(t)$ is usually not as close to the actual trajectory as is $\underline{\hat{x}}(t)$. However, it offers the computational advantage that the gains $\{K_k\}$ can be computed off-line and stored in a computer, since $\underline{\bar{x}}(t)$ is known *a priori*.

TABLE 6.1-3 SUMMARY OF CONTINUOUS-DISCRETE LINEARIZED KALMAN FILTER

System Model	$\dot{\underline{x}}(t) = \underline{f}(\underline{x}(t),t) + \underline{w}(t)$; $\underline{w}(t) \sim N(\underline{0}, Q(t))$
Measurement Model	$\underline{z}_k = \underline{h}_k(\underline{x}(t_k)) + \underline{v}_k$; $k = 1, 2, \ldots$; $\underline{v}_k \sim N(\underline{0}, R_k)$
Initial Conditions	$\underline{x}(0) \sim N(\underline{\hat{x}}_0, P_0)$
Other Assumptions	$E[\underline{w}(t)\,\underline{v}_k^T] = 0$ for all k and all t Nominal trajectory $\underline{\bar{x}}(t)$ is available
State Estimate Propagation	$\dot{\underline{\hat{x}}}(t) = \underline{f}(\underline{\bar{x}}(t),t) + F(\underline{\bar{x}}(t),t)[\underline{\hat{x}}(t) - \underline{\bar{x}}(t)]$
Error Covariance Propagation	$\dot{P}(t) = F(\underline{\bar{x}}(t),t)P(t) + P(t)F^T(\underline{\bar{x}}(t),t) + Q(t)$
State Estimate Update	$\underline{\hat{x}}_k(+) = \underline{\hat{x}}_k(-) + K_k[\underline{z}_k - \underline{h}_k(\underline{\bar{x}}(t_k)) - H_k(\underline{\bar{x}}(t))[\underline{\hat{x}}_k(-) - \underline{\bar{x}}(t_k)]]$
Error Covariance Update	$P_k(+) = [I - K_k H_k(\underline{\bar{x}}(t_k))]\,P_k(-)$
Gain Matrix	$K_k = P_k(-)H_k^T(\underline{\bar{x}}(t_k))\,[H_k(\underline{\bar{x}}(t_k))P_k(-)H_k^T(\underline{\bar{x}}(t_k)) + R_k]^{-1}$
Definitions	$F(\underline{\bar{x}}(t),t) = \left.\dfrac{\partial \underline{f}(\underline{x}(t),t)}{\partial \underline{x}(t)}\right\rvert_{\underline{x}(t) = \underline{\bar{x}}(t)}$ $H_k(\underline{\bar{x}}(t_k)) = \left.\dfrac{\partial \underline{h}_k(\underline{x}(t_k))}{\partial \underline{x}(t_k)}\right\rvert_{\underline{x}(t_k) = \underline{\bar{x}}(t_k)}$

Because the matrix P_k in Eq. (6.1-21) is only an approximation to the true covariance matrix, actual filter performance must be verified by monte carlo simulation. There is no guarantee that the actual estimate obtained will be close to the truly optimal estimate. Fortunately, the extended Kalman filter has been found to yield accurate estimates in a number of important practical applications. Because of this experience and its similarity to the conventional

Kalman filter, it is usually one of the first methods to be tried for any nonlinear filtering problem.

HIGHER-ORDER FILTERS

The Taylor series expansion method for treating the nonlinear filtering problem, outlined in the preceding section, can be extended to obtain higher-order nonlinear filters. One method of accomplishing this is to write the estimate, $\hat{\underline{x}}_k(+)$ in Eq. (6.1-9), as a higher-order power series in $\underline{z}_k$. Another more commonly used approach is to include more terms in the expansions for $\underline{f}(\underline{x},t)$ and $\underline{h}_k(\underline{x}_k)$ in Eqs. (6.1-6) and (6.1-19). These methods differ in that the latter seeks better approximations to the optimal filter whose structure is *constrained* by Eq. (6.1-9) – i.e., the measurement appears linearly; the former allows a more general dependence of the estimate upon the measurement data. If both techniques are applied simultaneously, a more general nonlinear filter will result than from either method alone. The discussion here is restricted to the case where the estimate is constrained to be linear in $\underline{z}_k$.

The Iterated Extended Kalman Filter – One method by which the estimate $\hat{\underline{x}}_k(+)$ given in Eq. (6.1-21) can be improved is by repeatedly calculating $\hat{\underline{x}}_k(+)$, K_k, and $P_k(+)$, each time linearizing about the most recent estimate. To develop this algorithm, denote the ith estimate of $\underline{x}_k(+)$ by $\hat{\underline{x}}_{k,i}(+)$, $i = 0, 1, \ldots$, with $\hat{\underline{x}}_{k,0}(+) = \hat{\underline{x}}_k(-)$ and expand $\underline{h}_k(\underline{x}_k)$ in Eqs. (6.1-13) and (6.1-17) in the form

$$\underline{h}_k(\underline{x}_{k,i}) = \underline{h}_k(\hat{\underline{x}}_{k,i}(+)) + H_k(\hat{\underline{x}}_{k,i}(+))(\underline{x}_k - \hat{\underline{x}}_{k,i}(+)) + \ldots$$

$$H_k(\hat{\underline{x}}_{k,i}(+)) \triangleq \left.\frac{\partial \underline{h}_k}{\partial \underline{x}_k}\right|_{\underline{x}_k = \hat{\underline{x}}_{k,i}(+)} \tag{6.1-22}$$

Observe that Eq. (6.1-22) reduces to Eq. (6.1-19) when i = 0. Truncating the series in Eq. (6.1-22) after the second term, substituting the result into Eqs. (6.1-13), (6.1-17), and (6.1-18), performing the indicated expectation operations, and observing that* $E[\underline{x}_k] = \hat{\underline{x}}_k(-)$, produces the following iterative expressions for the updated estimate:

$$\hat{\underline{x}}_{k,i+1}(+) = \hat{\underline{x}}_k(-) + K_{k,i}[\underline{z}_k - \underline{h}_k(\hat{\underline{x}}_{k,i}(+)) - H_k(\hat{\underline{x}}_{k,i}(+))(\hat{\underline{x}}_k(-) - \hat{\underline{x}}_{k,i}(+))]$$

$$K_{k,i} = P_k(-)H_k^T(\hat{\underline{x}}_{k,i}(+))\,[H_k(\hat{\underline{x}}_{k,i}(+))\,P_k(-)\,H_k^T(\hat{\underline{x}}_{k,i}(+)) + R_k]^{-1}$$

---

*This is true because the expectation $\hat{\underline{h}}_k$ is conditioned on all the measurements up to, but not including, time t_k.

$$P_{k,i+1}(+) = [I - K_{k,i}\, H_k(\hat{\underline{x}}_{k,i}(+))]\, P_k(-)$$

$$\hat{\underline{x}}_{k,0}(+) \triangleq \hat{\underline{x}}_k(-) \quad , \quad i = 0, 1, \ldots \tag{6.1-23}$$

As many calculations of $\hat{\underline{x}}_{k,i}(+)$ in Eq. (6.1-23), over the index i, can be performed as are necessary to reach the point where little further improvement is realized from additional iterations. However, it should be recognized that each iteration of Eq. (6.1-23) contributes to the computation time required to mechanize the filter.

A similar procedure can be devised for iterating over the nonlinear dynamics in Eq. (6.1-7); for details the reader is referred to Ref. 1. It has been found that such iteration techniques can significantly reduce that part of the extended Kalman filter estimation error which is caused by system nonlinearities (Ref. 5).

A Second-Order Filter – Another type of higher-order approximation to the minimum variance estimate is achieved by including second-order terms in the expansions for $\underline{h}_k(\underline{x}_k)$ and $\underline{f}(\underline{x}_k, t)$ in Eqs. (6.1-6) and (6.1-19). Thus, we write

$$\underline{f}(\underline{x}(t),t) = \underline{f}(\hat{\underline{x}}(t),t) - F(\hat{\underline{x}}(t),t)\, \tilde{\underline{x}}(t) + \frac{1}{2}\underline{\partial}^2\, (\underline{f}, \tilde{\underline{x}}(t)\, \tilde{\underline{x}}^T(t)) + \ldots$$

$$\underline{h}_k(\underline{x}_k) = \underline{h}_k(\hat{\underline{x}}_k(-)) - H(\hat{\underline{x}}_k(-))\, \tilde{\underline{x}}_k(-) + \frac{1}{2}\underline{\partial}^2\, (\underline{h}_k, \tilde{\underline{x}}_k(-)\, \tilde{\underline{x}}_k{}^T(-)) + \ldots$$

$$\tag{6.1-24}$$

where the operator $\underline{\partial}^2(\underline{f},B)$, for *any* function $\underline{f}(\underline{x},t)$ and *any* matrix B, is a vector whose i^{th} element is defined by

$$\partial_i{}^2(\underline{f},B) \triangleq \text{trace}\left\{\left[\frac{\partial^2 f_i}{\partial x_p\, \partial x_q}\right] B\right\} \tag{6.1-25}$$

The bracketed quantity in Eq. (6.1-25) is a matrix whose pq^{th} element is the quantity $\partial^2 f_i/\partial x_p \partial x_q$. Truncating the series in Eq. (6.1-24) after the third term, substituting the resulting approximations for $\underline{f}$ and $\underline{h}_k$ into Eqs. (6.1-3), (6.1-5), (6.1-13), (6.1-17), and (6.1-18), and carrying out the indicated expectation operations produces, after some manipulation,*

---

*An assumption used in deriving Eq. (6.1-26) is that third-order moments of $\tilde{\underline{x}}_k(-)$ are negligible – i.e., the probability density for $\tilde{\underline{x}}_k(-)$ is assumed to be symmetric about the mean.

$$\left.\begin{aligned}\dot{\hat{\underline{x}}}(t) &= \underline{f}(\hat{\underline{x}}(t),t) + \frac{1}{2}\underline{\partial}^2(\underline{f},P(t)) \\ \dot{P}(t) &= F(\hat{\underline{x}}(t),t)\,P(t) + P(t)\,F^T(\hat{\underline{x}}(t),t) + Q(t)\end{aligned}\right\}\quad t_{k-1} \leqslant t < t_k$$

$$\hat{\underline{x}}_k(+) = \hat{\underline{x}}_k(-) + K_k\left[\underline{z}_k - \underline{h}_k(\hat{\underline{x}}_k(-)) - \frac{1}{2}\underline{\partial}^2(\underline{h}_k, P_k(-))\right]$$

$$K_k = P_k(-)\,H_k^T(\hat{\underline{x}}_k(-))\left[H_k(\hat{\underline{x}}_k(-))\,P_k(-)\,H_k^T(\hat{\underline{x}}_k(-)) + R_k + A_k\right]^{-1}$$

$$P_k(+) = [I - K_k H_k(\hat{\underline{x}}_k(-))]\,P_k(-) \tag{6.1-26}$$

The matrix A_k is defined by the relation

$$A_k \triangleq \frac{1}{4}E\left[\underline{\partial}^2\left(\underline{h}_k, \tilde{\underline{x}}_k(-)\,\tilde{\underline{x}}_k^T(-)\right)\underline{\partial}^2\left(\underline{h}_k, \tilde{\underline{x}}_k(-)\,\tilde{\underline{x}}_k^T(-)\right)^T\right]$$
$$-\frac{1}{4}\underline{\partial}^2(\underline{h}_k, P_k(-))\,\underline{\partial}^2(\underline{h}_k, P_k(-))^T \tag{6.1-27}$$

The ijth element of A_k is given approximately by

$$a_{k_{ij}} \cong \frac{1}{4}\left[\sum_{p,q,m,n} \frac{\partial^2 h_i}{\partial x_p\,\partial x_q}(p_{pm}p_{qn} + p_{pn}p_{qm})\frac{\partial^2 h_j}{\partial x_m\,\partial x_n}\right]_{\underline{x}_k = \hat{\underline{x}}_k(-)} \tag{6.1-28}$$

where h_i denotes the ith element of $\underline{h}_k$ and the dependence of $\underline{h}$ and $\underline{x}$ on the time index k is suppressed for convenience.

Equation (6.1-28) is derived from the gaussian approximation

$$E[\tilde{x}_p\,\tilde{x}_q\,\tilde{x}_m\,\tilde{x}_n] = p_{pq}\,p_{mn} + p_{pm}\,p_{qn} + p_{pn}\,p_{qm}$$

where x_i and p_{ij} denote the ith and ijth elements of $\tilde{\underline{x}}_k(-)$ and $P_k(-)$, respectively. For this reason, Eq. (6.1-26) is sometimes referred to as a *gaussian second-order filter* (Ref. 1). It has a structure similar to the extended Kalman filter described in Table 6.1-1; however, there are additional terms to account for the effect of the nonlinearities. Consequently the improved performance is achieved at the expense of an increased computational burden.

EXAMPLES OF NONLINEAR FILTERING

In this section a number of examples are discussed which illustrate the comparative performance of various nonlinear filters. The first is a geometric example of a hypothetical navigation application; the second and third present computer simulation results of linearized and nonlinear filters applied to a tracking problem.

Example 6.1-1

The extended Kalman filter is a popular method for treating nonlinear estimation problems (e.g., see Ref. 18). However, if nonlinearities are sufficiently important, the estimation error can be significantly reduced through use of a higher-order estimation technique. As a specific illustration, consider an aircraft whose estimated location in x, y coordinates is $\hat{x}_0,\hat{y}_0$ in Fig. 6.1-1, with a gaussian distribution p(x,y) in position uncertainty whose shape is indicated by the narrow shaded elliptical area; the value of p(x,y) is constant on the circumference of the ellipse.* Now suppose the aircraft measures its range to a transponder that has a known position in x,y coordinates. Furthermore, assume that the range measurement errors are very small. If the nonlinear relationship of range to x, y position were taken into account exactly, a new estimate of position would be deduced approximately, as follows: Assuming the range measurement, r_m, is perfect, the aircraft lies somewhere on the circle of radius r_m shown in Fig. 6.1-1. Given the *a priori* distribution of x and y denoted by the shaded ellipse, the new minimum variance estimates, $\hat{x}_1$ and $\hat{y}_1$ of x and y will lie close to the peak value of p(x,y) evaluated along the range circle. Hence, $\hat{x}_1$ and $\hat{y}_1$ will lie approximately at the point indicated in Fig. 6.1-1.

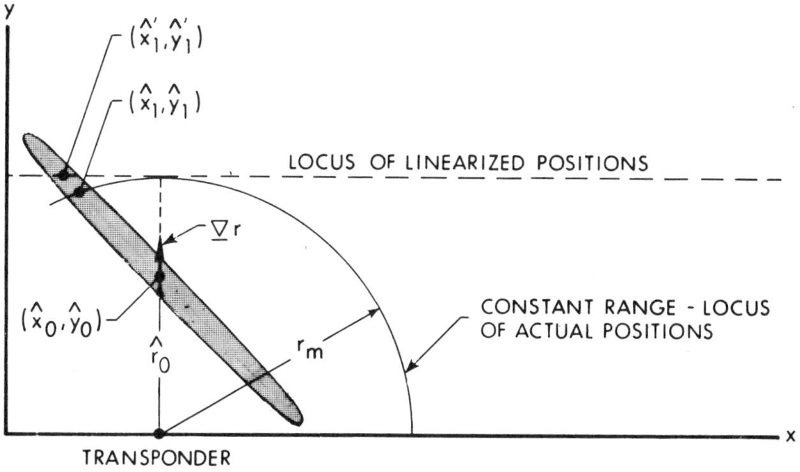

Figure 6.1-1 An Illustration of the Difference Between Linearized Estimates $(\hat{x}'_1, \hat{y}'_1)$ and Nonlinear Estimates $(\hat{x}_1, \hat{y}_1)$

---

*The elliptical contour is obtained by setting the magnitude of the exponent of the gaussian density function equal to a constant; that is, on the contour x and y satisfy

$$\begin{bmatrix} x-\hat{x}_0 \\ y-\hat{y}_0 \end{bmatrix}^T P^{-1} \begin{bmatrix} x-\hat{x}_0 \\ y-\hat{y}_0 \end{bmatrix} = \text{constant}$$

By contrast, if the range measurement is linearized as in an extended Kalman filter, the argument proceeds as follows: In terms of the initial range estimate $\hat{r}_0$ in Fig. 6.1-1, r_m is approximately expressed as

$$r_m \cong \hat{r}_0 + \left.\frac{\partial r}{\partial x}\right|_{\hat{x}_0,\hat{y}_0} (x-\hat{x}_0) + \left.\frac{\partial r}{\partial y}\right|_{\hat{x}_0,\hat{y}_0} (y-\hat{y}_0) \tag{6.1-29}$$

where r is the range to the transponder as a function of x and y. In other words

$$r_m - \hat{r}_0 \cong \underline{\nabla} r^T(\hat{x}_0,\hat{y}_0) \begin{bmatrix} x-\hat{x}_0 \\ y-\hat{y}_0 \end{bmatrix} \tag{6.1-30}$$

where $\underline{\nabla} r(\hat{x}_0,\hat{y}_0)$ is the gradient of r with respect to x and y evaluated at $\hat{r}_0$. The direction of the gradient is along the radius vector from the transponder. Equation (6.1-30) states that the new values of x and y deduced from r_m must lie along the straight dashed line shown in Fig. 6.1-1, which is normal to $\underline{\nabla} r(\hat{x}_0,\hat{y}_0)$. Choosing the new estimate of position at the maximum of p(x,y) along the straight line, linearized estimates, $\hat{x}'_1$ and $\hat{y}'_1$, are obtained at the point indicated in the figure. Depending upon the range accuracy desired, the difference between the nonlinear estimate $(\hat{x}_1,\hat{y}_1)$ and $(\hat{x}'_1,\hat{y}'_1)$ may be significant.

This example illustrates the point that nonlinearity can have an important effect on estimation accuracy. The degree of importance depends upon the estimation accuracy desired, the amount of nonlinearity, the shape of the density function p(x,y) and the strength of the measurement noise.

Example 6.1-2

In this example, we consider the problem of tracking a body falling freely through the atmosphere. The motion is modeled in one dimension by assuming the body falls in a straight line, directly toward a tracking radar, as illustrated in Fig. 6.1-2. The state variables for this problem are designated as

$$x_1 = x, \quad x_2 = \dot{x}, \quad x_3 = \beta \tag{6.1-31}$$

where β is the so-called ballistic coefficient of the falling body and x is its height above the earth.

The equations of motion for the body are given by

$$\underbrace{\begin{bmatrix} \dot{x}_1 \\ \dot{x}_2 \\ \dot{x}_3 \end{bmatrix}}_{\dot{\underline{x}}} = \underbrace{\begin{bmatrix} x_2 \\ d-g \\ 0 \end{bmatrix}}_{\underline{f}(\underline{x})}$$

$$d = \frac{\rho\, x_2^{\,2}}{2x_3}$$

$$\rho = \rho_0 e^{-\frac{x_1}{k_\rho}} \tag{6.1-32}$$

where d is drag deceleration, g is acceleration of gravity, ρ is atmospheric density (with ρ_0 the atmospheric density at sea level) and k_ρ is a decay constant. The differential equation

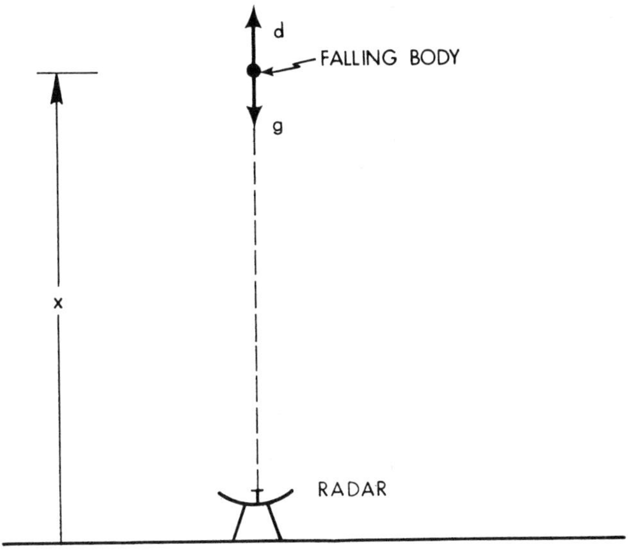

Figure 6.1-2 Geometry of the One-Dimensional Tracking Problem

for velocity, x_2, is nonlinear through the dependence of drag on velocity, air density and ballistic coefficient. Linear measurements are assumed available in the continuous form

$$z(t) = x_1(t) + v(t) \tag{6.1-33}$$

where $v(t) \sim N(0,r)$. Initial values of the state variables are assumed to have mean, $\underline{\mu}$, and a covariance matrix of the form

$$P_0 = \begin{bmatrix} p_{11_o} & 0 & 0 \\ 0 & p_{22_o} & 0 \\ 0 & 0 & p_{33_0} \end{bmatrix} \tag{6.1-34}$$

The problem of estimating all the state variables may be solved using both the extended and linearized Kalman filters discussed in Section 6.1. Recall that the extended Kalman filter is linearized about the current estimate of the state, whereas the general linearized filter utilizes a precomputed nominal trajectory $\underline{\bar{x}}(t)$. The latter is derived for this example by solving the differential equation

$$\dot{\underline{\bar{x}}} = \underline{f}(\underline{\bar{x}})$$
$$\underline{\bar{x}}(0) = \underline{\mu} \tag{6.1-35}$$

Comparative performance results from the two different filters are displayed in Figs. 6.1-3 and 6.1-4, which show the estimation errors achieved from a single monte carlo trial. The simulation parameter values are given below.

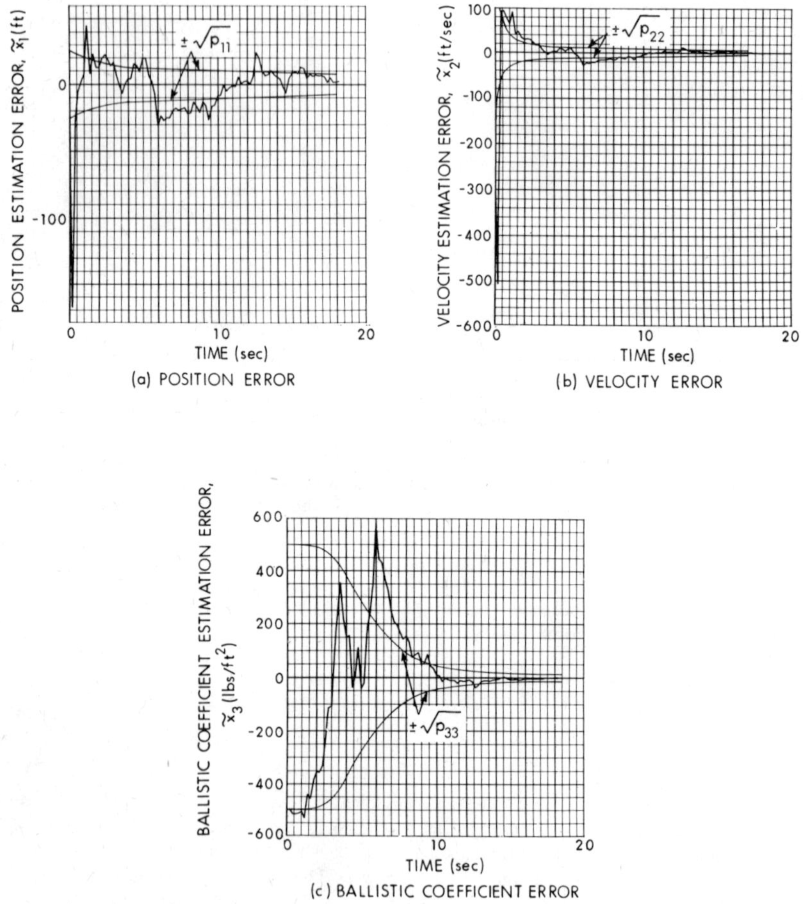

Figure 6.1-3 Performance of the Extended Kalman Filter

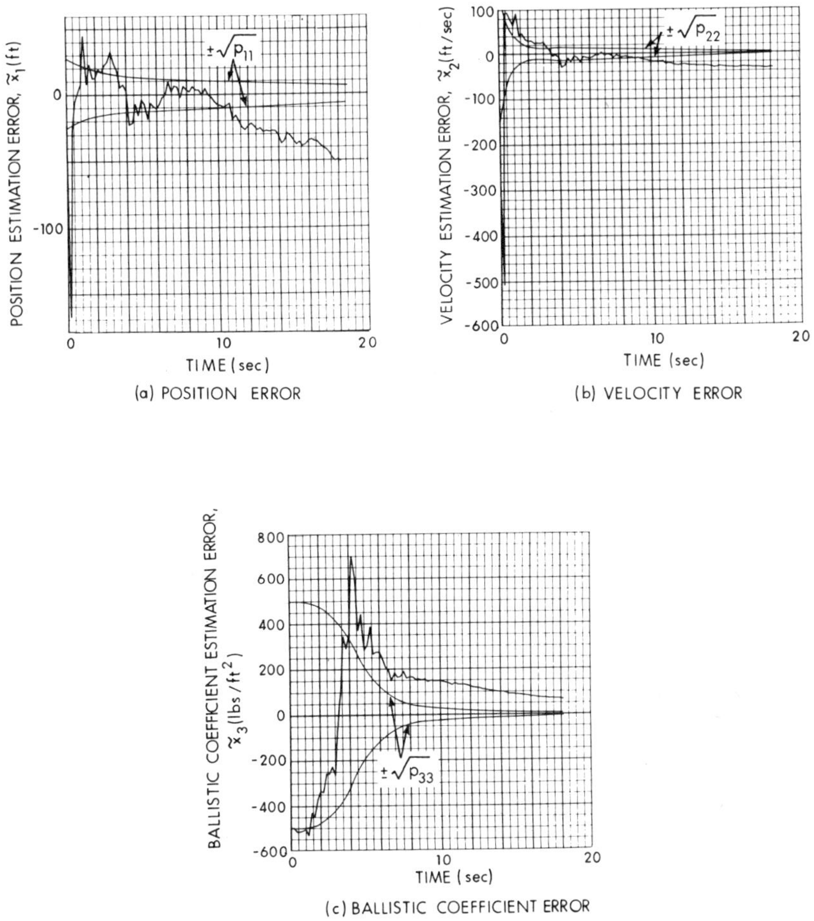

Figure 6.1-4 Performance of the Linearized Kalman Filter

PARAMETER VALUES FOR THE ONE-DIMENSIONAL TRACKING PROBLEM

$\rho_0 = 3.4 \times 10^{-3}$ lb sec^2/ft^4	$x(0) \sim N(10^5 \text{ ft}, 500 \text{ ft}^2)$
$g = 32.2$ ft/sec^2	$\dot{x}(0) \sim N(-6000 \text{ ft/sec}, 2 \times 10^4 \text{ ft}^2/\text{sec}^2)$
$k_\rho = 22000$ ft	$p_{11_0} = 500$ ft^2
$r = 100$ ft^2/Hz	$p_{22_0} = 2 \times 10^4$ ft^2/sec^2
$\beta \sim N(2000 \text{ lb/ft}^2, 2.5 \times 10^5 \text{ lb}^2/\text{ft}^4)$	$p_{33_0} = 2.5 \times 10^5$ lb^2/ft^4

Directing our attention to the errors in estimating the ballistic coefficient, we note that neither filter tracks β accurately early in the trajectory. Physically, this is due to the fact that the thin atmosphere at high altitude produces a small drag force on the body; consequently, the measurements contain little information about β. After the body enters the thicker atmosphere, the increased drag force enables both filters to achieve substantial reduction in the β estimation error; however, the extended Kalman filter (Fig. 6.1-3c) gives appreciably better performance. Consequently, the latter also gives better estimates of position and velocity as the body enters denser atmosphere. The actual performance observed for the extended Kalman filter is consistent with the behavior of its associated covariance matrix, computed from the algorithm in Table 6.1-2. This is demonstrated by the fact that the estimated states in Fig. 6.1-3 tend to remain between the smooth lines, which are the square roots of the corresponding diagonal elements in the computed P matrix. In contrast, the estimates for the linearized filter in Fig. 6.1-4 tend to have larger errors than predicted by the P matrix.

This example illustrates the fact that the extended Kalman filter generally performs better than a filter linearized about a precomputed nominal trajectory. However, the reader must remember that the latter is more easily mechanized because filter gains can be precomputed and stored. Consequently, there is a trade-off between filter complexity and estimation accuracy.

Example 6.1-3

As a further comparison of the types of nonlinear filters discussed in Section 6.1, we include a two-dimensional tracking problem in which the range measurement is made along the line-of-sight illustrated in Fig. 6.1-5. This example is taken from Ref. 5. The equations of motion are the same as in the one-dimensional tracking problem except that the third state variable is identified as $1/\beta$. Range measurements, taken at one second intervals, are now a nonlinear function of altitude given by

$$z_k = \sqrt{r_1^2 + (x_k - r_2)^2} + v_k$$
$$v_k \sim N(0,r) \tag{6.1-36}$$

The problem of estimating the three state variables was solved using the extended Kalman filter, a second-order filter, and an iterated filter. Monte carlo computer simulations were performed using the following parameter values:

PARAMETER VALUES FOR THE TWO-DIMENSIONAL TRACKING PROBLEM

$\frac{\rho_0 g}{2} = 0.05$ lb/ft^3	$x_3(0) = 2 \times 10^{-2}$ ft^2/lb
$x(0) = \hat{x}(0) = 3 \times 10^5$ ft	$\hat{x}_3(0) = 6 \times 10^{-4}$ ft^2/lb
$\dot{x}(0) = \hat{\dot{x}}(0) = 2 \times 10^4$ ft/sec	$r = 100$ ft^2/Hz

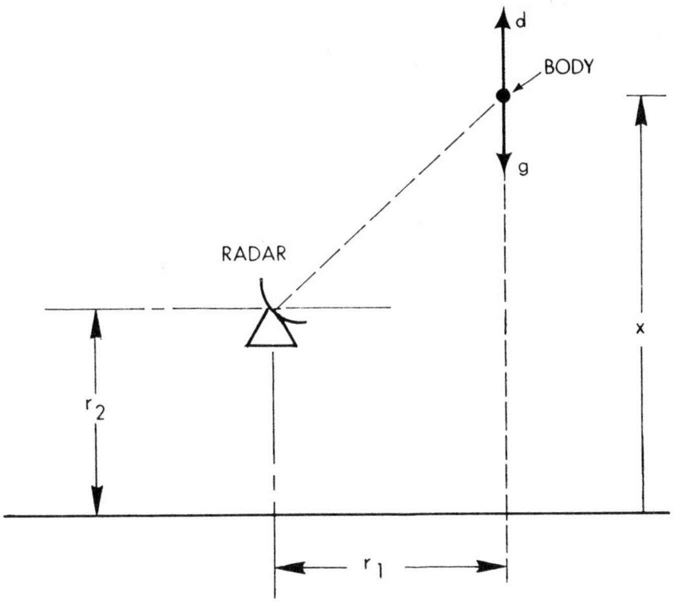

Figure 6.1-5 Geometry of the Two-Dimensional Tracking Problem

The rms errors incurred in estimating altitude and the inverse ballistic coefficient, $1/\beta$, using the three filters are shown in Fig. 6.1-6. These results have been obtained by computing the rms errors over 100 monte carlo runs.

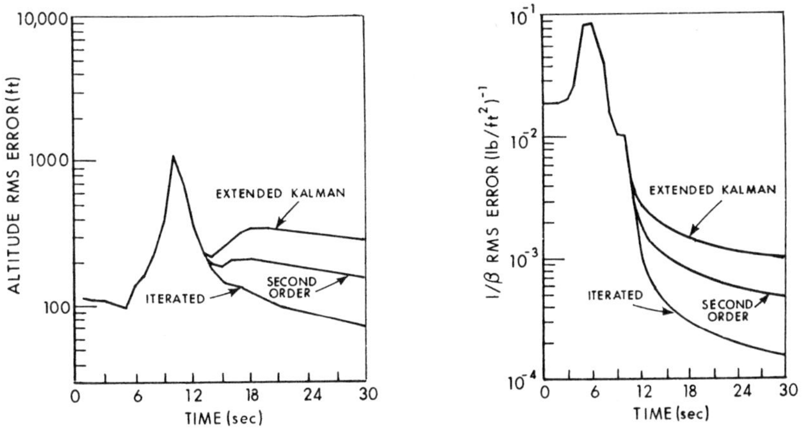

Figure 6.1-6 Comparative Performance of Several Tracking Algorithms (Ref. 5)

Near the beginning of the trajectory, all filters yield the same performance because they have the same initial estimation errors. Evidently, the higher-order methods – the second-order and iterated filters – yield substantially better estimates as the body falls into the earth's denser atmosphere. This occurs because both the measurement and dynamic nonlinearities become stronger as the altitude decreases.

This example also clearly demonstrates the trade-off between estimation accuracy and filter complexity. In terms of these criteria, the filters are ranked in the order of both increasing accuracy and complexity as follows: extended Kalman filter, second-order filter, iterated filter. Thus, better filter performance is achieved at the expense of a greater computational burden.

NONLINEAR SMOOTHING

The purpose of this section is to briefly indicate how smoothing algorithms for nonlinear systems can be derived using linearization techniques. The philosophy of smoothing for nonlinear systems is the same as for the linear case discussed in Chapter 5; namely, the filtered estimate of $\underline{x}(t)$ can be improved if future, as well as past, measurement data are processed. The associated estimation problem can be formulated as one of fixed-interval, fixed-point, or fixed-lag smoothing. The discussion here is confined to fixed-interval smoothing. Linear and second-order algorithms for the fixed-point and fixed-lag cases are described in Refs. 22 and 23. For a more theoretical treatment of this subject the reader is referred to Ref. 4.

Recall from Section 5.1 that the smoothed estimate $\hat{\underline{x}}(t|T)$ for a linear system, with a given data record ending at time T, can be expressed as

$$\hat{\underline{x}}(t|T) = P(t|T)[P^{-1}(t)\,\hat{\underline{x}}(t) + P_b^{-1}(t)\,\hat{\underline{x}}_b(t)]$$

$$P^{-1}(t|T) = P^{-1}(t) + P_b^{-1}(t) \tag{6.1-37}$$

where $\hat{\underline{x}}(t)$ and $\hat{\underline{x}}_b(t)$ are estimates provided by forward and backward filters, respectively, and $P(t)$ and $P_b(t)$ are their corresponding covariance matrices. If the same approach is taken in the nonlinear case – i.e., if it is assumed that the smoothed estimate is a linear combination of $\hat{\underline{x}}(t)$ and $\hat{\underline{x}}_b(t)$ – then Eq. (6.1-37) still holds (Ref. 3). The quantities $\hat{\underline{x}}(t)$ and $P(t)$, associated with the forward filter, can be obtained by any of the filtering algorithms given in the preceding sections. The subsequent discussion treats appropriate expressions for the backward filter.

The backward filter operates recursively on the measurement data, beginning at the terminal time and proceeding toward the desired smoothing point. Consequently, it is useful to redefine the equations of motion and the measurements in Eqs. (6.0-1) and (6.0-2) by making the change of variable $\tau = T - t$, resulting in

$$\frac{d}{d\tau}\,\underline{x}(T-\tau) = -\underline{f}(\underline{x}(T-\tau), T-\tau) + \underline{w}(T-\tau)$$

$$\underline{z}_{N-k} = \underline{h}_{N-k}(\underline{x}(T-\tau_k)) + \underline{v}_{N-k} \tag{6.1-38}$$

where

$$\tau_k \triangleq T - t_{N-k}$$

Using Eq. (6.1-38) as the model for the system dynamics, the linearization methods discussed in previous sections can be applied to obtain a backward filtering algorithm. In particular, it is convenient to assume that the forward extended Kalman filter, described in Table 6.1-1, is first applied to all the data; then the backward filter is derived by linearizing Eq. (6.1-38) about the forward filter estimate. This procedure results in a backward filter having the form given in Table 6.1-3, where the nominal trajectory $\bar{\underline{x}}(t)$ is taken to be $\hat{\underline{x}}(t) = \hat{\underline{x}}(T - \tau)$. The resulting expressions for $\hat{\underline{x}}_b(\tau)$ and $P_b(\tau)$ are as follows:

PROPAGATION EQUATIONS

$$\frac{d}{d\tau}\hat{\underline{x}}_b(\tau) = -\underline{f}(\hat{\underline{x}}(T-\tau), T-\tau) - F(\hat{\underline{x}}(T-\tau), T-\tau)(\hat{\underline{x}}_b(\tau) - \hat{\underline{x}}(T-\tau))$$

$$\frac{d}{d\tau}P_b(\tau) = -F(\hat{\underline{x}}(T-\tau), T-\tau)P_b(\tau) - P_b(\tau)F^T(\hat{\underline{x}}(T-\tau), T-\tau) + Q(T-\tau) \qquad (6.1\text{-}39)$$

MEASUREMENT UPDATE EQUATIONS

$$\hat{\underline{x}}_{b_k}(+) = \hat{\underline{x}}_{b_k}(-) + K_{b_k}\left[\underline{z}_{N-k} - \underline{h}_{N-k}(\hat{\underline{x}}_{N-k}) - H_{N-k}(\hat{\underline{x}}_{N-k})\left(\hat{\underline{x}}_{b_k}(-) - \hat{\underline{x}}_{N-k}\right)\right]$$

$$K_{b_k} = P_{b_k}(-)H_{N-k}^T(\hat{\underline{x}}_{N-k})\left[H_{N-k}(\hat{\underline{x}}_{N-k})P_{b_k}(-)H_{N-k}^T(\hat{\underline{x}}_{N-k}) + R_{N-k}\right]^{-1}$$

$$P_{b_k}(+) = \left[I - K_{b_k}H_{N-k}(\hat{\underline{x}}_{N-k})\right]P_{b_k}(-), \quad k=0,1,\ldots,N-1 \qquad (6.1\text{-}40)$$

where $\underline{h}_{N-k}$ and H_{N-k} are obtained from the forward filter. The quantities $\hat{\underline{x}}_{b_k}(-)$ and $P_{b_k}(-)$ are the solutions to Eq. (6.1-39) evaluated at time τ_k, just *before* $\underline{z}_{N-k}$ is processed.

The problem of obtaining initial conditions for the backward filter is circumvented by deriving algorithms for $P_b^{-1}(t)$ and $\underline{s}(t) \triangleq P_b^{-1}(t)\hat{\underline{x}}_b(t)$, both of which are initialized at zero, just as in the case of the linear smoother discussed in Chapter 5. Using Eqs. (6.1-39) and (6.1-40) and the matrix identities

$$\left(A^{-1} + B^TC^{-1}B\right)^{-1} = A - AB^T\left(BAB^T + C\right)^{-1}BA$$

$$\left(A^{-1} + B^TC^{-1}B\right)^{-1}B^TC^{-1} = AB^T\left(BAB^T + C\right)^{-1} \qquad (6.1\text{-}41)$$

and making the assumption that R_{N-k} has an inverse, the following, more convenient expressions for the backward filter are obtained:

PROPAGATION EQUATIONS

$$\frac{d}{d\tau}\,\underline{s}(\tau) = [F^T - P_b^{-1}(\tau)\,Q(T-\tau)]\;\underline{s}(\tau) - P_b^{-1}(\tau)\,[\underline{f} - F\hat{\underline{x}}(T-\tau)]$$

$$\frac{d}{d\tau}\,P_b^{-1}(\tau) = P_b^{-1}(\tau)\,F + F^T P_b^{-1}(\tau) - P_b^{-1}(\tau)\,Q(T-\tau)\,P_b^{-1}(\tau) \qquad (6.1\text{-}42)$$

MEASUREMENT UPDATE EQUATIONS

$$\underline{s}_k(+) = \underline{s}_k(-) + H_{N-k}{}^T R_{N-k}{}^{-1}\,[\underline{z}_{N-k} - \underline{h}_{N-k} + H_{N-k}\hat{\underline{x}}_k]$$

$$P_{b_k}{}^{-1}(+) = P_{b_k}{}^{-1}(-) + H_{N-k}{}^T R_{N-k}{}^{-1} H_{N-k}$$

$$P_{b_0}{}^{-1}(-) = 0, \quad \underline{s}_0(-) = \underline{0}, \quad k = 0, 1, \ldots, N-1 \qquad (6.1\text{-}43)$$

where F and $\underline{f}$ are understood to be functions of $\hat{\underline{x}}(T - \tau)$ and $(T - \tau)$. Substituting $\underline{s}(t)$ into Eq. (6.1-37), the smoothed estimate is obtained in the form

$$\hat{\underline{x}}(t|T) = P(t|T)\,[P^{-1}(t)\hat{\underline{x}}(t) + \underline{s}(t)] \qquad (6.1\text{-}44)$$

Again by analogy with linear systems, another form of the fixed-interval smoother is the Rauch-Tung-Striebel algorithm which expresses $\hat{\underline{x}}(t_k|T)$ and $P(t_k|T)$, denoted by $\hat{\underline{x}}_{k|N}$ and $P_{k|N}$, in terms of a forward filter estimate and the smoothed estimate at stage k + 1. This is more convenient to mechanize than the forward-backward filter inasmuch as it does not require inverting a covariance matrix. The relations for the smoothed estimate at successive measurement times are stated below without proof:

$$\hat{\underline{x}}_{k|N} = \hat{\underline{x}}_k + K_k\left(\hat{\underline{x}}_{k+1|N} - \Phi_k\hat{\underline{x}}_k - \underline{b}_k\right), \qquad \hat{\underline{x}}_{N|N} = \hat{\underline{x}}_N$$

$$K_k = P_k\Phi_k\left(\Phi_k P_k \Phi_k{}^T + Q_k\right)^{-1}$$

$$P_{k|N} = P_k + K_k\left(P_{k+1|N} - \Phi_k P_k \Phi_k{}^T - Q_k\right)K_k{}^T, \qquad P_{N|N} = P_N \qquad (6.1\text{-}45)$$

where P_k and $\hat{\underline{x}}_k$ are obtained from a forward filter and $P_{k|N}$ is (approximately) the error covariance matrix associated with $\hat{\underline{x}}_{k|N}$. The quantity $\underline{b}_k$ is the solution to the differential equation

$$\dot{\underline{b}}(t) = F(\hat{\underline{x}}(t), t)\,\underline{b}(t) + \underline{f}(\hat{\underline{x}}(t), t) - F(\hat{\underline{x}}(t), t)\hat{\underline{x}}(t)$$

$$\underline{b}(t_k) = \underline{0} \qquad (6.1\text{-}46)$$

at time t_{k+1}; Φ_k is the transition matrix from time t_k to t_{k+1} associated with the homogeneous differential equation

$$\dot{\underline{x}}(t) = F(\hat{\underline{x}}(t), t)\,\underline{x}(t) \qquad (6.1\text{-}47)$$

and Q_k is given by

$$Q_k = \int_{t_k}^{t_{k+1}} \Phi(\tau, t_k)\, Q(\tau)\, \Phi(\tau, t_k)^T\, d\tau \tag{6.1-48}$$

Observe that the initial conditions for Eq. (6.1-45) are the filter outputs $\hat{\underline{x}}_N$ and P_N; clearly the latter are identical with the corresponding smoothed quantities at the data endpoint. When the system dynamics and measurements are linear, $\underline{b}_k = \underline{0}$ and Eq. (6.1-45) reduces to the form given in Ref. 1.

6.2 NONLINEAR ESTIMATION BY STATISTICAL LINEARIZATION

In Section 6.1, a number of approximately optimal nonlinear filters have been derived using truncated Taylor series expansions to represent the system nonlinearities. An alternative approach, and one that is generally more accurate than the Taylor series expansion method, is referred to as *statistical approximation* (Refs. 6, 8, and 9). The basic principle of this technique is conveniently illustrated for a scalar function, f(x), of a random variable x.

Consider that f(x) is to be approximated by a series expansion of the form

$$f(x) \cong n_0 + n_1 x + n_2 x^2 + \ldots + n_m x^m \tag{6.2-1}$$

The problem of determining appropriate coefficients n_k is similar to the estimation problem where an estimate of a random variable is sought from given measurement data. Analogous to the concept of estimation error, we define a *function representation error*, e, of the form

$$e = f(x) - n_0 - \ldots - n_m x^m$$

It is desirable that the n_k's be chosen so that e is small in some "average" sense; any procedure that is used to accomplish this goal, which is based upon the statistical properties of x, can be thought of as a *statistical approximation technique*.

The most frequently used method for choosing the coefficients in Eq. (6.2-1) is to minimize the mean square value of e, $E[e^2]$. This is accomplished by forming

$$E\left[\left(f(x) - n_0 - n_1 x - \ldots - n_m x^m\right)^2\right] \tag{6.2-2}$$

and setting the partial derivatives of this quantity with respect to each n_k equal to zero. The result is a set of algebraic equations, linear in the n_k's, that can be solved in terms of the moments and cross-moments of x and f(x). Without carrying out the details of this procedure, we can see one distinct advantage that statistical approximation has over the Taylor series expansion; it does not require the existence of derivatives of f. Thus, a large number of nonlinearities – relay, saturation, threshold, etc. – can be treated by this method without having to approximate discontinuities or corners in f(x) by smooth functions. On the other hand, because of the expectation operation in Eq. (6.2-2), an apparent disadvantage of the method is that the probability density function for x must be known in order to compute the coefficients n_k, a requirement that does not exist when f(x) is expanded in a Taylor series about its mean value. However, it turns out that approximations can often be made for the probability density function used to calculate the coefficients, such that the resulting expansion for f(x) is considerably more accurate than the Taylor series, from a statistical point of view. Thus, statistical approximations for nonlinearities have potential performance advantages for designing nonlinear filters.

This section discusses the calculation of the first two terms in a series, having the form of Eq. (6.2-1), for a vector function $\underline{f}(\underline{x})$. This provides a *statistical linearization* of $\underline{f}$ that can be used to construct filter algorithms analogous to those provided in Section 6.1. An example is given that illustrates the comparative accuracy of the nonlinear filter algorithms derived from both the Taylor series and the statistical linearization approximations.

STATISTICAL LINEARIZATION

We seek a linear approximation for a vector function $\underline{f}(\underline{x})$ of a vector random variable $\underline{x}$, having probability density function $p(\underline{x})$. Following the statistical approximation technique outlined in the introduction, we propose to approximate $\underline{f}(\underline{x})$ by the linear expression

$$\underline{f}(\underline{x}) \cong \underline{a} + N_f \underline{x} \tag{6.2-3}$$

where $\underline{a}$ and N_f are a vector and a matrix to be determined. Defining the error

$$\underline{e} \triangleq \underline{f}(\underline{x}) - \underline{a} - N_f \underline{x} \tag{6.2-4}$$

we desire to choose $\underline{a}$ and N_f so that the quantity

$$J = E[\underline{e}^T A \underline{e}] \tag{6.2-5}$$

is minimized for some symmetric positive semidefinite matrix A. Substituting Eq. (6.2-4) into Eq. (6.2-5) and setting the partial derivative of J with respect to the elements of $\underline{a}$ equal to zero, we obtain

$$E[A(\underline{f}(\underline{x}) - \underline{a} - N_f \underline{x})] = \underline{0} \tag{6.2-6}$$

Therefore, $\underline{a}$ is given by

$$\underline{a} = \hat{\underline{f}}(\underline{x}) - N_f \hat{\underline{x}} \tag{6.2-7}$$

where the caret (ˆ) denotes the expectation operation. Substituting $\underline{a}$ from Eq. (6.2-7) into J and taking the partial derivative with respect to the elements of N_f, we obtain

$$E\left[A[N_f\underline{\tilde{x}}\underline{\tilde{x}}^T + (\hat{\underline{f}}(\underline{x}) - \underline{f}(\underline{x}))\underline{\tilde{x}}^T]\right] = 0 \tag{6.2-8}$$

where

$$\underline{\tilde{x}} = \underline{\hat{x}} - \underline{x}$$

Solving Eq. (6.2-8) produces

$$N_f = [\widehat{\underline{f}\underline{x}^T} - \hat{\underline{f}}\underline{\hat{x}}^T]\,P^{-1} \tag{6.2-9}$$

where P is the conventional covariance matrix for $\underline{x}$. Observe that both $\underline{a}$ and N_f, as given by Eqs. (6.2-7) and (6.2-9), are independent of the weighting matrix A; hence, they provide a generalized minimum mean square error approximation to $\underline{f}$.

Equation (6.2-9) has an important connection with *describing function theory* (Ref. 7) for approximating nonlinearities. In particular, if both $\underline{f}$ and $\underline{x}$ are scalars and their mean values are zero, N_f becomes the scalar quantity

$$n_f = \frac{E[fx]}{E[x^2]} \tag{6.2-10}$$

which is the *describing function gain* for an odd-function nonlinearity whose input is a zero mean random variable (e.g., see Fig. 6.2-1). The use of describing functions to approximate nonlinearities has found wide application in analyzing nonlinear control systems. Tables of expressions for n_f have been computed for many common types of nonlinearities having gaussian inputs.

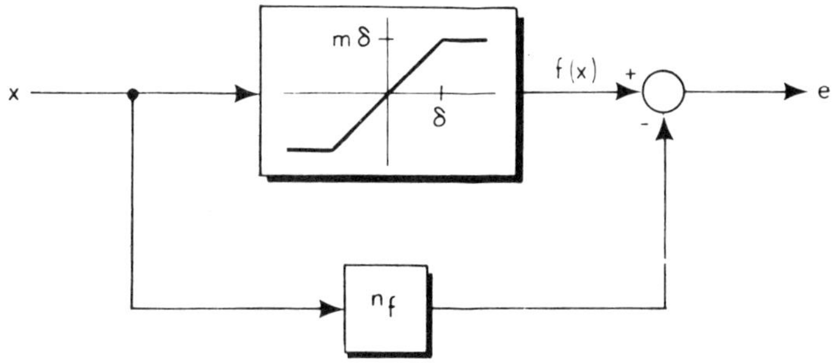

Figure 6.2-1 The Describing Function Approximation for a Scalar Odd Nonlinearity

At this point, it is worthwhile to note why statistical linearization tends to be more accurate than the Taylor series approximation method. Consider the example of the saturation nonlinearity in Fig. 6.2-1. If f(x) for this case is expanded in a Taylor series of any order about the origin ($\hat{x} = 0$), we obtain

$$f(x) \cong x \tag{6.2-11}$$

The effect of the saturation is completely lost because of the discontinuity in the first derivative of f. By contrast, if statistical linearization is used, we have

$$f(x) \cong n_f x \tag{6.2-12}$$

where n_f is the describing function gain defined by

$$n_f = \frac{\int_{-\infty}^{\infty} x\, f(x)\, p(x)\, dx}{\int_{-\infty}^{\infty} x^2\, p(x)\, dx} \tag{6.2-13}$$

and p(x) is the probability density function for x. If we now assume that x is a gaussian random variable, then

$$p(x) = \frac{1}{\sqrt{2\pi}\,\sigma} e^{-\frac{x^2}{2\sigma^2}} \tag{6.2-14}$$

Substituting Eq. (6.2-14) into Eq. (6.2-13) and evaluating n_f for the saturation function shown in Fig. 6.2-1, we obtain (from Ref. 7) the result shown in Fig. 6.2-2. Evidently, n_f is a function of the linear part of f(x), the point δ at which

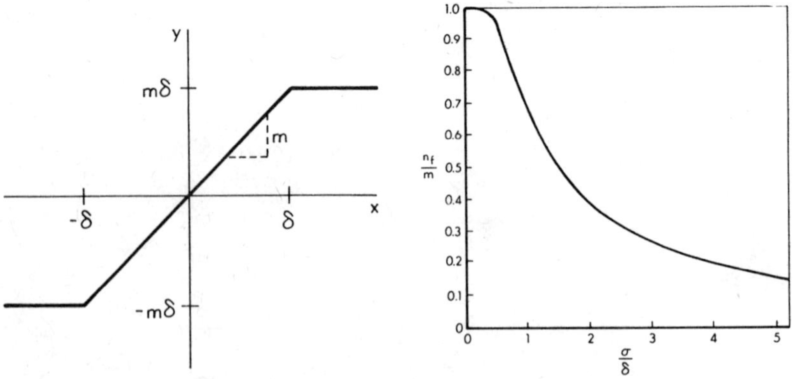

(a) Saturating Nonlinearity (b) Describing Function

Figure 6.2-2 The Describing Function for a Saturation Nonlinearity (Ref. 7) saturation occurs, and the standard deviation of x. The essential feature of the describing function is that it takes into account the probability that x can lie within the saturation region.

For values of σ which are small relative to δ, the probability of saturation is low and n_f is approximately equal to one – i.e., f is approximately equal to the Taylor series given in Eq. (6.2-11). For larger values of σ, n_f is significantly smaller than one because there is a higher probability of saturation. Thus, for a given saturation function, statistical linearization provides a series of σ-dependent linearizations illustrated in Fig. 6.2-3. In subsequent sections the usefulness of this approximation for designing nonlinear filters and analyzing nonlinear system performance is demonstrated.

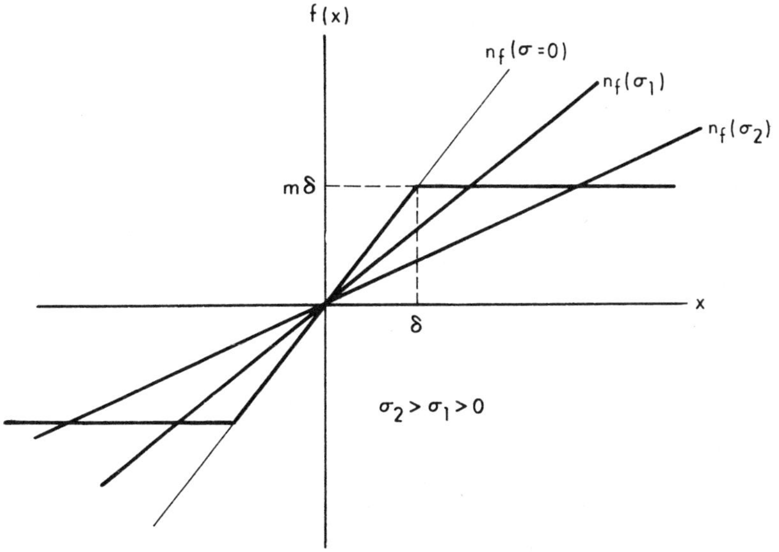

Figure 6.2-3 Illustration of Statistical Linearization as a σ-Dependent Gain

DESIGN OF NONLINEAR FILTERS

The approximation of a nonlinear function $\underline{f}(\underline{x})$ by a statistically optimized power series can be combined with results obtained in Section 6.1 to derive approximate minimum variance filtering algorithms. In particular, by substituting Eq. (6.2-7) into Eq. (6.2-3), we obtain

$$\underline{f}(\underline{x}) \cong \hat{\underline{f}}(\underline{x}) + N_f(\underline{x} - \hat{\underline{x}})$$

Because of the aforementioned connection with describing function theory, we shall refer to N_f as the *describing function gain matrix*. We now can identify $\underline{f}$ with the nonlinear function in the equations of motion [Eq. (6.0-1)] for a nonlinear stochastic system. The latter is in general a function of time as well, in which case N_f as computed from Eq. (6.2-9) is also time-dependent. Introducing the notation $N_f(t)$ to denote the describing function gain matrix for $\underline{f}(\underline{x},t)$, we write

$$\underline{f}(\underline{x},t) \cong \hat{\underline{f}}(\underline{x},t) + N_f(t)(\underline{x} - \hat{\underline{x}}) \tag{6.2-15}$$

Using similar notation for the measurement nonlinearity in Eq. (6.0-2) produces

$$\underline{h}_k(\underline{x}_k) \cong \hat{\underline{h}}_k(\underline{x}_k) + N_h(k)(\underline{x}_k - \hat{\underline{x}}_k) \tag{6.2-16}$$

Now the issue of computing $\hat{\underline{f}}$, $\hat{\underline{h}}_k$, N_f, and N_h arises. Each of these quantities depends upon the probability density function for $\underline{x}$, a quantity that is generally not readily available. In fact, the absence of the latter is the entire motivation for seeking power series expansions for $\underline{f}$ and $\underline{h}_k$. Consequently, an approximation is needed for $p(\underline{x})$ that permits the above quantities to be calculated. For

TABLE 6.2-1 SUMMARY OF THE STATISTICALLY LINEARIZED FILTER

System Model	$\dot{\underline{x}}(t) = \underline{f}(\underline{x}(t), t) + \underline{w}(t)$; $\underline{w}(t) \sim N(\underline{0}, Q(t))$
Measurement Model	$\underline{z}_k = \underline{h}_k(\underline{x}(t_k)) + \underline{v}_k$; $k = 1,2,\ldots$; $\underline{v}_k \sim N(\underline{0}, R_k)$
Initial Conditions	$\underline{x}(0) \sim N(\hat{\underline{x}}_0, P_0)$
Other Assumptions	$E[\underline{w}(t)\underline{v}_k^T] = 0$ for all k and all t
State Estimate Propagation	$\dot{\hat{\underline{x}}}(t) = \hat{\underline{f}}(\underline{x}(t), t)$
Error Covariance Propagation	$\dot{P}(t) = N_f(t)P(t) + P(t)N_f^T(t) + Q(t)$
Describing Function Calculations	$N_f(t) = [\widehat{\underline{f}\underline{x}^T} - \hat{\underline{f}}\hat{\underline{x}}^T]P^{-1}(t)$
State Estimate Update	$\hat{\underline{x}}_k(+) = \hat{\underline{x}}_k(-) + K_k[\underline{z}_k - \hat{\underline{h}}_k(\underline{x}_k)]$
Error Covariance Update	$P_k(+) = [I - K_k N_h(k)]P_k(-)$
Gain Matrix Calculation	$K_k = P_k(-)N_h(k)^T[N_h(k)P_k(-)N_h^T(k) + R_k]^{-1}$
Describing Function Calculations	$N_h(k) = [\widehat{\underline{h}_k(\underline{x}_k(-))\underline{x}_k^T(-)} - \hat{\underline{h}}_k(\underline{x}_k(-))\hat{\underline{x}}_k^T(-)]P_k^{-1}(-)$
Definitions	$\widehat{\underline{f}\underline{x}^T}$, $\hat{\underline{f}}$, $\hat{\underline{x}}$ are expectations calculated assuming $\underline{x} \sim N(\hat{\underline{x}}, P)$ $\hat{\underline{h}}_k$, $\hat{\underline{x}}_k(-)$ and $\widehat{\underline{h}_k\underline{x}_k^T(-)}$ are expectations calculated assuming $\underline{x}_k(-) \sim N[\hat{\underline{x}}_k(-), P_k(-)]$

this purpose, it is most frequently assumed that $\underline{x}$ is gaussian. Since the probability density function for a gaussian random variable is completely defined by its mean $\hat{\underline{x}}$ and its covariance matrix P, both of which are part of the computation in any filtering algorithm, it will be possible to compute all the averages in Eqs. (6.2-15) and (6.2-16).

Assuming that the right sides of Eqs. (6.2-15) and (6.2-16) are calculated by making the gaussian approximation for $\underline{x}$, the statistical linearization for $\underline{f}$ and $\underline{h}_k$ can be substituted into Eqs. (6.1-3), (6.1-5), (6.1-13), (6.1-17), and (6.1-18) for the nonlinear filter. Making this substitution and carrying out the indicated expectation operations produces the filter equations given in Table 6.2-1. Observe that the structure of the latter is similar to the extended Kalman filter in Table 6.1-1; however, N_f and N_h have replaced F and H_k. The computational requirements of the statistically linearized filter may be greater than for filters derived from Taylor series expansions of the nonlinearities because the expectations – $\hat{\underline{f}}$, $\hat{\underline{h}}_k$, etc. – must be performed over the assumed gaussian density for $\underline{x}$. However, as demonstrated below, the performance advantages offered by statistical linearization may make the additional computation worthwhile.

Example 6.2-1

To compare the filtering algorithm in Table 6.2-1 with those derived in Section 6.1 the following scalar example, due to Phaneuf (Ref. 6), is presented. The equation of motion is

$$\dot{x}(t) = -\sin(x(t)) + w(t)$$

$$x(0) \sim N(0,p_0), \qquad w(t) \sim N(0,q) \tag{6.2-17}$$

The unforced solution (w(t) = 0) to Eq. (6.2-17) is of interest, and is displayed in Fig. 6.2-4 for several different initial values of x. Observe that x(t) approaches a steady-state limit that depends upon the value of the initial condition.

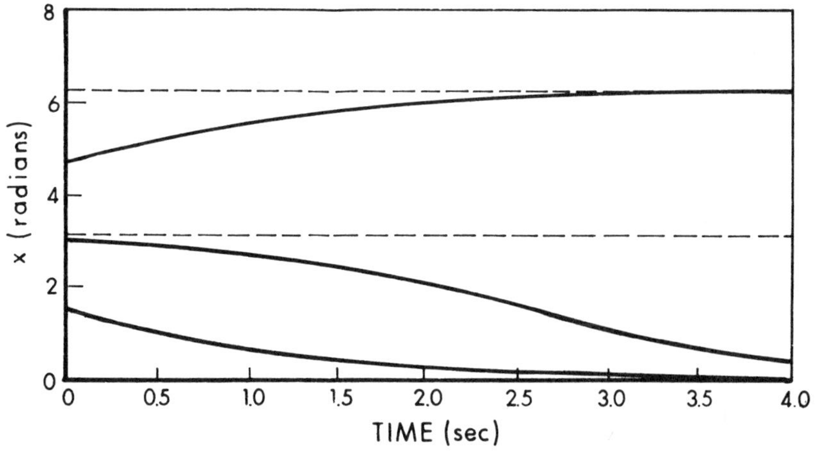

Figure 6.2-4 Unforced Solutions to Eq. (6.2-17)

A nonlinear discrete-time measurement equation of the form

$$z_k = \frac{1}{2}\sin(2x_k) + v_k, \quad v_k \sim N(0,r) \tag{6.2-18}$$

is assumed.

The dynamics of this example are sufficiently simple so that a variety of nonlinear filters can be constructed for estimating x(t). Algorithms were derived for the extended Kalman filter, for higher-order filters using second-, third-, and fourth-order Taylor series expansions of the nonlinearities, and for the statistically linearized (quasi-linear) filter discussed in this section. The results of monte carlo runs, performed with identical noise sequences for each of the above five filters, are compared in Fig. 6.2-5. Evidently the error associated with statistical linearization is much smaller than for the filters derived from Taylor series expansions, even up to fourth-order, during the transient period. This is explained on the basis that Taylor series expansions are least accurate when the estimation error is large.

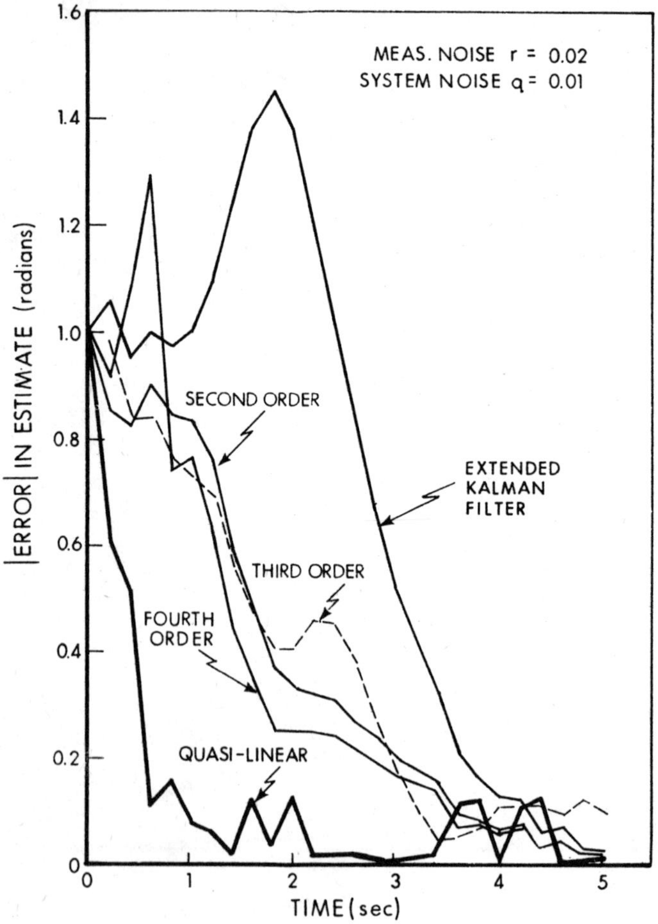

Figure 6.2-5 Estimation Errors for Five Filters (Ref. 6)

In the vector case, the statistically linearized filter requires computation of the matrices N_f and N_h in Eqs. (6.2-15) and (6.2-16). This, in general, involves the evaluation in real-time of multidimensional expectation integrals having the form

$$\hat{f}_i = \int_{-\infty}^{\infty} \cdots \int_{-\infty}^{\infty} f_i(\underline{x})p(\underline{x})d\underline{x}$$

$$\xi_{ij} = \int_{-\infty}^{\infty} \cdots \int_{-\infty}^{\infty} x_j f_i(\underline{x})p(\underline{x})d\underline{x} \tag{6.2-19}$$

Similar expressions must be evaluated for the components of $\underline{h}$. The quantity ξ_{ij} denotes the ijth element of the matrix $\widehat{\underline{f}(\underline{x})\underline{x}^T}$ in Eq. (6.2-9) and $p(\underline{x})$ is the assumed joint gaussian density function for $\underline{x}(t)$,

$$p(\underline{x}) = c \exp\left[-\frac{1}{2}(\underline{x} - \hat{\underline{x}})^T P^{-1}(\underline{x} - \hat{\underline{x}})\right] \tag{6.2-20}$$

where c is the appropriate normalizing constant. The degree of difficulty in calculating ξ_{ij} and $\hat{f}_i$ is an important consideration in judging whether to use statistical linearization for a particular application. The computational effort required is, of course, dependent upon the specific type of nonlinearity. Some systematic procedures for evaluating the expectation integrals are discussed next.

COMPUTATIONAL CONSIDERATIONS

Both integrals in Eq. (6.2-19) can be represented by the general expression

$$\hat{g} = \int_{-\infty}^{\infty} \cdots \int_{-\infty}^{\infty} c g(\underline{x}) \exp\left[-\frac{1}{2}(\underline{x} - \hat{\underline{x}})^T P^{-1}(\underline{x} - \hat{\underline{x}})\right] d\underline{x} \tag{6.2-21}$$

where $g(\underline{x})$ is a nonlinear function. In general, $\hat{g}$ will be difficult to evaluate; however, its computation can be facilitated by making a change of variables which simplifies the exponential term in Eq. (6.2-21) (Ref. 6). First, we define

$$\underline{r} = \underline{x} - \hat{\underline{x}} \tag{6.2-22}$$

so that $\hat{g}$ becomes

$$\hat{g} = \int_{-\infty}^{\infty} \cdots \int_{-\infty}^{\infty} c g(\underline{r} + \hat{\underline{x}}) \exp\left[-\frac{1}{2}\underline{r}^T P^{-1}\underline{r}\right] d\underline{r} \tag{6.2-23}$$

Next, recognizing that P^{-1} is a symmetric matrix, we can find a nonsingular transformation matrix, T, such that

$$T^T P^{-1} T = D \tag{6.2-24}$$

where D is a diagonal matrix whose diagonal elements are the eigenvalues of P^{-1} (Ref. 10). Now, if we define

$$\underline{s} \triangleq T^{-1}\, \underline{r} \tag{6.2-25}$$

it follows from the properties of T that $\hat{g}$ can be written in the form*

$$\hat{g} = \int_{-\infty}^{\infty} \cdots \int_{-\infty}^{\infty} c g(T\underline{s} + \hat{\underline{x}}) \exp\left[-\frac{1}{2}\underline{s}^T D \underline{s}\right] d\underline{s} \tag{6.2-26}$$

or alternatively,

$$\hat{g} = c \int_{-\infty}^{\infty} ds_n e^{-d_n s_n^2/2} \int_{-\infty}^{\infty} ds_{n-1} e^{-d_{n-1} s_{n-1}^2/2} \cdots$$

$$\cdots \int_{-\infty}^{\infty} ds_1\, g(T\underline{s} + \hat{\underline{x}}) e^{-d_1 s_1^2/2} \tag{6.2-27}$$

where d_i is the ith diagonal element of D. Consequently, $\hat{g}$ can be evaluated from Eq. (6.2-27) by successively computing integrals of the form

$$\hat{g}_{k+1}(s_{k+1}, \ldots, s_n) = \int_{-\infty}^{\infty} ds_k \hat{g}_k(s_k, \ldots, s_n) e^{-d_k s_k^2/2}, \qquad k=1, \ldots, n \tag{6.2-28}$$

where

$$\hat{g}_{n+1} \triangleq \hat{g} \qquad \text{and} \qquad g_1(s_1, \ldots, s_n) \triangleq g(T\underline{s} + \hat{\underline{x}})$$

The task of evaluating $\hat{g}$ from Eq. (6.2-27) may still be difficult; however, formulating the solution as a sequence of integrals of the form given in Eq. (6.2-28) permits the systematic application of approximate integration techniques, such as gaussian quadrature formulae. Other methods are suggested in Ref. 7.

Turning now to some important special cases of Eq. (6.2-21), we note that if $g(\underline{x})$ has the form

$$g(\underline{x}) = x_1^{m_1} x_2^{m_2} \ldots x_n^{m_n} \tag{6.2-29}$$

---

*In particular, T is an orthogonal matrix whose determinant is one. Consequently, after changing variables of integration from $\underline{r}$ to $\underline{s}$, $\hat{g}$ is given by Eq. (6.2-26).

then $\hat{g}$ can be expressed analytically in terms of the first and second moments of $\underline{x}$, assuming the latter is gaussian. For example, if $\underline{x}$ has zero mean and

$$g(\underline{x}) = x_1 x_2 x_3 x_4 \tag{6.2-30}$$

then it can be shown (Ref. 24) that

$$\hat{g} = \widehat{x_1 x_2}\,\widehat{x_3 x_4} + \widehat{x_1 x_3}\,\widehat{x_2 x_4} + \widehat{x_1 x_4}\,\widehat{x_2 x_3} \tag{6.2-31}$$

Product nonlinearities are an important class of functions because they can often be used to approximate more general nonlinearities.

A second special case of practical interest arises when the nonlinear part of the system dynamics is a function of only a few of the total number of state variables. For example, there may be a single saturation nonlinearity of the type shown in Fig. 6.2-1 embedded in a dynamical system which is otherwise linear. In this case, the system equation of motion could be expressed as

$$\dot{\underline{x}}(t) = F\underline{x}(t) + \begin{bmatrix} 0 \\ \vdots \\ 0 \\ f_i(x_j) \\ 0 \\ \vdots \\ 0 \end{bmatrix} + \underline{w}(t) \tag{6.2-32}$$

where $f_i(x_j)$ is a nonlinear function of only one state variable. It is demonstrated below that this type of structure generally simplifies the computational effort involved in computing the describing function gain matrix N_f in Eq. (6.2-9).

Consider the ijth element of N_f which requires calculating the quantity ξ_{ij} in Eq. (6.2-19). When the nonlinearities are functions of a limited number of state variables, there will be some elements of N_f that depend upon values of ξ_{ij} of the form

$$\xi_{ij} = \int_{-\infty}^{\infty} \cdots \int_{-\infty}^{\infty} x_j f_i(\underline{x}_s) p(\underline{x}) d\underline{x} \tag{6.2-33}$$

where $\underline{x}_s$ is a subset of the vector $\underline{x}$ which *excludes* x_j. If it is further assumed that $\underline{x}$ is a gaussian random variable, then with some manipulation the expression for ξ_{ij} in Eq. (6.2-33) can be simplified as follows*:

*This derivation was suggested by Prof. Wallace E. Vander Velde, Department of Aeronautics and Astronautics, Massachusetts Institute of Technology. Line three of Eq. (6.2-34) is obtained from line two by substituting the gaussian form for $p(\underline{x}_s, x_j)$ and $p(\underline{x}_s)$.

$$\xi_{ij} = \int_{-\infty}^{\infty} \cdots \int_{-\infty}^{\infty} f_i(\underline{x}_s)p(\underline{x}_s)d\underline{x}_s \int_{-\infty}^{\infty} x_j p(x_j|\underline{x}_s)dx_j$$

$$= \int_{-\infty}^{\infty} \cdots \int_{-\infty}^{\infty} f_i(\underline{x}_s)p(\underline{x}_s)d\underline{x}_s \int_{-\infty}^{\infty} x_j \frac{p(\underline{x}_s,x_j)}{p(\underline{x}_s)} dx_j$$

$$= E\left[\left\{(\underline{x}_s-\hat{\underline{x}}_s)^T \left(E[(\underline{x}_s-\hat{\underline{x}}_s)(\underline{x}_s-\hat{\underline{x}}_s)]^T\right)^{-1} E[(x_j-\hat{x}_j)(\underline{x}_s-\hat{\underline{x}}_s)] + \hat{x}_j\right\} f_i(\underline{x}_s)\right]$$

$$= \hat{x}_j\hat{f}_i + \underline{n}_{s_i}{}^T \underline{p}_{js} \tag{6.2-34}$$

where

$$\underline{p}_{js} \overset{\Delta}{=} E[(x_j - \hat{x}_j)(\underline{x}_s - \hat{\underline{x}}_s)] \tag{6.2-35}$$

and $\underline{n}_{s_i}$ is the *describing function vector* for $f_i(\underline{x}_s)$, defined by

$$\underline{n}_{s_i} \overset{\Delta}{=} E[f_i(\underline{x}_s)(\underline{x}_s - \hat{\underline{x}}_s)^T]\,(E[(\underline{x}_s - \hat{\underline{x}}_s)(\underline{x}_s - \hat{\underline{x}}_s)^T])^{-1} \tag{6.2-36}$$

The last equality in Eq. (6.2-34) verifies that ξ_{ij} in Eq. (6.2-33) can be calculated by first linearizing $f_i(\underline{x}_s)$ independently of x_j, viz:

$$f_i(\underline{x}_s) \cong \hat{f}_i(\underline{x}_s) + \underline{n}_{s_i}{}^T(\underline{x}_s - \hat{\underline{x}}_s)$$

and then carrying out the indicated expectation operation. This result simplifies the statistical linearization computations considerably, when nonlinearities are functions of only a few state variables.

SUMMARY

This section demonstrates a method of deriving approximately optimal nonlinear filters using statistical linearization. In many instances, this technique yields superior performance to filters developed from Taylor series expansion of the nonlinearities. The fact that practical application of the filter algorithm requires that a form be assumed for the density function of the state is not overly restrictive, since the same assumption is implicit in many of the existing Taylor series-type filters. The decision as to which of the several types of filters discussed in Sections 6.1 and 6.2 should be employed in a particular application ultimately depends upon their computational complexity and relative performance as observed from realistic computer simulations.

6.3 NONLINEAR LEAST-SQUARES ESTIMATION

As an alternative to the minimum variance estimation criterion employed throughout Sections 6.1 and 6.2, we *briefly* mention the subject of least-squares estimation. This approach requires no statistical assumptions about the sources of uncertainty in the problem, because the estimate of the state is chosen to provide a "best" fit to the observed measurement data in a deterministic, rather than a statistical sense. To illustrate, for the case with no process noise – i.e.,

$$\dot{\underline{x}}(t) = \underline{f}(\underline{x}(t),t) \tag{6.3-1}$$

$$\underline{z}_k = \underline{h}_k(\underline{x}_k) + \underline{v}_k \tag{6.3-2}$$

a weighted least-squares estimate of the state at stage k, $\hat{\underline{x}}_k$, is one which minimizes the deterministic performance index

$$J_k = \sum_{i=1}^{k} (\underline{z}_i - \underline{h}_i(\hat{\underline{x}}_i))^T W_i (\underline{z}_i - \underline{h}_i(\hat{\underline{x}}_i)) \tag{6.3-3}$$

where $\{W_i\}$ is a sequence of weighting matrices. Since Eq. (6.3-1) is unforced, $\underline{x}(t)$ is determined for all time by its initial value $\underline{x}_0$; hence, the estimation problem is equivalent to determining the value of $\underline{x}_0$ which minimizes J_k. The kth estimate of the initial condition, $\hat{\underline{x}}_{0k}$, can in principle be obtained by solving the familiar necessary conditions for the minimum of a function,

$$\frac{\partial J_k}{\partial \hat{\underline{x}}_{0k}} = \underline{0} \tag{6.3-4}$$

where $\hat{\underline{x}}_k$ in Eq. (6.3-3) is expressed in terms of $\hat{\underline{x}}_{0k}$ through the solution to Eq. (6.3-1).

The above formulation of the estimation problem is in the format of a classical parameter optimization problem, where J, the function to be minimized, is subject to the equality constraint conditions imposed by the solutions to Eq. (6.3-1) at the measurement times. The latter can be expressed as a nonlinear difference equation

$$\underline{x}_{k+1} = \underline{f}_k(\underline{x}_k) \tag{6.3-5}$$

where

$$\underline{f}_k(\underline{x}_k) \triangleq \underline{x}_k + \int_{t_k}^{t_{k+1}} \underline{f}(\underline{x}(t),t)\,dt \tag{6.3-6}$$

Consequently, the estimate $\hat{\underline{x}}_{0k}$ at each stage is typically determined as the solution to a two-point boundary problem (e.g., see Ref. 17). For general nonlinear functions $\underline{h}_k$ and $\underline{f}$, the boundary value problem cannot be solved in closed form; thus, approximate solution techniques must be employed to obtain practical algorithms. To obtain *smoothed* estimates of $\underline{x}_0$, after a fixed number of measurements have been collected, various iterative techniques that have been developed specifically for solving optimization problems can be employed – e.g., steepest descent (Ref. 17), conjugate gradient (Ref. 11) and quasi-linearization (Refs. 12 and 13). To obtain recursive estimates at each measurement stage – i.e., to mechanize a *filter* – the sequence of functions, $J_1, J_2, \ldots,$ defined in Eq. (6.3-3), must be minimized. Approximate recursive solutions have been obtained using both the invariant embedding (Ref. 14) and quasi-linearization methods. Extensions of the least-squares method to cases where Eq. (6.3-1) includes a driving noise term, and where the measurements are continuous are given in Refs. 15 and 16.

The above discussion presents a very brief summary of the philosophy of least-squares estimation. It is a useful alternative to minimum variance estimation in situations where the statistics of uncertain quantities are not well defined. To obtain specific least-squares data processing algorithms, the reader is referred to the cited works.

6.4 DIRECT STATISTICAL ANALYSIS OF NONLINEAR SYSTEMS – CADET™

One often encounters systems, for which the statistical behavior is sought, in which significant nonlinearities exist. These problems may include filters, linear or otherwise, or may simply involve statistical signal propagation. In either case, the existence of significant nonlinear behavior has traditionally necessitated the employment of monte carlo techniques – repeated simulation trials plus averaging – to arrive at a statistical description of system behavior. Hundreds, often thousands, of sample responses are needed to obtain statistically meaningful results; correspondingly, the computer burden can be exceptionally severe both in cost and time. Thus, one is led to search for alternative methods of analysis.

An exceptionally promising technique (Ref. 21), developed by The Analytic Sciences Corporation specifically for the direct statistical analysis of dynamical nonlinear systems, is presented herein. It is called the *Covariance Analysis DEscribing function Technique* (CADET). This technique employs the device of *statistical linearization* discussed in Section 6.2; however, the viewpoint here is statistical analysis rather than estimation.

The general form of the system to be considered is given by Eq. (6.0-1) and is illustrated in Fig. 6.4-1, where $\underline{x}$ is the system state vector, $\underline{w}$ is a white noise input $[\underline{w} \sim N(\underline{b},Q)]$, and $\underline{f}(\underline{x})$ is a vector nonlinear function. The objective is to determine the statistical properties – mean and covariance matrix – of $\underline{x}(t)$. The success of CADET in achieving this goal depends on how well $\underline{f}(\underline{x})$ can be approximated. The approximation criterion used here is the same as that employed in Section 6.2; however, the CADET algorithm is derived somewhat differently here, in order to expose the reader to additional properties of statistical linearization.

Consider approximation of the nonlinear function $\underline{f}(\underline{x})$ by a linear function, in the sense suggested by Fig. 6.4-2. The input to the nonlinear function, $\underline{x}$, is

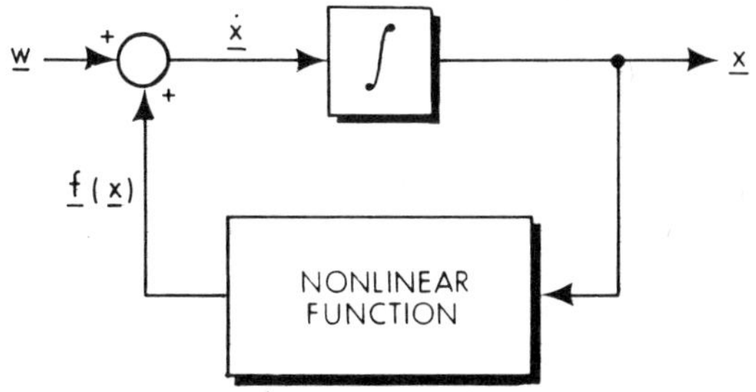

Figure 6.4-1 Nonlinear System Block Diagram

taken to be comprised of a mean, $\underline{m}$, plus a zero mean, independent, random process, $\underline{r}$. Thus,*

$$\underline{x} = \underline{m} + \underline{r} \qquad (6.4\text{-}1)$$

The mean can arise due to an average value of $\underline{w}$, a mean initial value of $\underline{x}$, a rectification in the nonlinearity, or a combination of these.

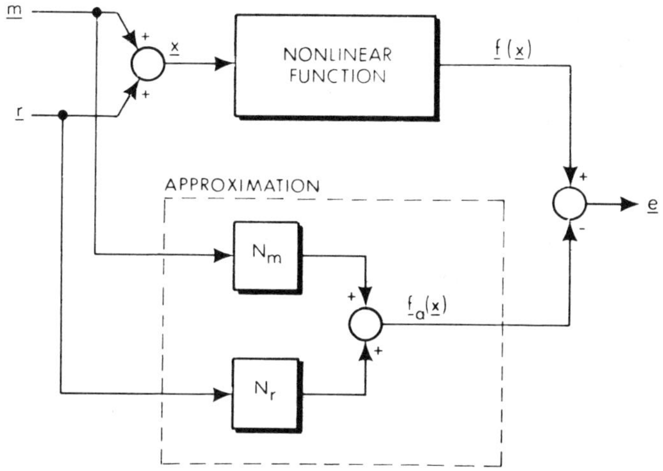

Figure 6.4-2 Describing Function Approximation

*Comparing subsequent notation with the development in Section 6.2, note that the following equivalences hold: $\underline{r} = -\underline{\tilde{x}}$, $N_m\underline{m} = \underline{\hat{f}}$, and $N_r = N_f$.

We think of the approximating output, $\underline{f}_a(\underline{x})$, as being comprised of the sum of two terms, one linearly related to $\underline{m}$ and the other linearly related to $\underline{r}$. The so-called *multiple-input describing function* gain matrices, N_m and N_r, are chosen to minimize the mean square error in representing the actual nonlinearity output $\underline{f}(\underline{x})$ by $\underline{f}_a(\underline{x})$.

Calculation of N_m and N_r is readily accomplished. Note first from Fig. 6.4-2 that

$$\begin{aligned} \underline{e} &= \underline{f}(\underline{x}) - \underline{f}_a(\underline{x}) \\ &= \underline{f}(\underline{x}) - N_m\underline{m} - N_r\underline{r} \end{aligned} \tag{6.4-2}$$

Forming the matrix $\underline{e}\underline{e}^T$, we minimize the mean square approximation error by computing

$$\frac{\partial}{\partial N_m}(\text{trace } E[\underline{e}\underline{e}^T]) = \frac{\partial}{\partial N_r}(\text{trace } E[\underline{e}\underline{e}^T]) = 0 \tag{6.4-3}$$

These computations result in the relationships

$$N_m\underline{m}\underline{m}^T = E\,[\underline{f}(\underline{x})]\,\underline{m}^T \tag{6.4-4}$$

and

$$N_r E[\underline{r}\underline{r}^T] = E[\underline{f}(\underline{x})\underline{r}^T] \tag{6.4-5}$$

since $E[\underline{r}\underline{m}^T] = E[\underline{m}\underline{r}^T] = 0$. Equations (6.4-4) and (6.4-5) define N_m and N_r. Denoting the random process covariance matrix by S(t), viz:

$$S(t) = E[\underline{r}(t)\underline{r}^T(t)] \tag{6.4-6}$$

and assuming that S is nonsingular, we find

$$N_r(\underline{m},S) = E[\underline{f}(\underline{x})\underline{r}^T]\,S^{-1} \tag{6.4-7}$$

Rather than attempt to solve for N_m (which requires a pseudoinverse calculation, since $\underline{m}\underline{m}^T$ is always singular), we simply require

$$N_m(\underline{m},S)\underline{m} = E[\underline{f}(\underline{x})] \tag{6.4-8}$$

This result is all we shall need to solve the problem at hand. Evaluation of the expectations in Eq. (6.4-7) and (6.4-8) requires an assumption about the probability density function of $\underline{r}(t)$. Most often a gaussian density is assumed, although this need not be the case.

Replacing the nonlinear function in Eq. (6.0-1) by the describing function approximation indicated in Fig. 6.4-2, we see directly that the differential equations of the resulting quasi-linear system are* (see Fig. 6.4-3)

$$\dot{\underline{m}} = N_m(\underline{m},S)\,\underline{m} + \underline{b}$$
$$\dot{\underline{r}} = N_r(\underline{m},S)\,\underline{r} + \underline{u} \qquad (6.4\text{-}9)$$

and the covariance matrix for $\underline{r}$ satisfies

$$\dot{S} = N_r(\underline{m},S)\,S + SN_r^T(\underline{m},S) + Q \qquad (6.4\text{-}10)$$

where Eq. (6.4-10) is simply the linear variance equation with F replaced by $N_r(\underline{m},S)$. These equations are initialized by associating the mean portion of $\underline{x}(0)$ with $\underline{m}(0)$, and the random portion with S(0), where

$$S(0) = E[(\underline{x}(0) - \underline{m}(0))\,(\underline{x}(0) - \underline{m}(0))^T]$$

A single forward integration of these coupled differential equations will then produce $\underline{m}(t)$ and S(t).

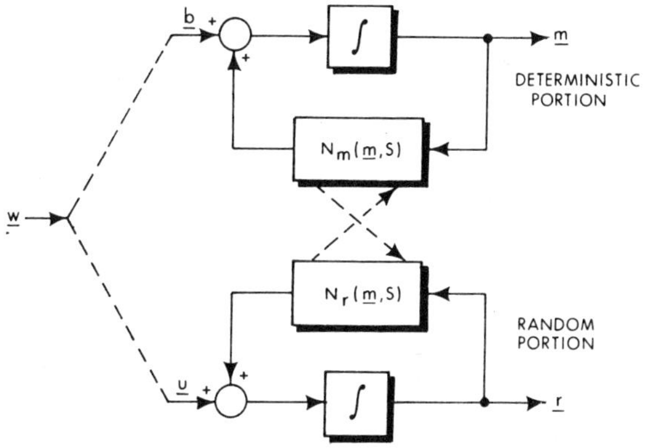

Figure 6.4-3 Quasi-linear System Model

A few special cases are worth noting. When the system is *linear*, $\underline{f}(\underline{x}) = F\underline{x}$ and Eqs. (6.4-7) and (6.4-8) immediately lead to the result $N_m = N_r = F$. Corresponding to Eqs. (6.4-9) and (6.4-10) are the familiar equations for mean and covariance propagation in linear systems. When the nonlinearity is *odd* [i.e., $\underline{f}(-\underline{x}) = -\underline{f}(\underline{x})$], the effective gain to a *small* mean $\underline{m}$ in the presence of a

*These equations also follow directly from Eqs. (6.1-3) and (6.1-5). Note that $\underline{u} = \underline{w} - \underline{b}$ is simply the zero mean value portion of the input, $\underline{w}$.

multidimensional gaussian process, $\underline{r}$, can be shown to be the same as the effective gain of that nonlinearity to $\underline{r}$ itself (Ref. 6), that is

$$\lim_{\underline{m} \to \underline{0}} N_m = N_r(\underline{m} = \underline{0}) \tag{6.4-11}$$

The same is true for the effective gain to a small sinusoid and other small signals. Discussion of this interesting result can be found in Ref. 7. It is also worth noting that *when $\underline{r}$ is gaussian*, N_r can be computed from the relationship

$$N_r(\underline{m},S) = \frac{d}{d\underline{m}} E[\underline{f}(\underline{x})] \tag{6.4-12}$$

Proof of this useful result is left as an exercise for the reader. Finally, in the case of *scalar* nonlinearities, the scalar describing function gains are computed from [Eqs. (6.4-7) and (6.4-8)]

$$n_r = \frac{1}{\sqrt{2\pi}\,\sigma^3} \int_{-\infty}^{\infty} f(r+m) r e^{-(r^2/2\sigma^2)} dr \tag{6.4-13}$$

and

$$n_m = \frac{1}{\sqrt{2\pi}\, m\sigma} \int_{-\infty}^{\infty} f(r+m) e^{-(r^2/2\sigma^2)} dr \tag{6.4-14}$$

where r has been assumed to be a gaussian random process. Tables of the result of this calculation for a wide variety of common nonlinearities can be found in Ref. 7. When the mean of x is known to be zero, we set m to zero and calculate only a single-input describing function,

$$n_r = \frac{1}{\sqrt{2\pi}\,\sigma^3} \int_{-\infty}^{\infty} f(r) r e^{-(r^2/2\sigma^2)} dr \tag{6.4-15}$$

These calculations are also extensively tabulated in Ref. 7.

Example 6.4-1

One issue of considerable importance is the degree of sensitivity which describing function gains display as a function of different input process probability densities. To get a feel for this sensitivity, we shall examine the limiter nonlinearity with a zero mean, random input whose probability density function is either *uniform, triangular* or *gaussian*.

Consider the uniform probability density function. For $a/2 \geqslant \delta$ (see Fig. 6.4-4) we first calculate σ^2 as

$$\sigma^2 = \int_{-\infty}^{\infty} r^2 p(r)dr = \frac{a^2}{12}$$

and then utilize this result to obtain n_r, viz:

$$n_r = \frac{1}{\sigma^2} \int_{-\infty}^{\infty} f(r) r p(r) dr$$

$$= \frac{2}{\sigma^2} \left[\int_0^{\delta} r^2 \frac{1}{a} dr + \int_{\delta}^{a/2} \delta r \frac{1}{a} dr \right]$$

$$= \frac{\sqrt{3}}{2} \frac{\delta}{\sigma} \left[1 - \frac{1}{9} \left(\frac{\delta}{\sigma} \right)^2 \right] \qquad \text{for} \qquad \frac{\sigma}{\delta} \geqslant \frac{1}{\sqrt{3}}$$

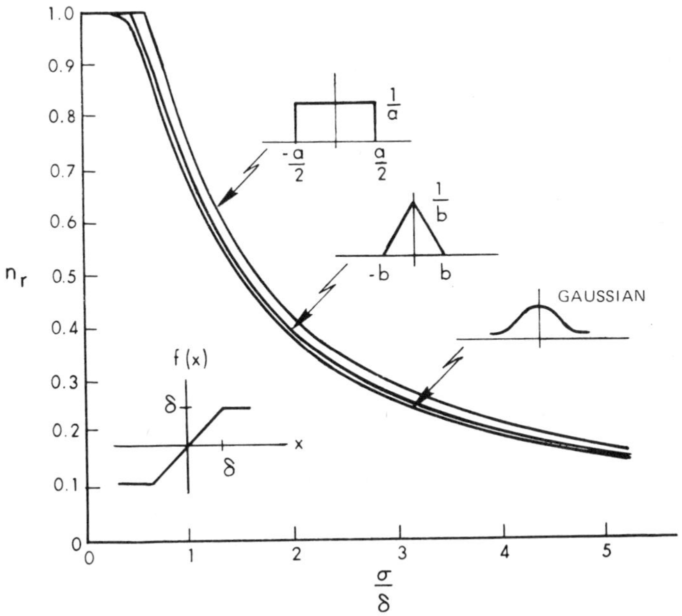

Figure 6.4-4 Describing Function Results With Different Input Probability Densities

For $a/2 < \delta$, it is clear that the input never causes output saturation to occur. Hence, in this event we find

$$n_r = 1 \quad \text{for} \ \frac{\sigma}{\delta} < \frac{1}{\sqrt{3}}$$

Similarly, for the triangular probability density function we calculate

$$\sigma^2 = \frac{b^2}{6}$$

and

$$n_r = \sqrt{\frac{2}{3}}\frac{\delta}{\sigma}\left[1 - \frac{1}{6}\left(\frac{\delta}{\sigma}\right)^2 + \frac{1}{12\sqrt{6}}\left(\frac{\delta}{\sigma}\right)^3\right] \qquad \text{for} \ \frac{\sigma}{\delta} \geqslant \frac{1}{\sqrt{6}}$$

$$= 1 \qquad \text{for} \ \frac{\sigma}{\delta} < \frac{1}{\sqrt{6}}$$

For the gaussian probability density function we obtain the form

$$n_r = \sqrt{\frac{2}{\pi}} \int_{-\infty}^{\delta/\sigma} e^{-r^2/2} dr - 1$$

in which the so-called probability integral, which is tabulated (Ref. 7), occurs. This result was depicted earlier in Fig. 6.2-2.

The results of these three calculations are plotted in Fig. 6.4-4. Qualitatively, at least, the relative *insensitivity* of the linearizing gain to the various input signal probability densities is apparent. This is obtained with other nonlinearities as well. It accounts, to some degree, for the success of CADET, given that the required probability densities are in fact never exactly known.

Example 6.4-2

Consider the problem of a pursuer and an evader, initially closing head-on in a plane. The evader has a random lateral acceleration maneuver, perpendicular to the initial line-of-sight, characterized by a first-order markov process with a standard deviation of 0.5g (16.1 ft/sec^2). The pursuer has first-order commanded lateral acceleration dynamics with lateral acceleration saturation. Pursuer lateral acceleration guidance commands are taken as proportional to the pursuer-evader line-of-sight rate (proportional guidance), with the guidance gain α set at 3. A block diagram of the system under consideration is shown in Fig. 6.4-5, where T_{GO} is the time-to-intercept. Total problem time is 10 seconds.

In the absence of information to the contrary, we assume x_1 to be a gaussian random process. If it is, the nonlinearity output will be significantly nongaussian for $\sigma > \delta$; but the action of linear filtering around the loop back to x_1 will be such as to reshape the density function, whatever it is, back towards gaussian (a result of the central limit theorem). This is the often cited *filter hypothesis*, common to most describing function analyses.

Making use of Eqs. (6.4-7) and (6.4-12), the quasi-linearized system can be represented in state vector notation as follows:

$$\begin{bmatrix} \dot{x}_1 \\ \dot{x}_2 \\ \dot{x}_3 \\ \dot{x}_4 \end{bmatrix} = \begin{bmatrix} -1 & \frac{\alpha}{T_{GO}^2} & \frac{\alpha}{T_{GO}} & 0 \\ 0 & 0 & 1 & 0 \\ -n_r\,(\sigma_{x_1}) & 0 & 0 & 1 \\ 0 & 0 & 0 & -1 \end{bmatrix} \begin{bmatrix} x_1 \\ x_2 \\ x_3 \\ x_4 \end{bmatrix} + \begin{bmatrix} 0 \\ 0 \\ 0 \\ w_4 \end{bmatrix}$$

CADET and 200 run ensemble monte carlo simulation results for the evader acceleration (x_4) and the relative separation (x_2) are presented in Fig. 6.4-6 for the linearized system ($\delta=\infty$) along with a 1g and a 0.1g pursuer lateral acceleration saturation level. The rms miss distance is given by the relative separation at 10 seconds ($T_{GO}=0$). The results clearly demonstrate good agreement between CADET and the monte carlo method, even when the nonlinear saturation effect is dominant, as in Figure 6.4-6d. The advantage of CADET, of course, is that it *analytically* computes the system statistics, thereby saving considerable computer time. More extensive applications of CADET are discussed in Ref. 25.

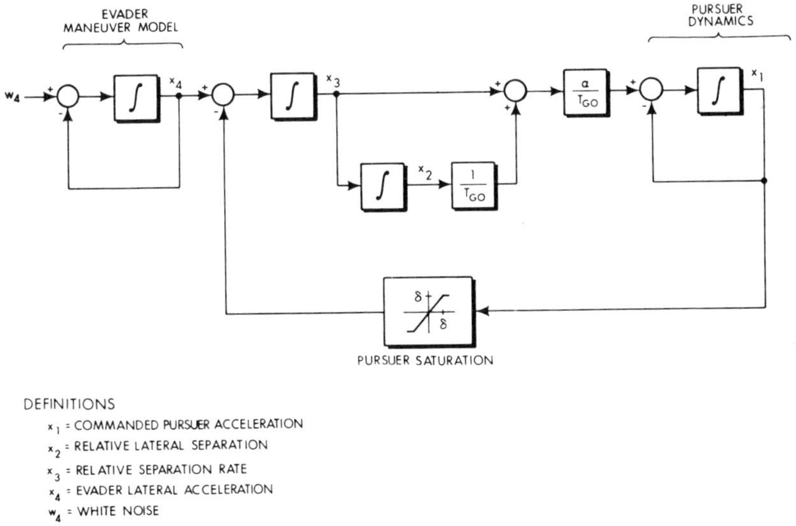

Figure 6.4-5 Kinematic Guidance Loop With Pursuer Lateral Acceleration Saturation

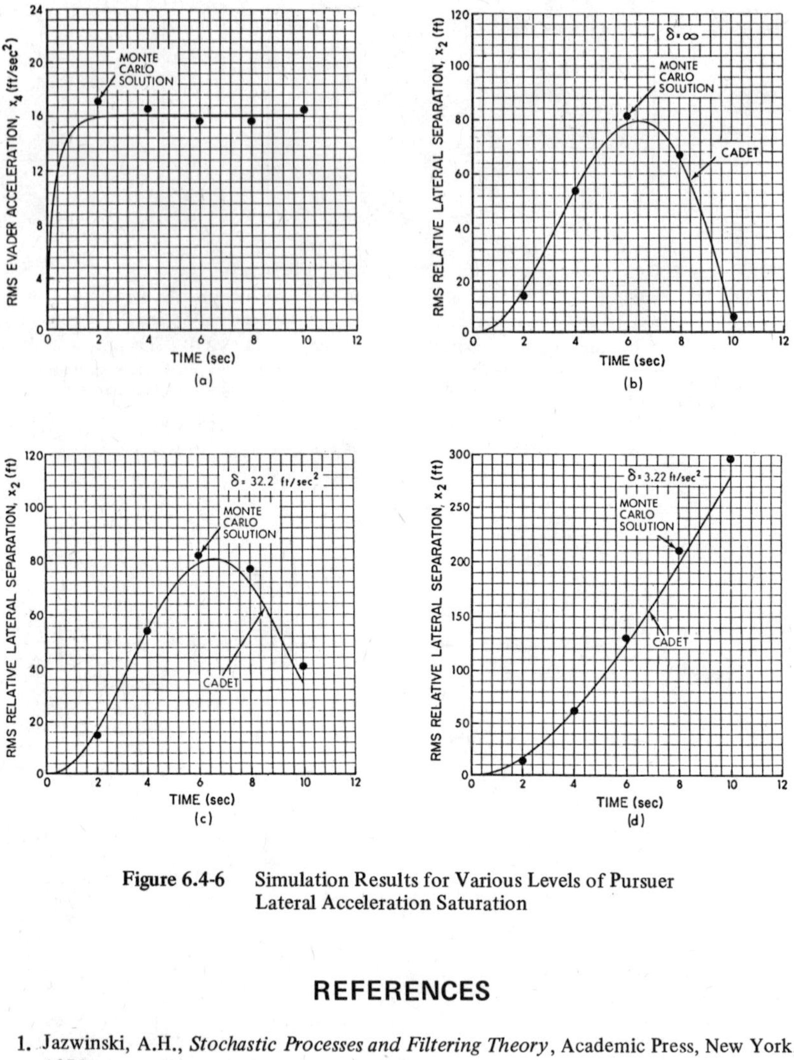

Figure 6.4-6 Simulation Results for Various Levels of Pursuer Lateral Acceleration Saturation

REFERENCES

1. Jazwinski, A.H., *Stochastic Processes and Filtering Theory*, Academic Press, New York, 1970.
2. Wong, E., *Stochastic Processes in Information and Dynamical Systems*, McGraw-Hill Book Co., Inc., New York, 1971.
3. Fraser, D.C., "A New Technique for the Optimal Smoothing of Data," Ph.D. Thesis, Massachusetts Institute of Technology, January 1967.
4. Leondes, C.T., Peller, J.B., and Stear, E.B., "Nonlinear Smoothing Theory," *IEEE Trans. Systems Science and Cybernetics*, Vol. SSC-6, No. 1, January 1970.
5. Wishner, R.P., Tabaczynski, J.A., and Athans, M., "A Comparison of Three Non-Linear Filters," *Automatica*, Vol. 5, 1969, pp. 487-496.
6. Phaneuf, R.J., "Approximate Nonlinear Estimation," Ph.D. Thesis, Massachusetts Institute of Technology, May 1968.

7. Gelb, A. and Vander Velde, W.E., *Multiple-Input Describing Functions and Nonlinear System Design*, McGraw-Hill Book Co., Inc., New York, 1968.

8. Sunahara, Y., "An Approximate Method of State Estimation for Nonlinear Dynamical Systems," Joint Automatic Control Conference, University of Colorado, 1969.

9. Mahalanabis, A.K. and Farooq, M., "A Second-Order Method for State Estimation of Nonlinear Dynamical Systems," *Int. J. Control*, Vol. 14, No. 4, 1971, pp. 631-639.

10. Hildebrand, F.B., *Methods of Applied Mathematics*, Prentice-Hall, Inc., Englewood Cliffs, N.J., 1952.

11. Fletcher, R. and Reeves, C.M., "Function Minimization by Conjugate Gradients," *The Computer Journal*, Vol. 7, 1964, p. 149.

12. Bellman, R.E. and Kalaba, R., *Quasilinearization and Boundary Value Problems*, American Elsevier Publishing Co., New York, 1965.

13. Chen, R.T.N., "A Recurrence Relationship for Parameter Estimation by the Method Quasilinearization and Its Connection with Kalman Filtering," Joint Automatic Control Conference, Atlanta, Georgia, June 1970.

14. Pearson, J.B., "On Nonlinear Least-Squares Filtering," *Automatica*, Vol. 4, 1967, pp. 97-105.

15. Sage, A.P., *Optimum Systems Control*, Prentice-Hall, Inc., Englewood Cliffs, N.J., 1968.

16. Sage, A.P. and Melsa, J.L., *System Identification*, Academic Press, New York, 1971.

17. Bryson, A.E., Jr. and Ho, Y.C., *Applied Optimal Control*, Blaisdell Publishing Co., Waltham, Mass., 1969.

18. Schmidt, S.F., Weinberg, J.D. and Lukesh, J.S., "Case Study of Kalman Filtering in the C-5 Aircraft Navigation System," *Case Studies in System Control*, University of Michigan, June 1968, pp. 57-109.

19. Sorenson, H.W. and Stubberud, A.R., "Nonlinear Filtering by Approximation of the Aposteriori Density," *Int. J. Control*, Vol. 18, 1968, pp. 33-51.

20. Sorenson, H.W. and Alspach, D.L., "Recursive Bayesian Estimation Using Gaussian Sums," *Automatica*, Vol. 7, 1971, pp. 465-479.

21. Gelb, A. and Warren, R.S., "Direct Statistical Analysis of Nonlinear Systems," *Proc. AIAA Guidance and Control Conf.*, Palo Alto, August 1972.

22. Sage, A.P. and Melsa, J.L., *Estimation Theory with Applications to Communications and Control*, McGraw-Hill Book Co., Inc., New York, 1971.

23. Bizwas, K.K. and Mahalanabis, A.K., "Suboptimal Algorithms for Nonlinear Smoothing," *IEEE Trans. on Aerospace and Electronic Systems*, Vol. AES-9, No. 4, July 1973, pp. 529-534.

24. Davenport, W.B., Jr. and Root, W.L., *An Introduction to the Theory of Random Signals and Noise*, McGraw-Hill Book Co., Inc., New York, 1958.

25. Price, C.F., Warren, R.S., Gelb, A., and Vander Velde, W.E., "Evaluation of Homing Guidance Laws Using the Covariance Analysis Describing Function Technique," *Trans. First NWC Symposium on the Application of Control Theory to Modern Weapons Systems*," China Lake, June 1973, pp. 73-94.

PROBLEMS

Problem 6-1

Suppose that a parameter x has the probability density function

$$p(x) = \lambda^2 x\, e^{-\lambda x} \quad \text{for } x \geqslant 0\,, \qquad = 0 \qquad \text{for } x < 0$$

with a specified value of λ. (a) Show that $E[x] = 2/\lambda$. (b) Show that the maximum value of $p(x)$ occurs at $x = 1/\lambda$. (c) Consider an arbitrary density function, $p(y)$, for a random variable. Determine restrictions on $p(y)$ such that $E[y]$ is the same as the value of y that maximizes $p(y)$.

Problem 6-2

Prove that Eq. (6.1-5) reduces to

$$\dot{P} = FP + PF^T + Q$$

when $\underline{f}(\underline{x}) = F\underline{x}$

Problem 6-3

Consider a scalar nonlinear system

$$\dot{x}(t) = f(x) + w(t); \qquad w(t) \sim N(0,q)$$

with measurements

$$z_k = h(x_k) + v_k; \qquad v_k \sim N(0,r)$$

Let the estimate of the state at time t_k be updated according to

$$\hat{x}_k(+) = a_k + b_k z_k + c_k z_k^2$$

Assuming that the conditional mean just before the update, $\hat{x}_k(-)$, is known, (a) show that $\hat{x}_k(+)$ is unbiased if

$$a_k + b_k\hat{h} + c_k(\widehat{h^2} + r) - \hat{x}_k(-) = 0$$

(b) Determine b_k and c_k, such that $\hat{x}_k(+)$ is a minimum variance estimate.

Problem 6-4

Derive the estimation equations given in Table 6.1-3 following the steps outlined in the text.

Problem 6-5

Defining

$$\underline{s}(t) = P_b^{-1}(t)\,\underline{\hat{x}}_b(t)$$

derive Eqs. (6.1-42) and (6.1-43) from Eqs. (6.1-39) and (6.1-40).

Problem 6-6

(a) Given a scalar nonlinear function $f(x)$ of a random variable x, show that the constants a, b, and c which minimize the mean square error

$$E[(f(x) - a - bx - cx^2)^2]$$

are given by

$$a = \hat{f} - b\hat{x} - c\widehat{x^2}$$

$$b = \frac{(4\hat{x}^2 m_2 + 4\hat{x} m_3 + m_4)(\widehat{fx} - \hat{f}\hat{x}) - (2\hat{x} m_2 + m_3)(\widehat{fx^2} - \hat{f}\widehat{x^2})}{m_2 m_4 - m_3^{\,2}}$$

$$c = \frac{-(2\hat{x} m_2 + m_3)(\widehat{fx} - \hat{f}\hat{x}) + m_2(\widehat{fx^2} - \hat{f}\widehat{x^2})}{m_2 m_4 - m_3^{\,2}}$$

where

$$m_i = E[(x - \hat{x})^i] \qquad \text{for } i = 2,3,4$$

(b) Show that if f(x) is an odd function of x, and if

$\hat{x} = m_3 = 0$, then $c = 0$ and

$$b = \frac{\widehat{fx} - \hat{f}\hat{x}}{m_2}$$

Compare these values for b and c with Eqs. (6.2-7) and (6.2-10).

Problem 6-7

Suppose x is a zero mean gaussian random variable with variance σ^2. Show that $E[x^n] = (1)\,(3)\,\ldots\,(n-1)\sigma^n$ for n even and zero for n odd. (Hint: Use the characteristic function (Ref. 24) of the probability density for x.)

Problem 6-8

Supply the missing details in deriving the third equality in Eq. (6.2-34).

Problem 6-9

Employ Eq. (6.4-3) to arrive at the relationships which define the multiple-input describing functions $N_m(\underline{m},S)$ and $N_r(\underline{m},S)$, Eqs. (6.4-4) and (6.4-5).

Problem 6-10

Demonstrate that the describing function approximation error, $\underline{e}$ (see Fig. 6.4-2), is uncorrelated with the nonlinearity input. That is, show that $E[\underline{e}\underline{x}^T] = \underline{0}$.

Problem 6-11

For a scalar nonlinearity described by

$$f(x) = x(1 + x^2)$$

show that the multiple-input describing function gains n_m and n_r are given by

$$n_m(m, \sigma_r^{\,2}) = 1 + m^2 + 3\sigma_r^{\,2}$$

$$n_r(m, \sigma_r^{\,2}) = 1 + 3m^2 + 3\sigma_r^{\,2}$$

(Hint: Use the result $E[r^4] = 3\sigma_r^{\,4}$, valid for a gaussian random variable r.)

Problem 6-12

For the ideal relay nonlinearity defined by

$$
\begin{aligned}
f(x) &= D \qquad \text{for } x > 0 \\
&= 0 \qquad \text{for } x = 0 \\
&= -D \qquad \text{for } x < 0
\end{aligned}
$$

show that the describing functions for gaussian, triangular and uniform input signal probability density functions are:

$$n_r\,(\text{gaussian}) = \sqrt{\frac{2}{\pi}}\,\frac{D}{\sigma} \approx 0.80\frac{D}{\sigma}$$

$$n_r\,(\text{triangular}) = \sqrt{\frac{2}{3}}\,\frac{D}{\sigma} \approx 0.82\frac{D}{\sigma}$$

$$n_r\,(\text{uniform}) = \sqrt{\frac{3}{4}}\,\frac{D}{\sigma} \approx 0.87\frac{D}{\sigma}$$

Problem 6-13

For the nonlinear differential equation

$$\dot{x} = a_1 x + a_2 x^2 + w$$

where $w \sim N(b,q)$, show that the covariance analysis describing function technique (CADET) yields the following equations for mean and covariance propagation:

$$\dot{m} = a_1 m + a_2 m^2 + a_2 p + b$$

$$\dot{p} = 2(a_1 + 2a_2 m)p + q$$

7. SUBOPTIMAL FILTER DESIGN AND SENSITIVITY ANALYSIS

After an excursion into the realm of nonlinear systems, measurements and filters in Chapter 6, we now direct our attention back to the more mathematically tractable subject of linear filtering, picking up where we left off at the conclusion of Chapter 5. The subject here, suboptimal linear filters and linear filter sensitivity, is of considerable practical importance to the present-day design and operation of multisensor systems.

The filter and smoother equations, developed in Chapters 4 and 5, provide a simple set of rules for designing optimal linear data processors. At first glance, the problem of filter design appears to have been solved. However, when the Kalman filter equations are applied to practical problems, several difficulties quickly become obvious. The truly *optimal* filter must model *all* error sources in the system at hand. This often places an impossible burden on the computational capabilities available. Also, it is assumed in the filter equations that exact descriptions of system dynamics, error statistics and the measurement process are known. Similar statements apply to the use of the optimal smoother equations. Because an unlimited computer capability is not usually available, the designer of a filter or smoother purposely ignores or simplifies certain effects when he represents them in his design equations; this results in a *suboptimal* data processor. For this reason, and because some of the information about the system behavior and statistics is not known precisely, the prudent engineer performs a separate set of analyses to determine the sensitivity of his design to any differences that might exist between his filter or smoother and one that fits the optimal mold exactly. This process is called *sensitivity analysis*. The sensitivity analysis discussed here is distinct from another procedure which may, from time to time, be given the same name in other writings – that of recomputing the optimal filter equations for several sets of assumptions, each time finding the accuracv which could be achieved if all the conditions of optimality were perfectly satisfied.

In addition to establishing the overall sensitivity of a particular linear data processing algorithm, the equations and procedures of sensitivity analysis can inform the filter designer of individual error source contributions to estimation errors. This type of source-by-source breakdown is valuable in assessing potential hardware improvements. For example if, for a given system and filter, the error contribution from a particular component is relatively small, specifications on that piece of equipment can be relaxed without seriously degrading system performance. Conversely, critical component specifications are revealed when errors in a particular device are found to be among the dominant sources of system error.

This chapter discusses various proven approaches to suboptimal filter design. The equations for sensitivity analysis of linear filters and smoothers are presented, with emphasis on the underlying development of relationships for analyzing systems employing "optimal" filters. Several valuable methods of utilizing information generated during sensitivity analyses are illustrated, and a computer program, organized to enable study of suboptimal filter design and sensitivity analysis, is described.

7.1 SUBOPTIMAL FILTER DESIGN

The data-combination algorithm, or filter, for a multisensor system is very often a deliberate simplification of, or approximation to, the optimal (Kalman) filter. One reason, already noted, for which the filter designer may choose to depart from the strict *Kalman* filter formula is that the latter may impose an unacceptable computer burden. Figure 7.1-1 illustrates the common experience that a judicious reduction in filter size, leading to a suboptimal filter, often provides a substantially smaller computer burden with little significant reduction in system accuracy (the "conventional" data processing algorithm is understood to be a fixed-gain filter, arrived at by some other means).

Since solution of the error covariance equations usually represents the major portion of computer burden in any Kalman filter, an attractive way to make the filter meet computer hardware limitations is to precompute the error covariance, and thus the filter gain. In particular, the behavior of the filter gain elements may be approximated by a time function which is easy to generate using simple electronic components. Useful choices for approximating the filter gain are: constants, "staircases" (piecewise constant functions), and decaying exponentials. Whenever an approximation to the optimal filter gain behavior is made, a sensitivity analysis is required to determine its effect on estimation errors.

The filter designer may also depart from the optimal design because it appears too sensitive to differences between the parameter values he uses to design the

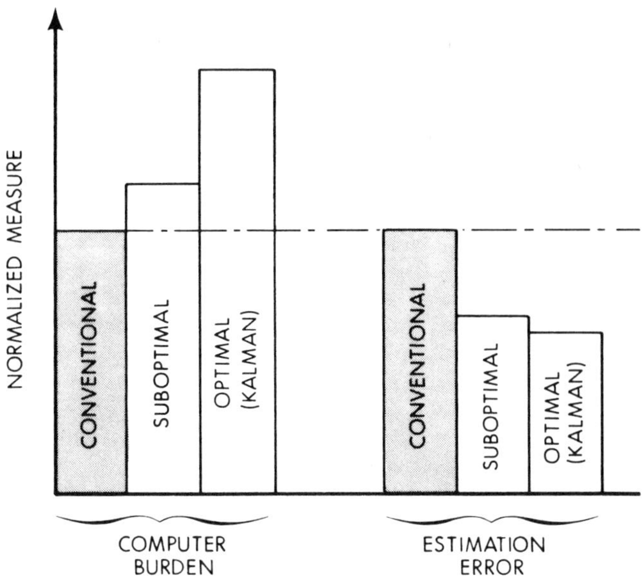

Figure 7.1-1 Illustration of Relative Computer Burden and Estimation Error for Three Data Processing Algorithms

filter and those which may exist in practice. By choosing an appropriate suboptimal design, it may be possible to reduce the sensitivity to uncertain parameters, as illustrated in Fig. 7.1-2. In the figure, it is assumed that there is a range of uncertainty in the value of σ_{true}, a certain parameter which is critical to the filter design. The value of the parameter that is used, σ_{design}, is chosen in the center of the interval in which σ_{true} is known to lie. Of course when $\sigma_{true} \neq \sigma_{design}$ the filter suffers some loss of accuracy, as indicated by the concave shape of the plots. In the hypothetical case shown, the theoretically optimal filter requires thirty states (n=30), but exhibits a wide variation in performance over the range of design uncertainty. The 10-state design is relatively insensitive but provides less accuracy than the 20-state filter shown, over the region of uncertainty. The minimum sensitivity filter of 20 states represents the "best" filter in this case. The best filter may be found, for example, by assuming a probability density function for σ_{true} and computing an expected value of the performance measure for each proposed filter, selecting the one with the lowest mean performance measure. It must be noted that reduced sensitivity is achieved at the price of a larger *minimum* error.

While the discussion of suboptimal filter design in this chapter centers on modifications to the basic Kalman filter procedure, observer theory (see Chapter 9) is also suggested as a viable technique for producing realizable filters.

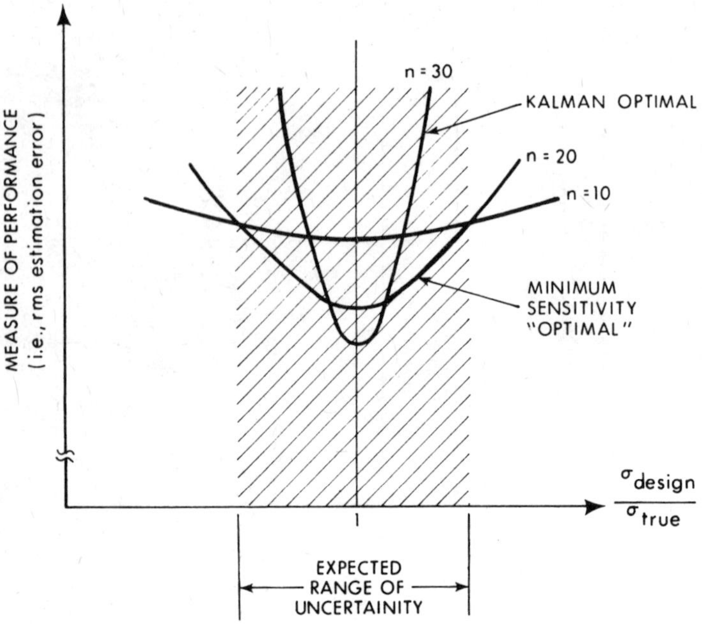

Figure 7.1-2 Conceptual Example of Designing for Minimum Sensitivity (σ_{true} held constant, σ_{design} is varied)

CHOOSING SIMPLIFIED SYSTEM MODELS

As just indicated, the filter designer may need to reduce the number of states modeled not only because a fully optimal filter imposes too great a burden on the computer available, but also because the optimal filter may be too sensitive to uncertainties in the statistical parameters (spectral densities of system noises, etc.) which must be provided. There is emerging a body of theory which can help the designer to reduce the sensitivity of his filter; this facet of the design problem is discussed later. By and large, however, the filter designer is left with few rules of general applicability to guide him in the process of eliminating those parts of the full description of the system that can be ignored, rearranged or replaced by a simpler mathematical description. He must depend, first and foremost, on his *physical* understanding of the system with which he is dealing. The common procedure is to make a simplified model based on insight, then analyze the accuracy of the resulting filter *in the presence of a complete set of system dynamics*, properly represented. Thus, the approach is generally one of analysis rather than synthesis, and a number of steps are necessary before a satisfactory result emerges. The equations for such analyses are given in Section 7.2. The present discussion concentrates on particular filter simplification approaches which have proven successful in the past, illustrated by examples.

Decoupling States – The number of multiplications necessary to compute the error covariance matrix used in Kalman-based linear filters generally varies as the third power of the state size (for more details see Chapter 8). Therefore, the main thrust of attempts to reduce the computer burden imposed by such filters is aimed at reducing the filter state size. Often the number of state variables reaches an irreducible minimum and the filter still makes excessive demands on the computer. Sometimes, an alternative technique is available that complements the more obvious approach of deleting state variables; if certain portions of the system are weakly coupled, it may be possible to break the relatively high-order filter into several mutually exclusive lower-order filters – each with separate covariance calculations, filter gains, etc. The advantage is evident if one considers breaking an n-state filter into three n/3-state filters. Using the rule stated above, the n-state filter requires kn^3 multiplications each time its covariance matrix is propagated. On the other hand the three n/3-state filters require a total of $3k(n^3/27) = kn^3/9$ multiplications to perform the same operation; the corresponding computer burden is thus reduced by about 90%.

Example 7.1-1

Consider the second-order coupled system, with continuous measurements (see Fig. 7.1-3), given by

$$\begin{bmatrix} \dot{x}_1 \\ \dot{x}_2 \end{bmatrix} = \begin{bmatrix} -\alpha_1 & 0 \\ \gamma & -\alpha_2 \end{bmatrix} \begin{bmatrix} x_1 \\ x_2 \end{bmatrix} + \begin{bmatrix} w_1 \\ w_2 \end{bmatrix}$$

$$\begin{bmatrix} z_1 \\ z_2 \end{bmatrix} = \begin{bmatrix} 1 & 0 \\ 0 & 1 \end{bmatrix} \begin{bmatrix} x_1 \\ x_2 \end{bmatrix} + \begin{bmatrix} v_1 \\ v_2 \end{bmatrix} \tag{7.1-1}$$

where w_1, w_2, v_1 and v_2 are uncorrelated white noises.

Notice that the subsystem whose state is x_1 receives no feedback from x_2. If γ is zero, the system is composed of two independent first-order markov processes. The estimation error covariance equations for a Kalman filter for the *coupled*, second-order system are

$$\dot{p}_{11} = -2\alpha_1 p_{11} + q_{11} - \frac{p_{11}^2}{r_{11}} - \frac{p_{12}^2}{r_{22}} \tag{7.1-2}$$

$$\dot{p}_{12} = -\left(\alpha_1 + \alpha_2 + \frac{p_{11}}{r_{11}} + \frac{p_{22}}{r_{22}}\right) p_{12} + \gamma p_{11} \tag{7.1-3}$$

$$\dot{p}_{22} = -2\alpha_2 p_{22} + q_{22} - \frac{p_{22}^2}{r_{22}} - \frac{p_{12}^2}{r_{11}} + 2\gamma p_{12} \tag{7.1-4}$$

where q_{11}, q_{22}, r_{11} and r_{22} are the spectral densities of w_1, w_2, v_1 and v_2, respectively.

If the coupling term γ is small, it is tempting to view the system as two separate, uncoupled systems, as follows (see Fig. 7.1-4):

$$\dot{x}_1 = -\alpha_1 x_1 + w_1$$

$$z_1 = x_1 + v_1 \tag{7.1-5a}$$

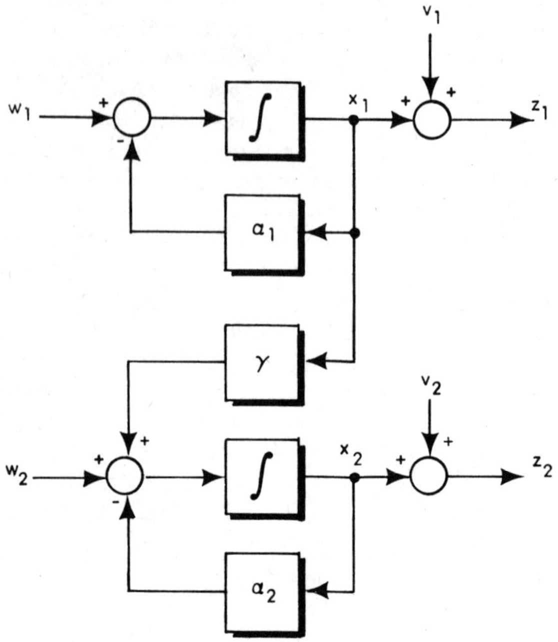

Figure 7.1-3 Block Diagram of Example System

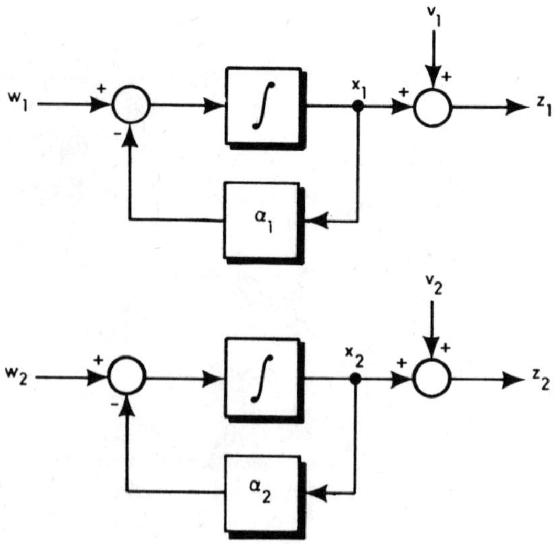

Figure 7.1-4 Block Diagram of Example System with γ Taken as Zero

and

$$\dot{x}_2 = -\alpha_2 x_2 + w_2$$
$$z_2 = x_2 + v_2 \tag{7.1-5b}$$

In this case the estimation error covariance equations are simply

$$\dot{p}_{11} = -2\alpha_1 p_{11} + q_{11} - \frac{p_{11}^2}{r_{11}} \tag{7.1-6}$$

$$\dot{p}_{22} = -2\alpha_2 p_{22} + q_{22} - \frac{p_{22}^2}{r_{22}} \tag{7.1-7}$$

Several qualitative observations can be drawn from a comparison of Eqs. (7.1-2), (7.1-3) and (7.1-4) with Eqs. (7.1-6) and (7.1-7). The expressions for $\dot{p}_{11}$ and $\dot{p}_{22}$ differ by terms proportional to p_{12} and p_{12}^2; if p_{12} is small, the decoupled filter error covariances are similar to those of the filter modeling the complete system. Furthermore, the terms in $\dot{p}_{11}$ that involve p_{12} always acts to *reduce* the error covariance in the complete filter; this suggests that the filter using the complete model of the system will always do a better job of estimating x_1. Finally, if we view the error covariances as varying very slowly (a quasi-static approximation), Eq. (7.1-3) shows that the estimation error correlation term p_{12} can be viewed as the output of a stable first-order system (assuming $\alpha_1 > 0$, $\alpha_2 > 0$) driven by a forcing term p_{11}. It can be seen that if the coupling coefficient is zero and $p_{12}(0) = 0$, the two sets of error covariance equations are identical.

Deleting States — When the computer burden indicates a need to delete states, it must be done very carefully, and always with some risk. Here the filter designer's knowledge of the physics of the problem and of *how the filter works* will assist him in selecting which states to eliminate first. Once the state vector and nominal dynamic and statistical parameters are selected, the resulting filter is subjected to a performance evaluation via sensitivity analysis techniques (see Section 7.2). If the performance is not acceptable, dominant error sources must be isolated and an effort made to reduce their effects. Where a dominant correlated error source was not estimated initially in the filter, it may be necessary to estimate it by adding extra states to the filter.* If it was too simply modeled in the filter, model complexity may have to be increased or model parameters adjusted to reduce the effect of this error source. Based on considerations of this type, the filter design is amended and the evaluation process repeated, thus starting a second cycle in this trial and error design approach.

Example 7.1-2

Consider the second-order system shown in Fig. 7.1-5 and given by Eq. (7.1-8),

*It should be emphasized that the act of estimating an additional important correlated error source does not, in itself, guarantee significantly improved filter performance.

$$\begin{bmatrix} \dot{x}_1 \\ \dot{x}_2 \end{bmatrix} = \begin{bmatrix} -\alpha_1 & 0 \\ 1 & -\alpha_2 \end{bmatrix} \begin{bmatrix} x_1 \\ x_2 \end{bmatrix} + \begin{bmatrix} w \\ 0 \end{bmatrix} \tag{7.1-8}$$

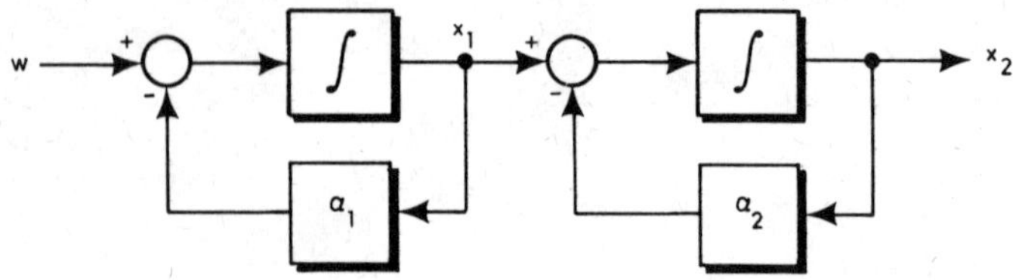

Figure 7.1-5 Block Diagram of Eq. (7.1-8)

Between measurements the error covariance equations for the state variables in Eq. (7.1-8) are given by

$$\dot{p}_{11} = -2\alpha_1 p_{11} + q \tag{7.1-9}$$

$$\dot{p}_{12} = p_{11} - (\alpha_1 + \alpha_2)\, p_{12} \tag{7.1-10}$$

$$\dot{p}_{22} = 2p_{12} - 2\alpha_2 p_{22} \tag{7.1-11}$$

where q is the spectral density of w. If the frequency content of x_1 is high, compared to the bandwidth of the first-order loop whose output is x_2, the filter designer may wish to model the entire system by a single first-order markov process driven by white noise. In that case, only one state variable, x_2, remains. The spectral density, q', of the "white noise" driving this system is given by $q' = 2p_{11\infty}/\alpha_1 = q/\alpha_1^2$ (Refs. 1 and 2), where $p_{11\infty}$ is the steady-state value of p_{11} calculated from Eq. (7.1-9). The error covariance equation for the lone state variable of the simplified system is

$$\dot{p} = -2\alpha_2 p + q' \tag{7.1-12}$$

In the steady state, the state error covariance of the simplified system can be found from Eq. (7.1-12) to be

$$p_\infty = \frac{q'}{2\alpha_2} = \frac{q}{2\alpha_1^2\alpha_2} \tag{7.1-13}$$

The steady-state error covariance of the corresponding state, x_2, in the full system is found from Eqs. (7.1-9), (7.1-10) and (7.1-11) to be

$$p_{22\infty} = \frac{q}{2\alpha_1\alpha_2(\alpha_1 + \alpha_2)} \tag{7.1-14}$$

It can be seen from Eqs. (7.1-13) and (7.1-14) that when the frequency content of x_1 is much larger than the bandwidth of the first-order system whose output is x_2 (i.e., $\alpha_1 \gg \alpha_2$), the steady-state covariance for the output of the simplified system closely approximates that of the full system. Reference 3 explores this technique in more generality. It was also employed in Example 4.6-1.

Verification of Filter Design – We conclude this section with an attempt to impress upon the reader the importance of checking, by suitable sensitivity analyses, any suboptimal filter design he may produce. The motivation is best illustrated by discussing an unusual type of filter behavior that has been observed in many apparently well-thought-out suboptimal filter designs, when they were subjected to analysis which correctly accounts for all error sources. This anomalous behavior is characterized by a significant *growth* in errors in the estimate of a particular variable when some measurements are incorporated by the filter. The reason for this unusual performance is usually traced to a difference between error correlations indicated by the filter covariance matrix, and those that are found to exist when the true error behavior is computed.

Figure 7.1-6 helps to illustrate how incorrect correlations can cause a measurement to be processed improperly. Before the measurement the state estimates are $\hat{x}_0$ and $\hat{y}_0$, with the estimation error distributions illustrated in the figure by ellipses which represent equal-error contours. The measurement indicates, within some narrow band of uncertainty not shown, that the state y is at y_m. Because the measurement is quite precise, the filter corrects the estimate of the y coordinate, $\hat{y}$, to a value that is essentially equal to the measurement, y_m. Because of the correlation that the filter thinks exists between the x and y coordinates, the x coordinate is corrected to $\hat{x}_1$. Observe that no direct measurement of x is made, but the correlation is used to imply x from a measurement of y. If the correct error correlations were known, the filter would

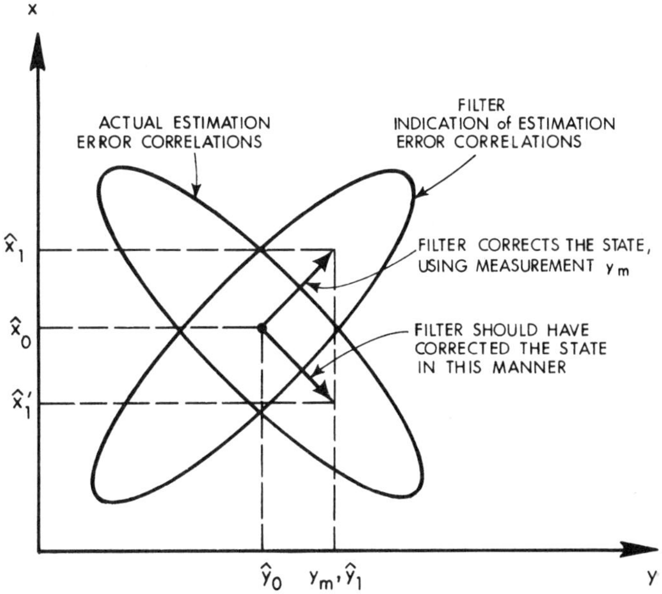

Figure 7.1-6 Illustration of Improper Use of a Measurement

correct the state to $\hat{y}_1 \cong y_m$ and $\hat{x}'_1$. Because the filter covariance matrix indicated the wrong correlation, the filter has "corrected" its estimate of x in the wrong direction, thus increasing the error in its estimate of that variable. This sort of improper behavior cannot usually be anticipated when choosing states, etc., for a suboptimal filter design. It can only be observed by performing sensitivity analyses.

CHOOSING SIMPLIFIED FILTER GAINS

It has been pointed out previously that the major portion of the computational burden imposed by the Kalman filter involves computing the filter error covariance matrix for use in determining the filter gains. While the previous discussions emphasized the role of reducing this effort by simplifying the system description and cutting the filter state size, it is sometimes possible and desirable to eliminate *altogether* the on-line covariance calculations. In these cases, the error covariance matrix is computed beforehand and the filter gain histories are observed. While a pre-recorded set of precise gain histories could be stored and used, a more attractive approach is to approximate the observed gain behavior by analytic functions of time, which can be easily computed in real

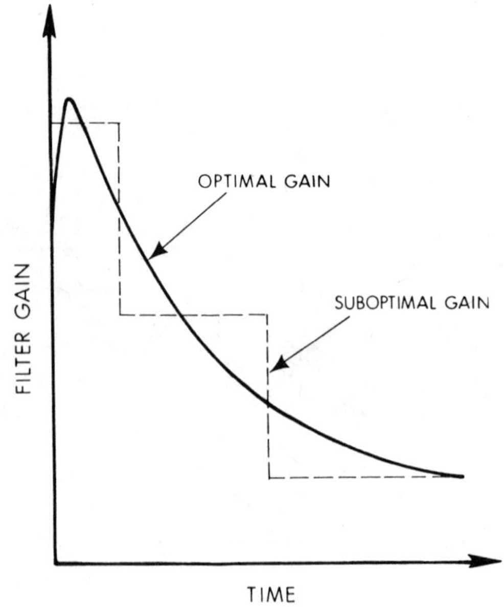

Figure 7.1-7 Piecewise Constant Suboptimal Gain

time. These tend to be exponentials, staircases and constants. Figure 7.1-7 illustrates a staircase, or piecewise constant, approximation to an optimal time history for a single element in a filter gain matrix. The same gain element could also be well approximated by a decaying exponential function of time.

Approximating the Optimal Gains – Reference 4 discusses a system for which, as a practical matter, an investigation was performed to determine a set of piecewise constant filter gains that will approximate the performance of a Kalman optimal filter. Figure 7.1-8 illustrates such a gain approximation used during initial operation of the system. The continuous optimal gain curve was approximated by a piecewise constant gain; Fig. 7.1-9 shows one resulting error in the system. As expected, the suboptimal gain history produces larger errors for a certain duration of time. The interesting observation here is that the *steady-state* error does not suffer as a consequence of the gain approximation. In many systems, the longer convergence time may be an acceptable price to pay for not having to compute the error covariance and gain matrices on-line or to store their detailed time histories.

While it is usually possible to select good approximations to the optimal filter gains simply by observing their time behavior and using subjective judgement, some analysis has been performed which could be brought to bear on the selection of an *optimal* set of piecewise constant gains with which to approximate time-varying gain elements. The work is described in Ref. 5, and deals with the linear regulator problem whose formulation is similar to that of

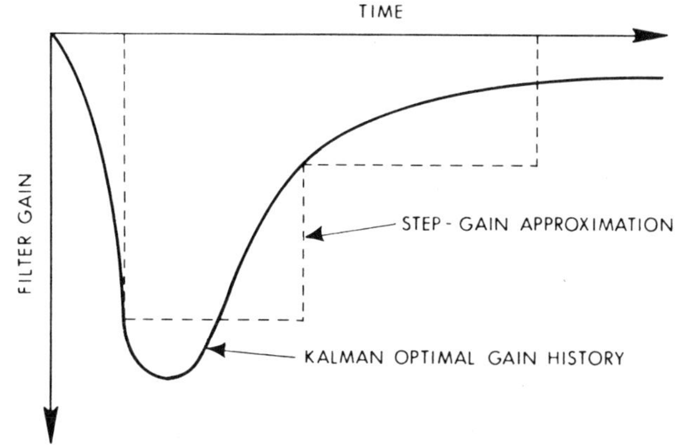

Figure 7.1-8 Fixed-Gain Approximations (Ref. 4)

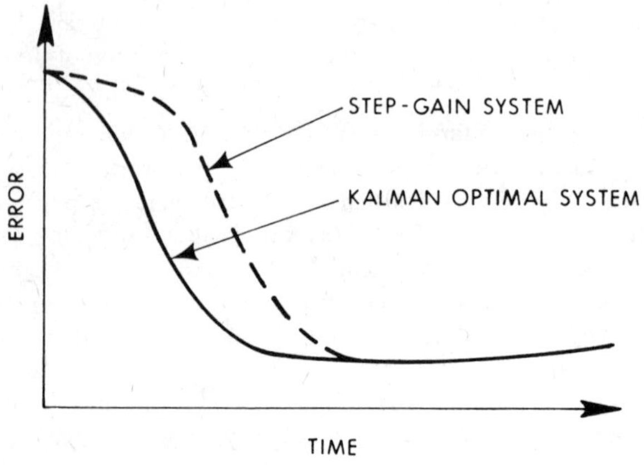

Figure 7.1-9 History of rms Error for Gain Choices Shown in Fig. 7.1-8 (Ref. 4)

optimal estimation. Briefly, a set of discrete time intervals – $t_i \leqslant t \leqslant t_{i+1}$; $i = 1, 2, \ldots N$ – is established and a constant controller gain matrix is sought for each of these intervals by minimizing the average quadratic cost function. A steepest descent technique is used to converge on the minimum. The same technique could be useful for selecting piecewise constant filter gains for a linear estimator. Figure 7.1-10 shows the one set of optimal piecewise constant gains, $k_1(t)$, chosen in Ref. 5 and the optimal continuous gain they replace. It can be seen that the problem was solved for several arbitrary subdivisions of the time scale. Note that the constant gains do not usually represent the average optimal continuous gain for the time interval they span.

Using Steady-State Gains – The limiting case of a set of piecewise constant gains is choosing each gain to be constant over all time; logical choices for the constants are the set of gains reached when the filter error covariance equations are allowed to achieve steady state. The gain matrix is simply given by

$$K = P_\infty H^T R^{-1}$$

where the H and R matrices, as well as the dynamics matrix F, must be constant for the steady-state error covariance, P_∞, to exist. More generally, $H^T R^{-1} H$ must be constant; this condition is seldom satisfied if H and R are not constant. In Chapter 4 it is shown that a Kalman filter which uses gains derived from the steady-state covariance is, in fact, identical with the Wiener filter.

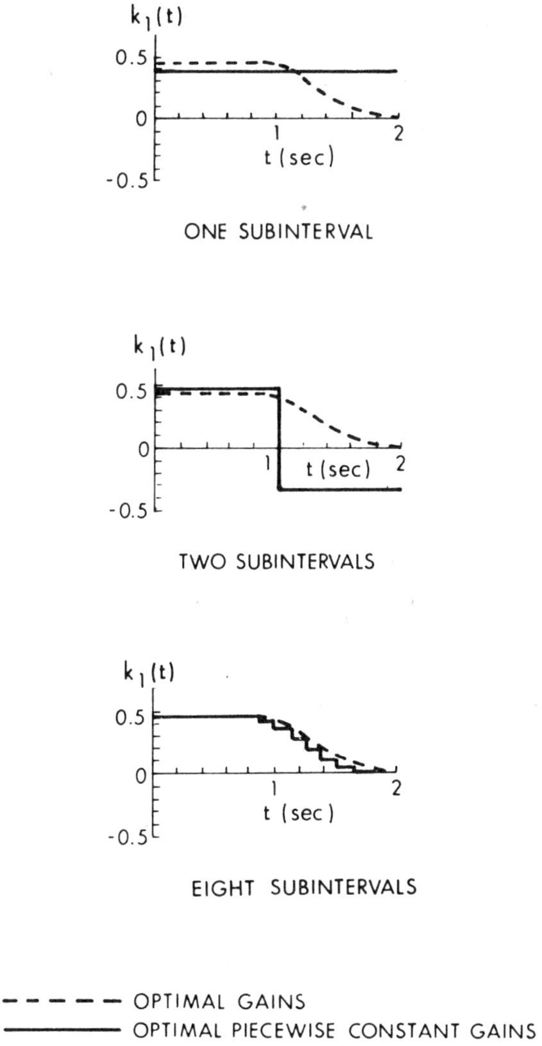

Figure 7.1-10 Optimal and Suboptimal Feedback Gains for Third-Order System (Ref. 5)

The use of steady-state gains assumes that measurements are available over a sufficiently long period of time such that a steady-state condition in the filter is achieved before critical (in terms of system accuracy) points in time are reached. This approach forfeits the rapid convergence capability of the filter, which depends largely on time-varying gains to weight the first few measurements heavily when initial uncertainty about the state value is high. How much time is required for the fixed gain filter errors to approach steady state (and therefore satisfy conditions for which the filter was designed) is, of course, a function of the particular problem. It is sufficient to say that, in most cases, the constant gain filter will be considerably slower in reaching steady state than the time-varying filter whose steady-state gains the former may be using. A lower bound on the convergence time of the fixed-gain filter can be determined by noting how long it takes for the steady-state covariance matrix, on which those fixed gains are based, to develop.

Example 7.1-3

Consider the problem of estimating a random walk from a noisy measurement

$$\dot{x} = w, \qquad w \sim N(0,q)$$

$$z = x + v, \qquad v \sim N(0,r)$$

The estimation error covariance equation for the optimal filter is simply

$$\dot{p} = -\frac{p^2}{r} + q$$

In the steady state $\dot{p}_\infty = 0$ and

$$p_\infty = \sqrt{rq}$$

$$k_\infty = \sqrt{\frac{q}{r}}$$

Observe that the steady-state gain weights the residuals, $(z - \hat{x})$, highly when the process noise is high and the measurement noise is low. Also, the steady-state error is driven by the process and measurement noises alone.

Consider the error covariance matrix equations for a continuous Kalman filter and a Wiener filter

$$\dot{P}_K = FP_K + P_K F^T + GQG^T - P_K H^T R^{-1} HP_K \qquad \text{(Kalman filter)}$$

$$\dot{P}_W = (F - K_\infty H)P_W + P_W(F - K_\infty H)^T + GQG^T + K_\infty R K_\infty{}^T \qquad \text{(Wiener filter)*}$$

where

$$K_\infty = P_\infty H^T R^{-1}$$

---

*An error covariance differential equation for a filter with the *structure* of a Kalman filter but an arbitrary gain K is

$$\dot{P} = (F - KH)P + P(F - KH)^T + GQG^T + KRK^T$$

Proof of this is left as an exercise for the reader (see Problem 7-7).

and P_∞ is the steady-state Kalman filter error covariance. Reference 6 shows that an upper bound on the Hilbert or spectral norm (Ref. 7)* of the difference between the Kalman and Wiener error covariances is given by

$$\| P_W(t) - P_K(t) \| \leqslant \frac{\| P(0) - P_\infty \|^2 \; \| H^T R^{-1} H \|}{8 \, | \alpha_{max} |} \tag{7.1-15}$$

where P(0) is the initial error covariance and α_{max} is the maximum real part of the eigenvalues of $(F - K_\infty H)$. Equation (7.1-15) indicates that an upper bound on the difference between the Kalman and Wiener filter errors is large when the difference between the initial and steady-state errors, $\delta P(0) = P(0) - P_\infty$, is large. Also, α_{max} is the inverse of the largest time constant in the Wiener filter; the bound on δP varies with that time constant. The matrix $H^T R^{-1} H$ is recognized as being related to the information contained in each measurement, from the matrix inversion relationship, Eq. (4.2-19), of Chapter 4.

Example 7.1-4

Consider the Kalman and Wiener filters for the scalar system

$$\dot{x} = ax + w, \qquad w \sim N(0,q)$$

$$z = bx + v, \qquad v \sim N(0,r)$$

The Kalman filter error covariance equation is

$$\dot{p}_K(t) = 2ap_K(t) + q - \frac{b^2 p_K{}^2(t)}{r}, \qquad p(0) = p_0$$

The solution of this equation is, from Problem 4-11,

$$p_K(t) = \frac{(ap_0 + q)\sinh\beta t + \beta p_0 \cosh\beta t}{\left(\frac{b^2}{r} p_0 - a\right)\sinh\beta t + \beta\cosh\beta t}$$

where

$$\beta = a\sqrt{1 + \frac{b^2 q}{a^2 r}}$$

---

*The Hilbert norm of a matrix M, denoted $\|M\|$, is given by

$$\|M\| = \sqrt{\lambda_{max}(M^T M)}$$

where $\lambda_{max}(M^T M)$ is the largest eigenvalue of $M^T M$.

It can be shown by limiting arguments that

$$p_\infty = \frac{ar}{b^2}\left(1 + \frac{\beta}{a}\right)$$

The gain of the Wiener filter is thus

$$k_\infty = \frac{a + \beta}{b}$$

Using Eq. (7.1-15), the upper bound on $\delta p(t) = p_W(t) - p_K(t)$ is

$$\|\delta p(t)\| \leqslant \frac{\|p_0 - p_\infty\|^2 \, \|\frac{b^2}{r}\|}{8|\alpha_{max}|}$$

which, given the properties of the Hilbert norm,* becomes

$$\delta p(t) \leqslant \frac{(p_0 - p_\infty)^2 \, b^2}{8r\beta}$$

MINIMUM SENSITIVITY DESIGN

In many problems dynamic and statistical parameters are known only to lie in certain bounded ranges. A game-theoretic approach to filter design in the presence of these parameter uncertainties has been formulated and studied in Refs. 8 and 9. Using any of the previously described state reduction techniques, a suboptimal filter form is selected *a priori*, but the filter parameters are left unspecified. Any filter performance measure is now a function of *uncertain* real world system parameters and the *unspecified* filter parameters. Let us denote the uncertain real world system parameters by a vector $\underline{\alpha}$ and the unspecified filter parameters by the vector $\underline{\beta}$. It is assumed that $\underline{\alpha}$ and $\underline{\beta}$ lie in closed bounded sets, A and B, respectively. A convenient *scalar* measure for any suboptimal filter performance is then

$$J(\underline{\alpha}, \underline{\beta}) = \text{Trace}\,[MP] \tag{7.1-16}$$

where M is a positive definite weighting matrix, included to balance the importance one assigns to *each* system error and P is the filter error covariance matrix. $J(\underline{\alpha}, \underline{\beta})$ is then simply a weighted sum of all system errors. For a given $\underline{\alpha}$, the minimum value of J, denoted $J_0(\underline{\alpha})$, is attained by the Kalman filter and is a function of $\underline{\alpha}$ alone. Clearly, $J(\underline{\alpha}, \underline{\beta}) \geqslant J_0(\underline{\alpha})$ for all $\underline{\alpha} \in A$ and all $\underline{\beta} \in B$. Since $\underline{\alpha}$ is unknown and $\underline{\beta}$ alone is available for selection by the designer, it seems most natural to view $\underline{\alpha}$ and $\underline{\beta}$ as adversaries in the game-theoretic sense. With this in

*From the definition of the Hilbert norm, when M is a scalar, m,

$$\|m\| = m$$

mind, three sensitivity measures and their associated rules of synthesis are appropriate. They are:

$$S_1 = \min_{\underline{\beta} \epsilon B} \max_{\underline{\alpha} \epsilon A} J(\underline{\alpha}, \underline{\beta})$$

$$S_2 = \min_{\underline{\beta} \epsilon B} \max_{\underline{\alpha} \epsilon A} [J(\underline{\alpha}, \underline{\beta}) - J_0(\underline{\alpha})]$$

$$S_3 = \min_{\underline{\beta} \epsilon B} \max_{\underline{\alpha} \epsilon A} \left[\frac{J(\underline{\alpha}, \underline{\beta}) - J_0(\underline{\alpha})}{J_0(\underline{\alpha})}\right]$$

The S_1 design simply minimizes the maximum value of J over the parameter set $\underline{\alpha}$. This places an *upper bound* on the cost, and might be interpreted as a "worst case" design. The second and third criteria minimize the maximum absolute and relative deviations, respectively, of the filter error from optimum, over the parameter set $\underline{\alpha}$. Thus, the S_2 and S_3 criteria force the filter error to track the optimum error within some tolerance over the entire set $\underline{\alpha} \epsilon A$. In each case the above procedures yield a fixed value of $\underline{\beta}$ and, therefore, a fixed filter design good for all values of the uncertain parameters $\underline{\alpha}$.

Research to date has concentrated on the design of optimally insensitive filters in the presence of uncertain system and measurement noise statistics. Specifically, it has been assumed that elements of the system and measurement noise covariance matrices, Q and R respectively, were unknown. These unknown elements then constitute the vector $\underline{\alpha}$. For the S_1 filter, which is the easiest to find, a rule of thumb is available: A good initial guess for $\underline{\alpha}$ is that value which maximizes $J_0(\underline{\alpha})$, denoted $\underline{\alpha}'$ – i.e.,

$$J_0(\underline{\alpha}') = \max_{\underline{\alpha} \epsilon A} J_0(\underline{\alpha})$$

When the filter state and the real world state are of the same dimension, this result is exact and the S_1 filter is simply the Kalman filter for $\underline{\alpha}'$.

Example 7.1-5

Consider a first-order plant with a noisy measurement

$$\dot{x} = -x + w, \qquad w \sim N(0,q), \qquad 0 \leqslant q \leqslant 1$$

$$z = x + v, \qquad v \sim N(0,r), \qquad 0 \leqslant r \leqslant 1$$

Given the filter equation

$$\dot{\hat{x}} = -\hat{x} + k(z - \hat{x}); \qquad \text{k unspecified}$$

select $J = p_\infty$, where p_∞ is the steady-state value of the filter error covariance, and perform the minimization over the range $\beta = k$, $k > -1$ for stability and the maximization over the range of q and r. [Note that this is the same measure of filter performance expressed in Eq. (7.1-16)]. Then

$$J = \frac{k^2 r + q}{2(1+k)} \; ; \; J_0 = (r^2 + rq)^{1/2} - r \tag{7.1-17}$$

The S_1 value of J, $S_1 = \min(k > -1) \max(q,r) [J(k,q,r)]$, occurs at $q = r = 1$ with $k = \sqrt{2} - 1$. The S_2 criterion is satisfied when $q = 0$, $r = 1$ and $q = 1$, $r = 0$ with $k = 1$. Since $J_0 = 0$ when r or q are zero, the argument of the S_3 criterion is infinite at those boundaries of the r,q space under consideration, and the filter satisfying the S_3 criterion does not exist in this example.

Example 7.1-6

Consider the continuous first-order markov process with noisy measurements ($\beta > 0$)

$$\dot{x} = -\beta x + w, \qquad w \sim N(0,q)$$

$$z = x + v, \qquad v \sim N(0,r)$$

Assuming that the value of β is uncertain, a filter identical in form to the Kalman filter is chosen to estimate x, namely

$$\dot{\hat{x}} = -\beta_f \hat{x} + k(z - \hat{x})$$

where it is required for stability that $\beta_f + k > 0$. The covariance of the estimation error for this simple problem has a steady-state value, denoted here as p_∞ and given by

$$p_\infty = \frac{k^2 r + q}{2(\beta_f + k)} + \frac{(\beta_f^2 - \beta^2) q}{2\beta(\beta_f + k)(\beta + \beta_f + k)}$$

Observe that p_∞ is a function of both the unknown β and the unspecified filter parameters β_f and k. The S_1, S_2 and S_3 filters described above were determined for this example for the case where $q = 10$, $r = 1$ and $0.1 \leqslant \beta \leqslant 1$.

Figure 7.1-11 shows the error performance of the optimal S_1, S_2, and S_3 filters versus the true value of β. The performance of a Kalman filter designed for $\beta = 0.5$ is also illustrated. Observe that the S_1 filter has the smallest maximum error, while the S_2 and S_3 filters tend to track the optimal performance more closely. The S_3 error covariance, for example, is everywhere less than 8% above optimum and is only 0.4% from optimum at $\beta = 0.3$. By comparison, the Kalman filter designed for a nominal value of β equal to 0.5 degrades rapidly as the true value of β drops below 0.5.

This example shows that the minimax filters can achieve near optimal performance over wide parameter variations. The filter implemented is identical in form to the Kalman filter and thus avoids the additional mechanization complexity of adaptive schemes sometimes suggested.

7.2 SENSITIVITY ANALYSIS: KALMAN FILTER

The statistical (or covariance) analysis which determines the true behavior of estimation errors in a suboptimal linear filter is necessarily more complex than the work presented in previous chapters. The equations are developed below. Emphasis here is placed on the technical aspects of the analysis, rather than the

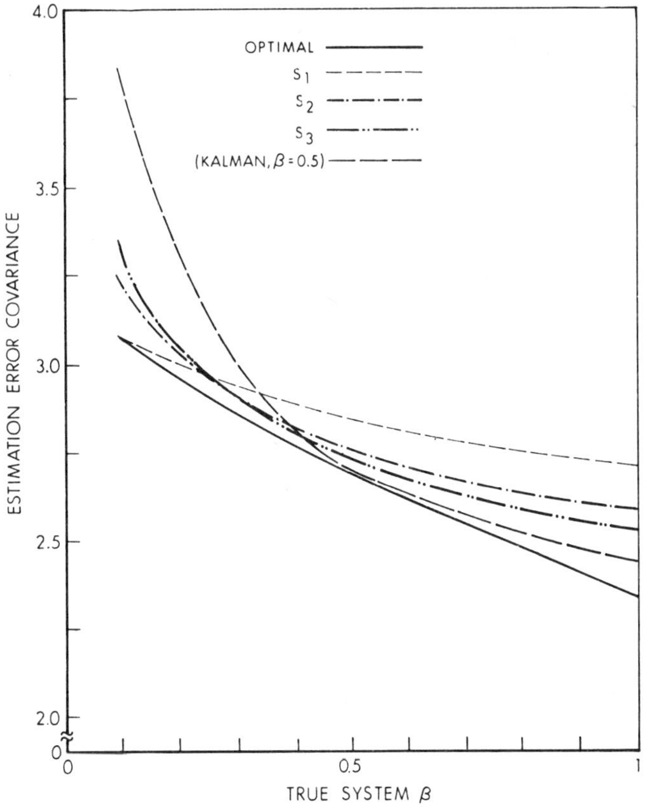

Figure 7.1-11 Estimation Error Covariance Comparison as a Function of True System Bandwidth (Ref. 9)

motivation. Also, the equations developed, while correct, are not necessarily most efficient, because emphasis here is placed on the tutorial presentation of the subject. References 10 and 11 provide alternative sensitivity equations which may require less computer space and time.

Observe the structure of the Kalman filter, illustrated in Fig. 7.2-1, for the continuous case; the filter contains an *exact model* of the system dynamics and measurement process. Additionally, note that the filter gain matrix is calculated using the exact models of dynamics and measurement and exact knowledge of the process noise covariance (and the influence matrix G), measurement error covariance, and initial estimation error covariance.

There are two broad questions we can ask with respect to the sensitivity of the filter: "How does the error covariance behave if we make approximations in computing the gain matrix K, but use the *correct* values of F and H in the implemented filter?" and "How does the error covariance behave if we compute

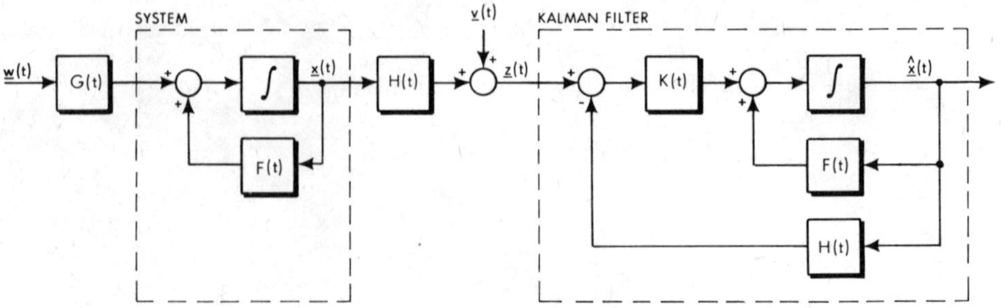

Figure 7.2-1 Block Diagram of the Continuous Filter Equations

the gain matrix in some manner (optimal or otherwise), and use *wrong* values of F and H in the implemented filter?" The first question is relatively easy to answer, while the second question requires a considerable amount of extra calculation.

EXACT IMPLEMENTATION OF DYNAMICS AND MEASUREMENTS

The error covariance relationships for a discrete filter with the same structure as the Kalman filter, but with an *arbitrary* gain matrix, are (see Problem 7-6):

$$P_k(+) = (I-K_kH_k)\,P_k(-)\,(I-K_kH_k)^T + K_kR_kK_k^T \tag{7.2-1}$$

$$P_{k+1}(-) = \Phi_kP_k(+)\,\Phi_k^T + Q_k \tag{7.2-2}$$

A single equation describes the corresponding error propagation for the continuous filter (see Problem 7-7), viz:

$$\dot{P} = (F-KH)P + P(F-KH)^T + GQG^T + KRK^T \tag{7.2-3}$$

Under the assumptions with which we are presently dealing, Fig. 7.2-1 (which illustrates the similarity between the filter and the system) can be rearranged to provide a corresponding block diagram for estimation *error* dynamics, shown in Fig. 7.2-2. A similar error block diagram for the discrete filter is shown in Fig. 7.2-3. The error equations illustrated in these two figures are used in the derivation of Eqs. (7.2-1), (7.2-2) and (7.2-3).

If the expressions for the optimal gain matrix provided earlier are substituted into Eqs. (7.2-1) and (7.2-3), they reduce to the previously stated covariance relations for the optimal filter, in which the K matrix does not appear explicitly.

Equations (7.2-3) or (7.2-1) and (7.2-2), together with the covariance equations for the optimal filter and the definition of the gain matrix, can be used to determine the effect of using incorrect values of F, H, G, R or P(0) in the calculation of the Kalman gain matrix. The procedure involves two steps which can be performed either simultaneously or in sequence. In the latter case K is computed and stored for later use in Eqs. (7.2-1) or (7.2-3).

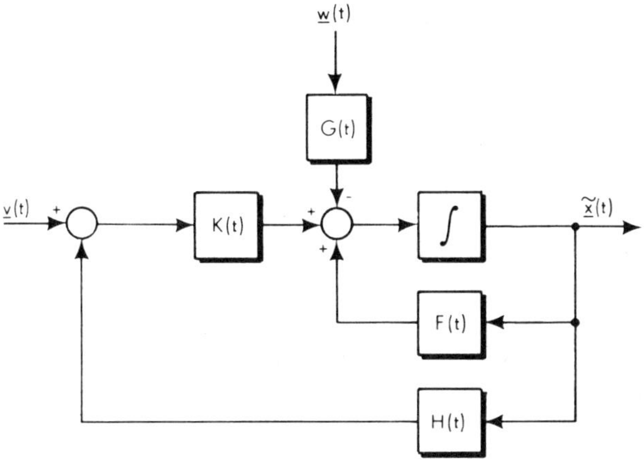

Figure 7.2-2 Block Diagram of Estimation Error Dynamics of a Continuous Filter: System Dynamics and Measurement Process Perfectly Modeled in the Filter

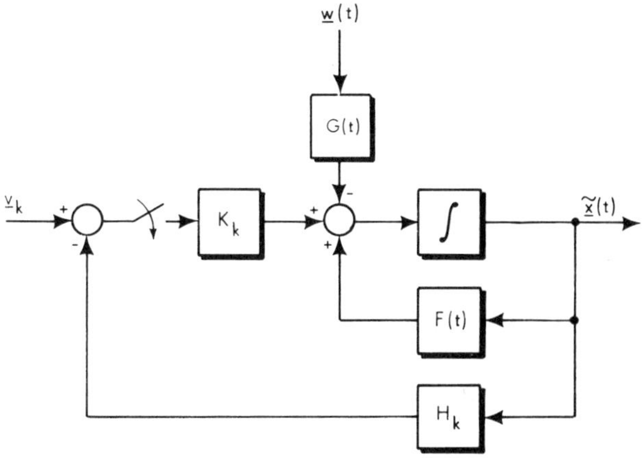

Figure 7.2-3 Block Diagram of Estimation Error Dynamics of a Discrete Filter: System Dynamics and Measurement Process Perfectly Modeled in the Filter

Step 1: In previous chapters the following error covariance equations, in which the filter gain matrix, K, does not appear explicitly, were derived. For the discrete-time case,

$$P_k(+) = P_k(-) - P_k(-) H_k^T (H_k P_k(-) H_k^T + R_k)^{-1} H_k P_k(-) \tag{7.2-4}$$

$$P_{k+1}(-) = \Phi_k P_k(+) \Phi_k^T + Q_k \tag{7.2-5}$$

or, for the continuous-time case,

$$\dot{P} = FP + PF^T - PH^T R^{-1} HP + GQG^T \tag{7.2-6}$$

Using Eqs. (7.2-4) and (7.2-5), or Eq. (7.2-6), and *design** values of F, H, G, Q, R and P(0), compute the error covariance history. Also, using the equations from previous discussions, compute the filter gain matrix which would be optimal *if the design values were correct.*

Step 2: Inserting the gain matrix computed in Step 1 into Eq. (7.2-1) or (7.2-3), and using the correct values of F, H, G, Q, R and P(0) (which are to be implemented in the filter), compute the "actual" error covariance history.

Because Eqs. (7.2-1), (7.2-2) and (7.2-3) are based only on the *structure* of the Kalman filter, and not on any assumption that the optimal gain matrix is employed, they can be used to analyze the filter error covariance for any set of filter gains. This permits investigation of proposed sets of precomputed gains or simplified filter gains, such as decaying exponentials, etc., assuming the correct F (or Φ) and H matrices are implemented in the filter. In that case, the gain matrix is simply inserted in Eqs. (7.2-1) or (7.2-3) and, using correct values for F, H, G, Q, R and P(0), Eqs. (7.2-1) and (7.2-2), or (7.2-3), are used to compute the error covariance.

INCORRECT IMPLEMENTATION OF DYNAMICS AND MEASUREMENTS

Answering the second question posed in the introduction to this section is more difficult. It is tempting to compute a filter gain matrix based on Eqs. (7.2-4) and (7.2-5) or (7.2-6) and a set of design values, and insert it into Eq. (7.2-1) or (7.2-3); essentially, this is following the same procedure as outlined above. The fallacy of this approach is that Eqs. (7.2-1) through (7.2-6) are all based on the assumption that the system dynamics and measurement process are *identical* in the Kalman filter implementation and the real world – the set of circumstances treated in the previous section.

---

*Design values are those used to derive the filter gain matrix. The filter *implementation* requires specification of system dynamics and measurement matrices, which do not necessarily have to be the same as the corresponding matrices used in filter gain matrix design.

Continuous Filter – The procedure for deriving error covariance equations which can be used to answer the second question posed is quite similar to that used in the earlier derivations in Chapters 3 and 4. The error sensitivity equations for the continuous filter are derived as follows: From Chapter 4 we know that the equations for the state and the estimate are given by

$$\dot{\underline{x}} = F\underline{x} + G\underline{w}$$

$$\underline{z} = H\underline{x} + \underline{v}$$

$$\dot{\hat{\underline{x}}} = F^*\hat{\underline{x}} + K^*\,[\underline{z} - H^*\hat{\underline{x}}] \tag{7.2-7}$$

where the asterisked quantities K*, H* and F* represent the filter gain and the measurement process and system dynamics *implemented* in the filter; H and F represent the actual measurement process and system dynamics. In this derivation, it is assumed that the state variables estimated and those needed to completelv describe the process are *identical*. The sensitivity equations for the case when the estimated state is a subset of the entire state are provided at the end of the derivation. Equation (7.2-7) illustrates the fact that the actual system dynamics and measurement process, represented by the matrices F and H, are not faithfully reproduced in the filter – i.e., $F \neq F^*$, $H \neq H^*$. The error in the estimate, $\tilde{\underline{x}} = \hat{\underline{x}} - \underline{x}$, obeys, from Eq. (7.2-7),

$$\dot{\tilde{\underline{x}}} = (F^* - K^*H^*)\hat{\underline{x}} - (F - K^*H)\underline{x} - G\underline{w} + K^*\underline{v} \tag{7.2-8}$$

Letting

$$\Delta F \triangleq F^* - F \qquad \text{and} \qquad \Delta H \triangleq H^* - H$$

and recalling the relation between $\underline{x}$, $\tilde{\underline{x}}$ and $\hat{\underline{x}}$, Eq. (7.2-8) can be written as

$$\dot{\tilde{\underline{x}}} = (F^* - K^*H^*)\tilde{\underline{x}} + (\Delta F - K^*\Delta H)\underline{x} - G\underline{w} + K^*\underline{v} \tag{7.2-9}$$

A new vector, $\underline{x}'$, is defined by

$$\underline{x}' \triangleq \begin{bmatrix} \tilde{\underline{x}} \\ \hline \underline{x} \end{bmatrix}$$

The differential equation for $\underline{x}'$ in vector-matrix form is, from Eqs. (7.2-7) and (7.2-9),

$$\overbrace{\begin{bmatrix} \dot{\tilde{\underline{x}}} \\ \hline \dot{\underline{x}} \end{bmatrix}}^{\dot{\underline{x}}'} = \overbrace{\left[\begin{array}{c|c} F^* - K^*H^* & \Delta F - K^*\Delta H \\ \hline 0 & F \end{array}\right]}^{F'} \overbrace{\begin{bmatrix} \tilde{\underline{x}} \\ \hline \underline{x} \end{bmatrix}}^{\underline{x}'} + \overbrace{\begin{bmatrix} K^*\underline{v} - G\underline{w} \\ \hline G\underline{w} \end{bmatrix}}^{\underline{w}'} \tag{7.2-10}$$

Note that Eq. (7.2-10) is in the form of the differential equation for the state of a linear system driven by white noise. From Chapter 4, the covariance equation for $\underline{x}'$ is

$$\frac{d(E[\underline{x}'\underline{x}'^T])}{dt} = F'E[\underline{x}'\underline{x}'^T] + E[\underline{x}'\underline{x}'^T]\,F'^T + E[\underline{w}'\underline{w}'^T] \tag{7.2-11}$$

The quantity of interest, the covariance of $\tilde{\underline{x}}$, is the upper left corner of the covariance matrix of $\underline{x}'$:

$$E[\underline{x}'\underline{x}'^T] = \left[\begin{array}{c|c} P & V^T \\ \hline V & U \end{array}\right] \tag{7.2-12}$$

where

$$P \triangleq E[\tilde{\underline{x}}\tilde{\underline{x}}^T] \quad V \triangleq E[\underline{x}\tilde{\underline{x}}^T] \quad U \triangleq E[\underline{x}\underline{x}^T]$$

Combining Eqs. (7.2-11) and (7.2-12) and expressing the expected value of $\underline{w}'\underline{w}'^T$ in terms of the spectral density matrices Q and R,

$$\begin{aligned}\left[\begin{array}{c|c} \dot{P} & \dot{V}^T \\ \hline \dot{V} & \dot{U} \end{array}\right] &= \left[\begin{array}{c|c} F^* - K^*H^* & \Delta F - K^*\Delta H \\ \hline O & F \end{array}\right] \left[\begin{array}{c|c} P & V^T \\ \hline V & U \end{array}\right] \\ &+ \left[\begin{array}{c|c} P & V^T \\ \hline V & U \end{array}\right] \left[\begin{array}{c|c} (F^* - K^*H^*)^T & O \\ \hline (\Delta F - K^*\Delta H)^T & F^T \end{array}\right] \\ &+ \left[\begin{array}{c|c} GQG^T + K^*RK^{*T} & -GQG^T \\ \hline -GQG^T & GQG^T \end{array}\right]\end{aligned} \tag{7.2-13}$$

Breaking Eq. (7.2-13) into its component parts, the error sensitivity equations become

$$\begin{aligned}\dot{P} = {} & (F^*-K^*H^*)P + P(F^*-K^*H^*)^T + (\Delta F - K^*\,\Delta H)V \\ & + V^T(\Delta F - K^*\,\Delta H)^T + GQG^T + K^*RK^{*T}\end{aligned} \tag{7.2-14}$$

$$\dot{V} = FV + V(F^*-K^*H^*)^T + U(\Delta F - K^*\,\Delta H)^T - GQG^T \tag{7.2-15}$$

$$\dot{U} = FU + UF^T + GQG^T \tag{7.2-16}$$

Since the initial uncertainty in the estimate is identical with the uncertainty in the state,

$$P(0) = V(0) = U(0) = E[\underline{x}(0)\underline{x}(0)^T]$$

Note that when the *actual* system dynamics and measurement process are implemented ($\Delta F = 0$, $\Delta H = 0$), Eqs. (7.2-14), (7.2-15) and (7.2-16) reduce to Eq. (7.2-3), as expected. There can be only one gain matrix of interest here, the one implemented in the filter, consequently, $K^* = K$.

If the state vector of the implemented filter contains fewer elements than the correct state vector, F and F* and H and H* will not be compatible in terms of their dimensions. They can be made compatible by the use of an appropriate nonsquare matrix. For example, if the state vector implemented consists of the first four elements in a six-element true state vector, F is 6×6 and F* is 4×4. The difference matrix ΔF can be defined by

$$\Delta F = W^T F^* W - F \tag{7.2-17}$$

where the matrix W accounts for the dimensional incompatibility between F and F*,

$$W = \begin{bmatrix} 1 & 0 & 0 & 0 & 0 & 0 \\ 0 & 1 & 0 & 0 & 0 & 0 \\ 0 & 0 & 1 & 0 & 0 & 0 \\ 0 & 0 & 0 & 1 & 0 & 0 \end{bmatrix}$$

The matrix W is, in fact, the transformation between the true state and that implemented,

$$\underline{x}_{\text{implemented}} = W\underline{x}_{\text{true}}$$

A similar transformation can make H* compatible with H,

$$\Delta H = H^* W - H \tag{7.2-18}$$

The sensitivity equations for a continuous filter which does not estimate the entire state vector are

$$\begin{aligned} \dot{P} &= W^T(F^*-K^*H^*)WP + PW^T(F^*-K^*H^*)^T W + (\Delta F - W^T K^* \Delta H)V \\ &\quad + V^T(\Delta F - W^T K^* \Delta H)^T + GQG^T + W^T K^* R K^{*T} W \\ \dot{V} &= FV + VW^T(F^*-K^*H^*)^T W + U(\Delta F - W^T K^* \Delta H)^T - GQG^T \\ \dot{U} &= FU + UF^T + GQG^T \end{aligned} \tag{7.2-19}$$

where ΔF and ΔH are defined in Eqs. (7.2-17) and (7.2-18). A more complex set of sensitivity equations, covering the case where the filter state is a general linear combination of the true states, is given in Refs. 10 and 11 for both continuous and discrete linear filters.

It should be emphasized that the only purpose of the transformation W is to account for the elimination of certain state variables in the implementation. The approach illustrated here will not cover the situation where *entirely different* sets of state variables are used.

Discrete Filter – The sensitivity equations for the discrete filter which correspond to Eqs. (7.2-14) through (7.2-16) are, *between* measurements,

$$
\begin{aligned}
P_{k+1}(-) &= \Phi_k{}^* P_k(+)\, \Phi_k{}^{*T} + \Phi_k{}^* V_k(+) \Delta\Phi_k{}^T \\
&\quad + \Delta\Phi_k V_k(+)\, \Phi_k{}^{*T} + \Delta\Phi_k U_k \Delta\Phi_k{}^T + Q_k \\
V_{k+1}(-) &= \Phi_k V_k(+)\, \Phi_k{}^{*T} + \Phi_k U_k \Delta\Phi_k{}^T - Q_k \\
U_{k+1} &= \Phi_k U_k \Phi_k{}^T + Q_k
\end{aligned}
\tag{7.2-20}
$$

and *across* a measurement,

$$
\begin{aligned}
P_k(+) &= (I - K_k{}^* H_k{}^*) P_k(-)\, (I - K_k{}^* H_k{}^*)^T - (I - K_k{}^* H_k{}^*) V_k{}^T(-) \Delta H_k{}^T K_k{}^{*T} \\
&\quad - K_k{}^* \Delta H_k V_k(-)\, (I - K_k{}^* H_k{}^*)^T + K_k{}^* \Delta H_k U_k \Delta H_k{}^T K_k{}^{*T} \\
&\quad + K_k{}^* R_k K_k{}^{*T} \\
V_k(+) &= V_k(-)\, (I - K_k{}^* H_k{}^*)^T - U_k \Delta H_k{}^T K_k{}^{*T}
\end{aligned}
\tag{7.2-21}
$$

where

$$
\Delta\Phi \triangleq \Phi^* - \Phi
$$

When $\Delta\Phi$ and ΔH are zero, Eqs. (7.2-20) and (7.2-21) become identical with Eqs. (7.2-2) and (7.2-1), respectively.

It is worth noting here that if all the states whose values are estimated by the filter are instantaneously adjusted after each measurement according to

$$
\underline{x}_k(+) = \underline{x}_k(-) - \hat{\underline{x}}_k(+) \tag{7.2-22}
$$

the estimate of the state following such a correction must be zero. In that case, the equation for propagating the estimate between states and incorporating the next estimate is

$$
\hat{\underline{x}}_{k+1}(+) = K_k \underline{z}_k \tag{7.2-23}
$$

and there is no need for specifying filter matrices Φ^* and H^*. Equations (7.2-1) and (7.2-2) are the equations which apply in this case. Instantaneous adjustments can be made if the states being estimated are formed from variables stored in a digital computer.

7.3 SENSITIVITY ANALYSIS EXAMPLES

This section provides the reader with several examples of the application of sensitivity analysis.

Example 7.3-1

Consider a stationary, first-order markov process where the state is to be identified from a noisy measurement:

$$\dot{x}(t) = -\beta x + w, \qquad w \sim N(0,q)$$

$$z(t) = x + v, \qquad v \sim N(0,r) \tag{7.3-1}$$

Equation (7.2-3) becomes the scalar equation

$$\dot{p} = (-\beta - k)\, p + p(-\beta - k) + q + k^2 r \tag{7.3-2}$$

In the steady state, we set $\dot{p}_\infty = 0$, and thus find

$$p_\infty = \frac{q + k^2 r}{2(\beta + k)} \tag{7.3-3}$$

The optimal k(denoted k_0) is computed as $p_\infty h/r$ or, equivalently, from

$$\frac{\partial p_\infty}{\partial k} = 0$$

both of which yield

$$k_0 = -\beta + \sqrt{\beta^2 + q/r} \tag{7.3-4}$$

The variation of p_∞ as a function of k is illustrated in Fig. 7.3-1.

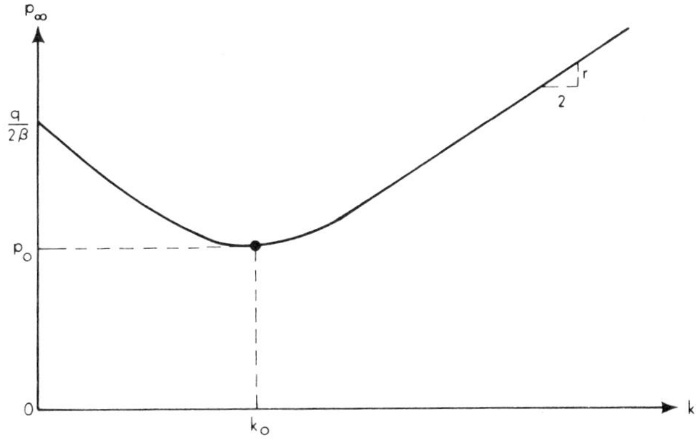

Figure 7.3-1 Sensitivity Curve, p_∞ vs k

If the filter is designed for particular values of r and q, k is fixed according to Eq. (7.3-4). It can be seen from Eq. (7.3-3) that, with k *fixed*, the steady-state error covariance varies linearly with the actual process noise spectral density or measurement error spectral density. This is illustrated in Fig. 7.3-2.

If the true noise variances are assumed fixed and the *design* values of q and r are varied, quite different curves result. Any deviation of the design variances, and consequently k, from the correct values will cause an increase in the filter error variance. This is a consequence of the optimality of the filter, and is illustrated in Fig. 7.3-3. The sensitivities of the first-order process shown in Figs. 7.3-2 and 7.3-3 are similar to those observed in higher-order systems.

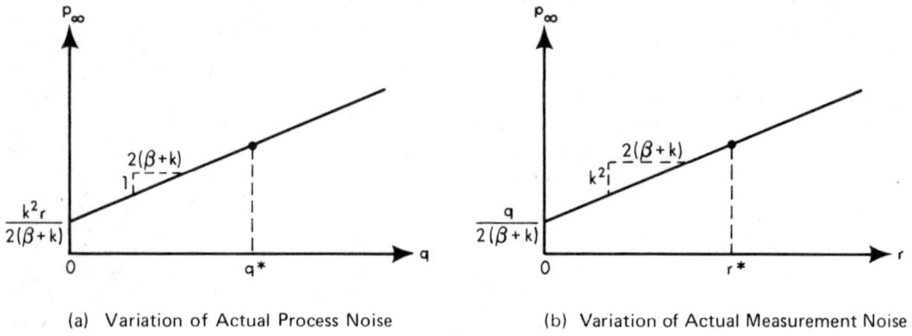

Figure 7.3-2 Effect of Changing Actual Noise Variances

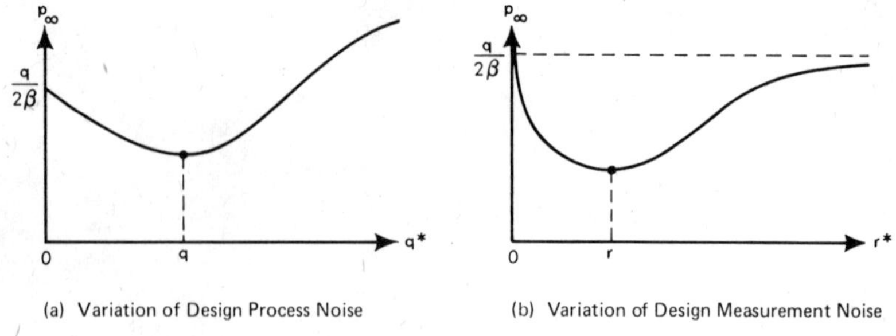

Figure 7.3-3 Effects of Changing Design Values of Noise Variances

Example 7.3-2

Continuing with the example, let us suppose that there is an uncertainty in the value of the coefficient β in the system dynamics. Furthermore, let us suppose that the filter designer selects a design value β^* and is certain of the mean square magnitude, σ^2, of the state x. Consequently, the designer always picks $q^* = 2\beta^*\sigma^2$. In this case, the steady-state estimation error covariance can be shown to be

$$p_\infty = \frac{1}{\lambda(\beta+\lambda)}\left\{\sigma^2\left[(\beta+\beta^*)\lambda+(\beta-\beta^*)\beta^*\right]+r\beta^*(\lambda-\beta^*)(\beta^*-\beta)\right\} \qquad (7.3\text{-}5)$$

where $\lambda = \sqrt{\beta^{*2} + 2\beta^*\sigma^2/r}$

Figure 7.3-4 is a plot of p_∞, as a function of β^*, for the case $\beta = r = \sigma^2 = 1$. Notice the similarity with Fig. 7.3-3.

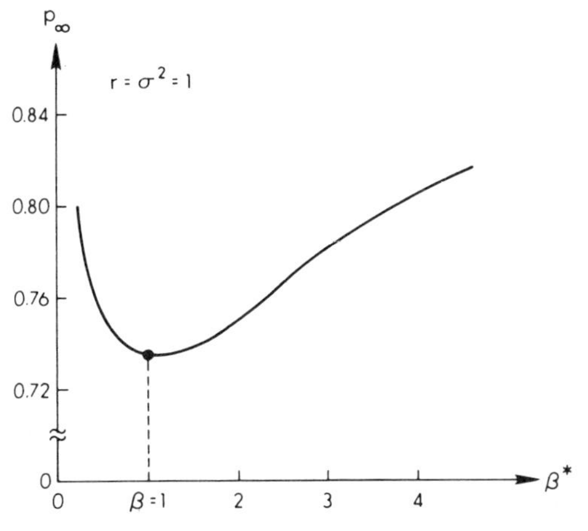

Figure 7.3-4 Effect of Varying Design Value of System Dynamics Parameter

Example 7.3-3

To illustrate the effect of different system dynamics models, we consider the system shown in Figure 7.3-5a. The state of interest is influenced by the forcing function. A measurement, several integrations removed, is used to estimate the system state. The two possible models for the forcing function considered are: a first-order markov process and a random ramp, both of which are illustrated in Fig. 7.3-5b. Figures 7.3-6, 7.3-7, and 7.3-8 illustrate the results of research into the importance of that model. In Fig. 7.3-6, the filter has been designed with the forcing function modeled as a first-order markov process but the actual forcing function is a random ramp. The rms error in the estimate of the state of interest is shown as a function of the rms ramp slope γ. In Fig. 7.3-7, the filter has been redesigned with the forcing function modeled as a random ramp *and* the value of γ in the filter design is assumed to be always correct (i.e., the filter and actual values of γ coincide for each value of γ shown; in this sense the filter is always optimal).

It can be seen by comparing Figs. 7.3-6 and 7.3-7 that having the correct system model makes a dramatic improvement in filter performance. However, the situation depicted in Fig. 7.3-7 is optimistic since the designer cannot hope to have perfect information about the rms slope of the random ramp. Figure 7.3-8 illustrates a more realistic situation in which the designer has picked a nominal value for γ (i.e., 10^{-6}). The effect of the existence of other values of γ is shown. It is evident that some deterioration in performance takes place when the actual forcing function is greater than that presumed in the filter design. However, it is interesting to note, by comparing Figs. 7.3-6, 7.3-7 and 7.3-8, that the *form* chosen for the system model has a much greater impact on accuracy in this case than the numerical parameters used to describe the magnitude of the forcing function.

Example 7.3-4

We conclude this section with an illustration of how the sensitivity analysis equation, developed in Section 7.2, can be used to help choose *which* states will provide the most efficient suboptimal filter and what deterioration from the optimal filter will be experienced. Figure 7.3-9 is drawn from a study of a multisensor system in which the complete (optimal) filter would have 28 state variables. It shows the effect of deleting state variables from the optimal filter. A judicious reduction from 28 to 16 variables produces less than a 1% increase in the error in the estimate of an important parameter, suggesting that the 12 states deleted were not significant. Overlooking for the moment the case of 13 state variables, we can see that further deletions of variables result in increasing estimation errors. But even the removal of 19 states can give rise to only a 5% increase in estimation error.

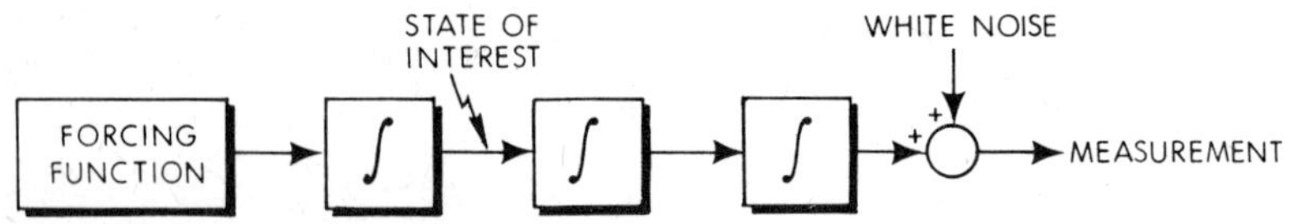

(a) The System

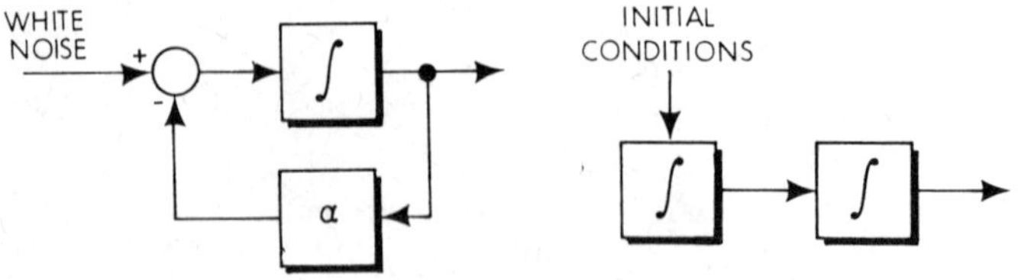

(b) Forcing Functions Considered

Figure 7.3-5 System Used to Illustrate the Effect of Different System Dynamics Modeled

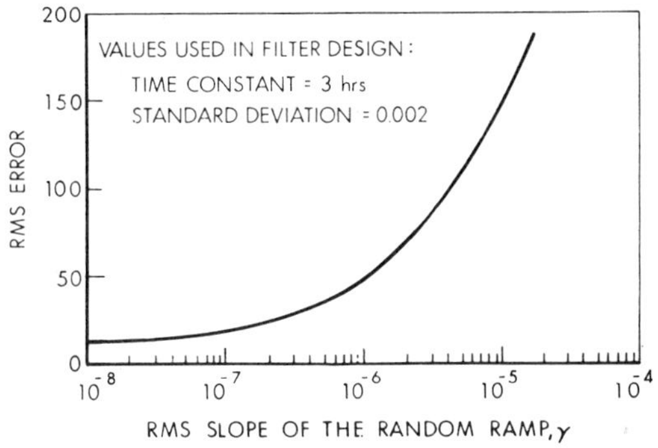

Figure 7.3-6 Filter Based on Exponential Correlation

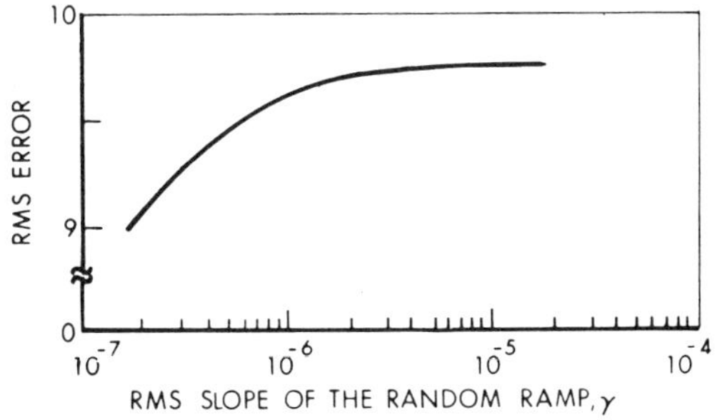

Figure 7.3-7 Filter Based on Ramp with Correct rms Slope

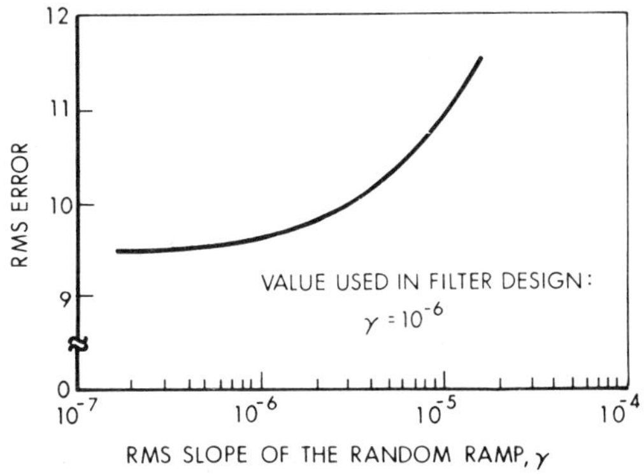

Figure 7.3-8 Filter Based on Ramp with Fixed rms Slope

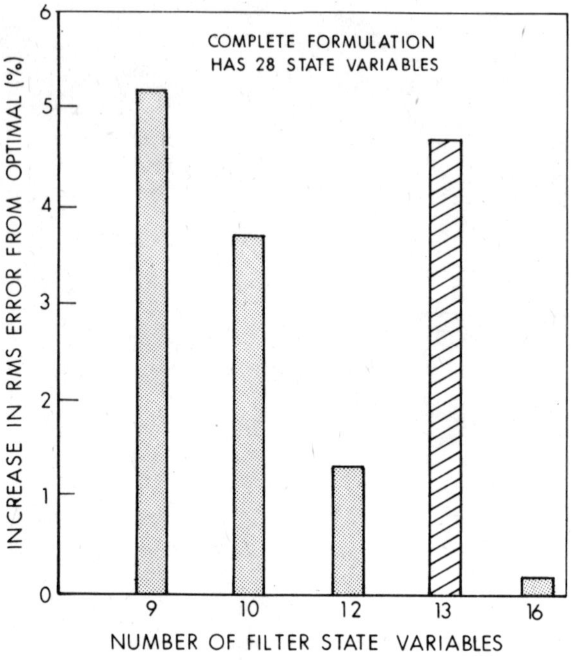

Figure 7.3-9 Effect of Reduction in Number of Estimated State Variables

Of course, the case of 13 variables does not fit the general pattern. The point illustrated here is that *care must be exercised in removing states.* The 13 variables represented by this bar are not the twelve variables represented by the bar to its left, with one addition. (If they were, a lower error than that shown for 12 states could generally be expected.) They are a different subset of the original 28 states and they have obviously not been well chosen. This kind of information comes to the filter designer only through the careful exercise of the sensitivity analysis equations. It is behavior such as this that keeps good filter design a mixture of both art and science.

7.4 DEVELOPING AN ERROR BUDGET

Error budget calculations are a specialized form of sensitivity analysis. They determine the separate effects of individual error sources, or groups of error sources, which are thought to have potential influence on system accuracy. The underlying assumption is that a Kalman-like linear filter is designed, based on some choice of state variables, measurement process, noise spectra etc. ("filter model"). When that filter is employed, all of the error sources ("truth model"), whether modeled in the filter or not, contribute in their own manner to errors in the filter estimate. *The error budget is a catalog of those contributions.* This section describes how the error budget calculations are performed. The discussion treats the discrete filter in some detail, but the approach is valid for continuous filters as well.

Briefly, the steps required to evaluate a proposed filter design by producing an error budget are as follows: *First*, using the filter designer's rules, determine the time history of the filter gain matrix. *Then*, using the complete model of error sources, evaluate system errors. Developing an error budget involves determining the individual effects of a single error source, or group of error sources. These steps are illustrated in Fig. 7.4-1. Their implementation leads to a set of time histories of the contributions of the sets of error sources treated. The error budget is a snapshot of the effects of the error sources at a particular point in time. Any number of error budgets can be composed from the time traces developed. In this way, information can be summarized at key points in the system operation. As illustrated in Fig. 7.4-1, each column of the error budget can represent the contributions to a particular system error of interest, while a row can represent the effects of a particular error source or group of error sources. For example, the columns might be position errors in a multisensor navigation system, while one of the rows might be the effects on those position errors of a noise in a particular measurement. Assuming the error sources are uncorrelated, the total system errors can be found by taking the root-sum-square of all the contributions in each column. Error sources correlated with each other can also be treated, either by including them within the same group or by providing a separate line on the error budget to account for correlation.

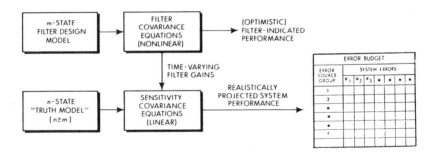

Figure 7.4-1 Diagram Illustrating the Steps in Error Budget Development

Notice that Fig. 7.4-1 indicates that the (optimal) filter error covariance equations are *nonlinear* [see Eqs. (7.2-4) through (7.2-6)] while the sensitivity equations [Eqs. (7.2-19) through (7.2-21)] are *linear*. That is, the sensitivity equations are linear in F(or Φ), H, R, Q, and initial conditions once the filter (F*, H*, R*, Q* and K*) is fixed. As a consequence of the linearity of the sensitivity equations, it is a simple matter to develop sensitivity curves once the error budget has been assembled. This will be illustrated in a later example. The figure also points out that the filter covariance calculations are generally an optimistic indication of the accuracy of the filter. When all the error sources are considered the true filter accuracy is usually seen to be somewhat poorer than the filter covariance, *per se*, would lead one to believe. Certain steps in the generation of an error budget are now discussed in some detail.

Filter Gain Calculations – Each Kalman filter mechanization specifies a particular set of matrices – a filter representation of state dynamics (symbolically, the matrix F* or Φ*), a filter representation of the measurement process (H*), a process noise covariance matrix (Q*), and a measurement noise covariance matrix (R*). These matrices, combined with an initial estimation error covariance [P*(0)], are used to compute the time history of filter error covariance and, subsequently, the filter gain matrix, K*. The filter then proceeds to propagate its estimate and process incoming measurements according to the Kalman filter equations, using K*, H* and F* (or Φ*). All of the above-listed matrices [F*, H*, Q*, R*, P*(0)] must be provided in complete detail.

True Error Covariance Calculations – If all of the pertinent error effects are properly modeled in the implemented filter, the error covariance matrix computed as described above would be a correct measure of the system accuracy. However, for reasons discussed previously, many sources of error are not properly accounted for in the design of most filters, and the consequences of ignoring or approximating these error effects are to be investigated. The behavior of each element of the so-called "*truth model*", whether or not included in the filter design under evaluation, must be described by a linear differential equation. For example, an individual error, α, which is presented as a first-order markov process obeys the differential equation

$$\dot{\alpha} = -\frac{1}{\tau_\alpha}\alpha + w$$

where τ_α is a time constant and w is a white noise. Information concerning the rms magnitude of such a variable is provided by its initial covariance and in the spectral density ascribed to w. When a constant error source such as a measurement bias error, b, is considered, it obeys the linear differential equation

$$\dot{b} = 0$$

In this case, the rms value of the error is entirely specified by an initial covariance parameter. All of the untreated sources of error are added to the filter states to form an augmented "state" vector for performing the error budget calculations.* The dynamics of the augmented state are represented by a matrix F or its counterpart in the discrete representation, Φ. An initial covariance matrix, P(0), is also formed for the augmented state.

*The augmented state vector is not to be confused with the vector $\underline{x}'$ whose dynamics are given in Eq. (7.2-10). The augmented state vector discussed here is always represented by the $\underline{x}$ component of $\underline{x}'$ in Section 7.2.

Of course, it is essential to the systems under consideration that measurements be incorporated to reduce errors in estimates of the variables of interest. The gain matrix designed for each filter only contains enough rows to account for the number of states modeled in the filter. But the augmented state is generally much larger and a convenient device is used to make the filter gain matrix, K*, conformable for multiplications with the covariance matrix for the new state. A matrix W is defined and a new gain matrix is formed as follows:

$$K' = WK^*$$

If the augmented state is formed by simply adding new state variables to those used in the filter, we can write

$$W^T = [I \mid 0]$$

where the identity matrix has dimensions equal to that of the filter state. Clearly, filter states which are more complex (but linear) combinations of the augmented state can be accommodated by proper definition of W.

The covariance calculations for the augmented state yield, at the appropriate locations in the matrix P, the true mean square errors in the estimates of interest. The difficulty involved in performing those calculations can be greatly reduced if it is assumed that all of the state variables are properly corrected [see the discussion surrounding Eqs. (7.2-22) and (7.2-23)] for each measurement; this assumption is made in what follows. The covariance equation at the time a measurement is taken is similar to Eq. (7.2-1), viz:

$$P_k(+) = (I - WK_k{}^*H_k)\, P_k(-)\, (I - WK_k{}^*H_k)^T + WK_k{}^*R_kK_k{}^{*T}W^T \qquad (7.4\text{-}1)$$

In the absence of measurements the augmented state (truth model) covariance changes in time according to Eq. (7.2-2), which is repeated here for convenience:

$$P_{k+1}(+) = \Phi_k P_k(-)\, \Phi_k + Q_k \qquad (7.4\text{-}2)$$

In Eq. (7.4-2) the matrix Q_k represents the effects of uncorrelated forcing functions over the interval t_k to t_{k+1}. When the state variables are not corrected, Eqs. (7.2-20) and (7.2-21) apply, with the appropriate substitution of $WK_k{}^*$ for $K_k{}^*$.

Equations (7.4-1) and (7.4-2) can be used to determine the effects of the various items in the truth model. In order to separate the effects, the equations must be exercised many times, with different initial conditions and forcing functions. For example, to investigate the effects of measurement bias errors alone, all elements of P(0) not corresponding to these errors will be set to zero, along with the Q and R matrices. The error covariance elements generated by this procedure only result from measurement bias errors. To look at first-order markov errors the appropriate elements of P(0) and Q are entered (all others set to zero) and the equations rerun. This procedure must be repeated many times to generate the error budget. The root-sum-square of all the effects provides the true measure of total system performance. Section 7.6 describes the organization of a computer program to perform these covariance calculations.

When *bias* error sources are under consideration, their separate effects can also be computed using "simulation" equations such as

$$\underline{x}_{k+1}(-) = \Phi_k \underline{x}_k(+) \tag{7.4-3}$$

$$\underline{x}_k(+) = (I - K_k{}^* H_k{}^*) \underline{x}_k(-) \tag{7.4-4}$$

Equations (7.4-3) and (7.4-4) are *vector* rather than *matrix* equations, and offer some reduction in computer complexity over the covariance equations described above [Eqs. (7.4-1) and (7.4-2)]. However, their use precludes calculating the effects of groups of error sources in a single computer run. Each bias error source must be treated separately with Eqs. (7.4-3) and (7.4-4) and their effects root-sum-squared, while the covariance equations permit the calculation of the effects of a set of bias error sources in one operation. Also, the "simulation" approach cannot handle correlations between error sources.

Example 7.4-1

From Example 7.3-1 we can form an error budget which displays individual contributions to the steady-state estimation error. The sources of error in the estimate of the single state, x, are: initial errors, $\tilde{x}(0)$, process noise, w, and measurement noise, v.

Effect of Initial Errors – By inspection of Eq. (7.3-3) we can see that the initial errors, represented by p(0), do not contribute to the steady-state estimation error.

Effect of Process Noise – Equation (7.3-3) permits us to calculate the effect of process noise, characterized by the spectral density matrix q, by setting r = 0:

$$p_\infty(q) = \frac{q}{2(\beta + k)}$$

Effect of Measurement Noise – We can find the effect of measurement noise on the steady-state error covariance by setting q = 0 in Eq. (7.3-3):

$$p_\infty(r) = \frac{k^2 r}{2(\beta + k)}$$

If we assign values to β, q and r, we can construct an error budget from the above equations. Since the error budget commonly gives the root-mean-square contribution of each error source we take the square root of p_∞ in finding entries in the error budget. Set $\beta = q = r = 1$. Then, from Eq. (7.3-4)

$$k = -1 + \sqrt{2} = 0.414$$

and

$$\sqrt{p_\infty(q)} = \sqrt{\frac{1}{2(1.414)}} = 0.595$$

$$\sqrt{p_\infty(r)} = \sqrt{\frac{(0.414)^2}{2(1.414)}} = 0.245$$

The error budget generated is given in Table 7.4-1. The root-sum-square error can be computed easily.

TABLE 7.4-1 EXAMPLE ERROR BUDGET

Error Sources	Contributions to Steady-State Errors in Estimate of x
Initial Error	0
Process Noise	0.595
Measurement Noise	0.245
Total (Root-Sum-Square) Error	0.643

The linearity of the equations used in forming error budgets permits easy development of sensitivity curves which illustrate the effects of different values of the error sources on the estimation errors, *as long as the filter design is unchanged.* The procedure for developing a set of sensitivity curves for a particular error source is as follows: First, subtract the contribution of the error source under consideration from the mean square total system error. Then, to compute the effect of changing the error source by a factor of y, multiply its contributions to the mean square system errors by y^2. Next replace the original contribution to mean square error by the one computed above, and finally, take the square root of the newly computed mean square error to obtain the new rss system error. Sensitivity curves developed in this manner can be used to establish the effect of incorrectly prescribed values of error sources, to identify critical error sources, and to explore the effects of substituting into the system under study alternative hardware devices which have different error magnitudes.

Example 7.4-2

Continuing Example 7.4-1, we can develop a sensitivity curve showing the effect of different values of measurement noise. If the rms measurement noise is halved (implying a 1/4 reduction in r), while k, β and q are held constant, the entry on the third line of the error budget becomes

$$\sqrt{\frac{(0.245)^2}{4}} = 0.123$$

All other entries are unchanged. The new total (rss) error is

$$\sqrt{(0.595)^2 + (0.123)^2} = 0.607$$

When the rms measurement noise is doubled (factor of 4 increase in r), the entry is doubled and the total error is

$$\sqrt{(0.595)^2 + (0.490)^2} = 0.770$$

The sensitivity curve constructed from the three points now available is shown in Fig. 7.4-2.

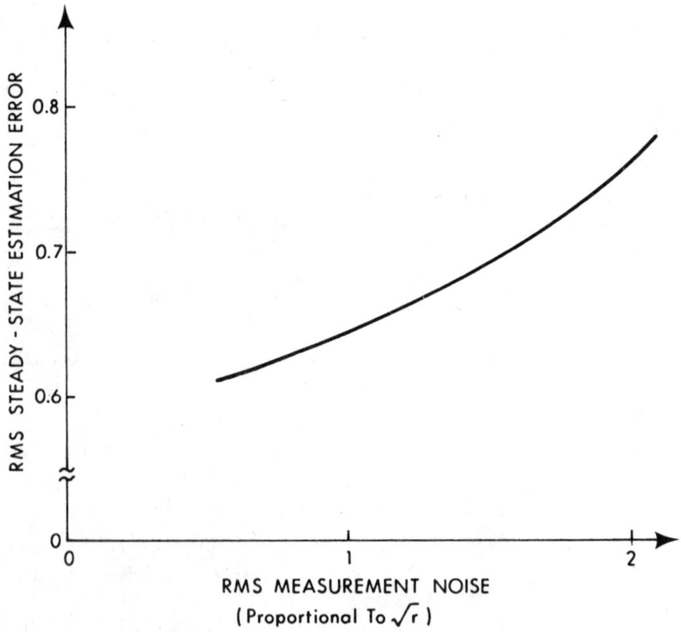

Figure 7.4-2 Example Sensitivity Curve

7.5 SENSITIVITY ANALYSIS: OPTIMAL SMOOTHER

Sensitivity analyses of optimal smoothers provide the same insights, and are motivated by the same concerns as those discussed in Section 7.2 with regard to optimal filters. The derivation of the sensitivity equations proceeds along lines similar to that detailed in Section 7.2. The two cases treated in Section 7.2 – when the system dynamics and measurement process are accurately modeled and when they are not – also arise in the sensitivity analysis of smoothers. They are discussed separately below.

EXACT IMPLEMENTATION OF DYNAMICS AND MEASUREMENTS

Using the same notation as developed in Section 5.2 for the Rauch-Tung-Striebel form of the optimal smoother, the optimal filter covariance is computed using Eq. (7.2-3),

$$\dot{P}(t) = (F-KH)\,P(t) + P(t)\,(F-KH)^T + KRK^T + GQG^T \tag{7.5-1}$$

Then, using the end condition $P(T|T) = P(T)$, the fixed-interval optimal smoother error covariance is computed from the relation [Eq. (5.2-15)]

$$\begin{aligned}\dot{P}(t|T) = {} & [F + GQG^T P^{-1}(t)]\; P(t|T) \\ & + P(t|T)\; [F + GQG^T P^{-1}(t)]^T - GQG^T\end{aligned} \tag{7.5-2}$$

If we define a "smoother gain", K_s, from the first equation in Table 5.2-2,

$$K_s = GQG^T P^{-1}(t) \tag{7.5-3}$$

Eq. (7.5-2) can be written

$$\dot{P}(t|T) = (F + K_s)\,P(t|T) + P(t|T)\,(F + K_s)^T - GQG^T \tag{7.5-4}$$

To determine the effects of differences between design values of F, H, G, Q, R, and P(0) and those that may actually exist, *assuming that correct values of F and H are used in the filter implementation equations*, the following steps are taken: First, compute P(t) using the design values. Next compute the filter gain K, as before, and K_s as in Eq. (7.5-3). Finally, using Eqs. (7.5-1), (7.5-4), the end condition given above, and the actual values of F, H, G, Q, R, and P(0), compute the actual error covariance history.

INCORRECT IMPLEMENTATION OF DYNAMICS AND MEASUREMENTS

When incorrect values of F and H are used in filter and smoother implementation, a more complex set of relations must be used to perform the sensitivity analysis. As before we designate implemented values with an asterisk. It is also necessary to define a number of new matrices, some with statistical meaning as covariances between familiar vectors (e.g., such as true state, estimate, smoothed estimate, error in smoothed estimate, etc.) and others which serve only as mathematical intermediaries. The equations for sensitivity analysis of the linear smoother are

$$\begin{aligned}\dot{P}(t) = {} & (F^*-K^*H^*)\,P(t) + P(t)\,(F^*-K^*H^*)^T + (\Delta F-K^*\Delta H)V \\ & + V^T(\Delta F-K^*\Delta H)^T + GQG^T + K^*RK^{*T}, \qquad P(0) = P^*(0)\end{aligned} \tag{7.5-5}$$

$$\dot{V} = FV + V(F^* - K^*H^*)^T + U(\Delta F - K^*\Delta H)^T - GQG^T, \quad V(0) = E[\underline{x}(0)\tilde{\underline{x}}^T(0)] \tag{7.5-6}$$

$$\dot{U} = FU + UF^T + GQG^T, \qquad U(0) = E[\underline{x}(0)\underline{x}^T(0)] \tag{7.5-7}$$

$$\begin{aligned}\dot{P}(t|T) = {} & (F^* + G^*Q^*G^{*T}B^{-1})\,P(t|T) + P(t|T)\,(F^* + G^*Q^*G^{*T}B^{-1})^T \\ & + G^*Q^*G^{*T}B^{-1}\,[V^TD^T - P(t)C^T] + [DV - CP(t)]\,B^{-1}G^*Q^*G^{*T} \\ & + \Delta F(VC^T - UD^T) + (CV^T - DU)\Delta F^T + (C + D)\,GQG^T \\ & + GQG^T(C + D)^T - GQG^T, \qquad P(T|T) = P(T)\end{aligned} \tag{7.5-8}$$

$$\dot{B} = F^*B + BF^{*T} - BH^*R^{*-1}H^{*T}B + G^*Q^*G^{*T}, \qquad B(0) = P^*(0) \tag{7.5-9}$$

$$\dot{C} = (F^* + G^*Q^*G^{*T}B^{-1})C - G^*Q^*G^{*T}B^{-1} - C(F^* - K^*H^*), \qquad C(T) = I \tag{7.5-10}$$

$$\dot{D} = (F^* + G^*Q^*G^{*T}B^{-1})\,D - DF - \Delta F + C(\Delta F - K^*\Delta H), \qquad D(T) = 0 \tag{7.5-11}$$

Note that P(t), V, U, ΔF and ΔH are the same as P, V, U, ΔF and ΔH in Section 7.2. The matrix B is the error covariance of the optimal filter if the set of "implemented" matrices are correct. Note also that when ΔF and ΔH are zero, B = P(t), D = 0, and, when the implemented matrices are the same as the design matrices ($F^* = F$, $Q^* = Q$, etc.) the equation for the actual smoother error covariance [Eq. (7.5-8)] reduces to Eq. (7.5-2) for the optimal smoother.

7.6 ORGANIZATION OF A COMPUTER PROGRAM FOR COVARIANCE ANALYSIS

This section describes the organization of a computer program which has been found useful in the evaluation and design of linear filters. The program description illustrates the practical application of the covariance analysis methods developed in previous sections. The description is presented in terms of algebraic equations and a flow chart and, therefore, is independent of any particular programming language.

EVALUATION OF AN n-STATE FILTER OPERATING IN AN m-STATE WORLD

The program described here is suited to a fairly broad class of filter evaluation problems. The overall program *scheme* is applicable, with minor modifications, to an even broader class of problems involving suboptimal or optimal linear filters. The general situation being considered is one where a linear filter design has been proposed; it is to receive measurements at discrete points in time and estimate states of a system which is continuously evolving. The filter design is based on an incomplete and/or incorrect model of the actual system dynamics. It is desired to obtain a measure of the filter's performance based on a *truth model* which is different (usually more complex) than the model contained in the filter itself.

The problem is formulated in terms of *two* covariance matrices: P, representing the truth model or *system* estimation error covariance, and P*, representing the *filter's* internal calculation of the estimation error covariance. P is an $m \times m$ matrix, where m is the dimension of the truth model. P* is an $n \times n$ matrix, where n is the number of estimated states. Usually, m is larger than n; sometimes much larger. The principal program output is the time history of P; its value before and after each update is computed. A useful graphical output is a collection of plots of the square roots of the diagonal elements of P versus time. The time histories of the filter gain matrix K and covariance P* are also computed, and may be printed or plotted if desired.

The principal program inputs are the following collection of system and filter matrices:

System Inputs

P_0	the initial truth model covariance matrix $(m \times m)$
F	the truth model dynamics matrix $(m \times m)$
H	the truth model measurement matrix $(\ell \times m)$, where ℓ is the measurement vector dimension
Q	the truth model process noise matrix $(m \times m)$
R	the truth model measurement noise matrix $(\ell \times \ell)$

Filter Inputs

P_0^*	the initial filter covariance matrix $(n \times n)$
F*	the filter dynamics matrix $(n \times n)$
H*	the filter measurement matrix $(\ell \times n)$
Q*	the filter process noise matrix $(n \times n)$
R*	the filter measurement noise matrix $(\ell \times \ell)$

The five system matrices represent a linearized description of the entire physical situation, as understood by the person performing the evaluation. The five filter matrices represent, usually, a purposely simplified model, which it is hoped will produce adequate results. In the general case, F, H, Q, R, F*, H*, Q*, and R* are time-varying matrices, whose elements are computed during the course of the problem solution. For the special case of constant dynamics and stationary noise statistics these eight matrices are simply held constant at their input values.

The immediate objective in running the program is to produce:

- A quantitative measure of overall system *performance*, showing how the n-state filter design performs in the m-state representation of the real world.
- A detailed *error budget*, showing the contribution of each system state or driving noise (both those that are modeled and those that are unmodeled in the filter) to estimation errors.

An underlying objective is to gain insight into the various error mechanisms involved. Such insight often leads the way to improved filter design.

TRUTH MODEL AND FILTER COVARIANCE EQUATION RELATIONSHIPS

Two sets of covariance equations, involving P and P*, are presented below. The relationship between these two sets is explained with the aid of an information flow diagram. This relationship is a key element in determining the organization of the covariance analysis computer program.

Both sets of equations are recursion relations which alternately propagate the covariance between measurements and update the covariance at measurement times. The truth model covariance equations used in the example program are a special case of equations given in Section 7.2. For *propagation* between measurements

$$P_{k+1}(-) = \Phi_k P_k(+) \Phi_k^T + Q_k \qquad (7.6\text{-}1)$$

For *updating*

$$P_k(+) = (I - W^T K_k{}^* H_k) P(-) (I - W^T K_k{}^* H_k)^T + W^T K_k{}^* R K_k{}^{*T} W \qquad (7.6\text{-}2)$$

where W is an m × n transformation matrix $[I \mid 0]$ as defined in Section 7.4. The transition matrix Φ_k and noise matrix Q_k are found using matrix series solutions to

$$\Phi_k = e^{F \Delta t}$$

and

$$Q_k = \int_0^{\Delta t} \Phi_k(\Delta t, \tau)\, Q\, \Phi_k(\Delta t, \tau)^T\, d\tau$$

where Δt is the time interval between measurements. Equations (7.6-1) and (7.6-2) are appropriate forms of Eqs. (7.2-20) and (7.2-21), accounting for the difference between filter state size and actual state size, when one of the two following situations obtains: (1) when $\Delta F = \Delta H = 0$ – i.e., when the part of the truth model state vector that is estimated by the filter is correctly modeled; or (2) when a *feedback* filter is mechanized and the estimated states are *immediately corrected*, in accordance with the filter update equation. When neither of the above conditions hold, the more complicated equations involving the V and U matrices should be used.

Except for one missing ingredient, the five system matrix inputs along with Eqs. (7.6-1) and (7.6-2) would completely determine the discrete time history of P_k, the estimation error covariance. The missing ingredient is the sequence of filter gain matrices K_k^*. The filter covariance equations must be solved in order to produce this sequence. These equations are, for *propagation*

$$P_{k+1}^*(-) = \Phi_k^* P_k^*(+) \Phi_k^{*T} + Q_k^* \tag{7.6-3}$$

for *gain calculation*

$$K_k^* = P_k^*(-) H_k^{*T} [H_k^* P_k^*(-) H_k^{*T} + R_k^*]^{-1} \tag{7.6-4}$$

and for *updating*

$$P_k^*(+) = (I - K_k^* H_k^*) P_k^*(-) (I - K_k^* H_k^*)^T + K_k^* R_k^* K_k^{*T} \tag{7.6-5}$$

where

$$\Phi_k^* = e^{F^* \Delta t}$$

and

$$Q_k^* = \int_0^{\Delta t} \Phi_k^*(\Delta t, \tau)\, Q^* \Phi_k^{*T}(\Delta t, \tau)\, d\tau$$

Figure 7.6-1 is an information flow diagram illustrating the relationship between these two sets of recursion formulae. The upper half of the diagram represents the iterative solution of the filter covariance equations. These are solved in order to generate the sequence of filter gains, K_k^*, which is a necessary input to the lower half of the diagram, representing the iterative solution of the truth model covariance equations.

PROGRAM ARCHITECTURE

A "macro flow chart" of a main program and its communication with two subroutines, is shown in Figure 7.6-2. The function of the main program is to update and propagate both system and filter covariances, P and P*, in a single loop. A single time step is taken with each passage around the loop. Note that the filter gain and update calculations are performed before the system update calculation because the current filter gain matrix must be available as an input to

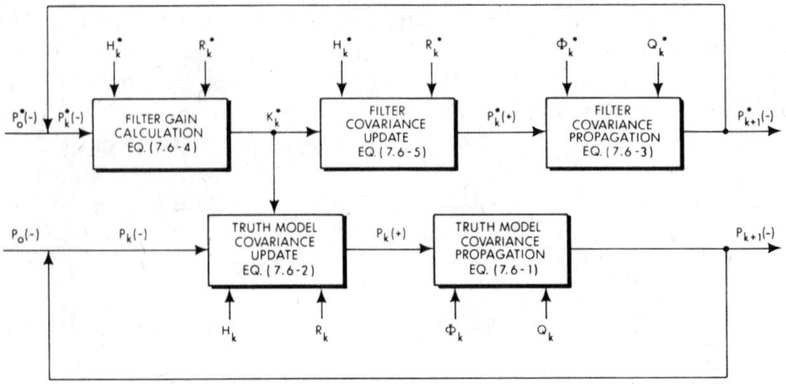

Figure 7.6-1 Covariance Analysis Information Flow

the latter. For constant dynamics problems, no subroutines are needed; the values of F, F*, Q, Q*, H, H*, R, and R* are read in as inputs and do not change thereafter. These constant matrices along with the initial values, P_0 and P_0*, completely determine the solution. For time-varying problems a subroutine TVM is called once during each passage around the loop. Its function is to generate the time-varying elements of the system and filter matrices.

While the main program is generally applicable to a broad class of problems, TVM is a special purpose subroutine which is tailored to a particular time-varying problem. TVM can be designed in a variety of ways. For example, in the case of problems involving maneuvering vehicles a useful feature, corresponding to the organization shown in Figure 7.6-2, is the inclusion of a subsidiary subroutine TRAJ, which provides trajectory information. Once each time step TRAJ passes position, velocity and acceleration vectors ($\overline{r}$, $\overline{v}$, and $\overline{a}$) to TVM. TVM generates various matrix elements, which are expressed in terms of these trajectory variables. This modular organization is useful in a number of ways. For example, once TVM is written for a particular truth model, corresponding to a given filter evaluation problem, the evaluation can be repeated for different trajectories by inserting different versions of TRAJ and leaving TVM untouched. Similarly, if two or more filter designs are to be compared over a given trajectory, different versions of TVM can be inserted while leaving TRAJ untouched. TRAJ can be organized as a simple table-look-up procedure, or as a logical grouping of expressions representing a sequence of trajectory phases. Individual phases can be constant velocity or constant acceleration straight line segments, circular arc segments, spiral climbs, etc. The particular group of expressions to be used during a given pass depends on the current value of time.

An overall system performance projection for a given trajectory can be generated in a single run by inputting appropriate elements of P_0, Q, and R, corresponding to the entire list of truth model error sources. The effects of

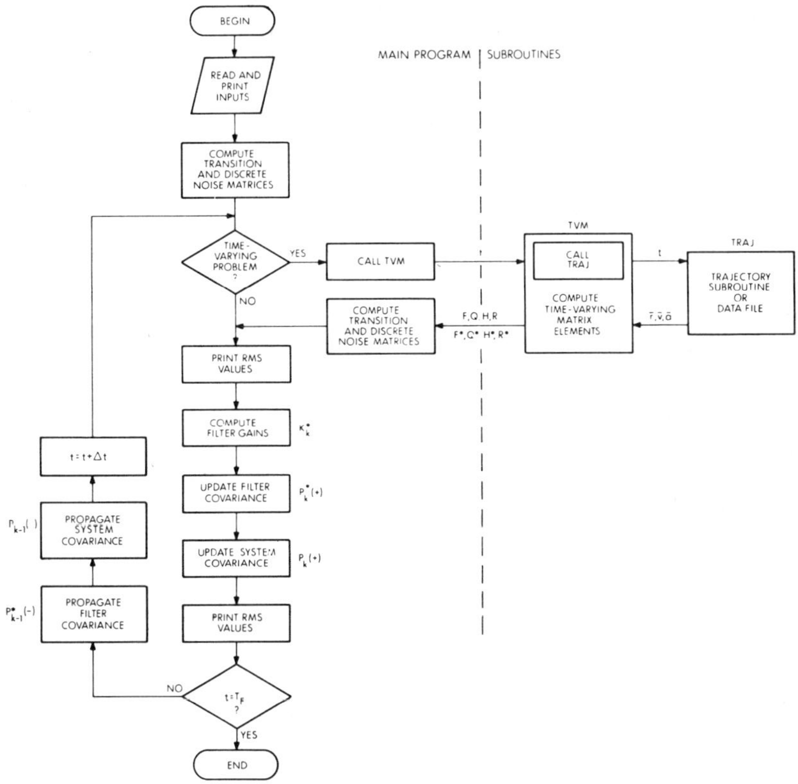

Figure 7.6-2 Covariance Program Macro Flow Chart

changing the trajectory, the update interval or the measurement schedule can be found using repeat runs with appropriate changes in the trajectory subroutine or the parameters, such as Δt, which control the sequencing of the main program loop. Contributions to system error of individual error sources, or small groups of error sources, can be found by inputting individual or small groups of elements of P_0, Q, and R. A sequence of such runs can be used to generate a *system error budget* of the type discussed in Section 7.4, which tabulates these individual contributions separately. Such a tabulation reveals the major error contributors for a given design. The time-history plots produced in conjunction with these error budget runs are also useful in lending insight into the important error mechanisms involved. This information and these insights are very helpful in suggesting filter design improvements.

REFERENCES

1. Bryson, A.E., Jr. and Ho, Y.C., *Applied Optimal Control*, Blaisdell Publishing Co., Waltham, Massachusetts, 1969.
2. Sutherland, A.A., Jr. and Gelb, A., "Application of the Kalman Filter to Aided Inertial Systems," Naval Weapons Center, NWC TP 4652, August 1968.
3. Huddle, J.R., "On Suboptimal Linear Filter Design," *IEEE Trans. on Automatic Control*, Vol. AC-12, No. 3, pp. 317-318, June 1967.
4. Downey, J.J., Hollander, S and Rottach, E.A., "A Loran Inertial Navigation System," *Sperry Engineering Review*, Vol. 21, No. 1, pp. 36-43, 1968.
5. Fortmann, T., Kleinman, D.L., and Athans, M., "On the Design of Linear Systems with Piecewise-Constant Feedback Gains," *IEEE Trans. on Automatic Control*, Vol. AC-13, No. 4, pp. 354-361, August 1968.
6. Singer, R.A. and Frost, P.A., "On the Relative Performance of the Kalman and Wiener Filters," *IEEE Trans. on Automatic Control*, Vol. AC-14, No. 4, pp. 391-394, August 1969.
7. Isaacson, E. and Keller, H.B., *Analysis of Numerical Methods*, John Wiley and Sons, Inc., New York, 1966.
8. Hutchinson, C.E. and D'Appolito, J.A., "Design of Low-Sensitivity Kalman Filters for Hybrid Navigation Systems," *AGARD Conference Proceedings No. 54 on Hybrid Navigation Systems*, AGARD CP No. 54, January 1970.
9. D'Appolito, J.A. and Hutchinson, C.E., "Low Sensitivity Filters for State Estimation in the Presence of Large Parameter Uncertainties," *IEEE Trans. on Automatic Control*, Vol. AC-14, No. 3, pp. 310-312, June 1969.
10. D'Appolito, J.A., "The Evaluation of Kalman Filter Designs for Multisensor Integrated Navigation Systems," The Analytic Sciences Corp., AFAL-TR-70-271, (AD 881286), January 1971.
11. Nash, R.A., D'Appolito, J.A. and Roy, K.J., "Error Analysis of Hybrid Inertial Navigation Systems," AIAA Guidance and Control Conference, Paper No. 72-848, Stanford, California, August 1972.

PROBLEMS

Problem 7-1

Using Eq. (7.2-3) show that in the steady state, the error covariance for the filter using the fixed gain $k_\infty = \sqrt{q/r}$ described in Example 7.1-3 is equal to $\sqrt{rq}$, the steady-state error covariance determined for a Kalman filter in the same example.

Problem 7-2

Using the relations

$$\underline{x}_{k+1} = \Phi_k \underline{x}_k + \underline{w}_k$$

$$\underline{z}_k = H_k \underline{x}_k + \underline{v}_k$$

$$\hat{\underline{x}}_{k+1}(-) = \Phi_k^* \hat{\underline{x}}_k(+)$$

$$\hat{\underline{x}}_{k+1}(-) = \hat{\underline{x}}_{k+1}(-) + K_k[\underline{z}_k - H_k^* \hat{\underline{x}}_{k+1}(-)]$$

and the definitions provided for V, U, $\Delta\Phi_k$ and ΔH, derive Eqs. (7.2-20) and (7.2-21).

Problem 7-3

Using Eq. (7.1-1) and the equation

$$\dot{P} = FP + PF^T + GQG^T - PH^TR^{-1}H$$

derive Eqs. (7.1-2) through (7.1-4).

Problem 7-4

Using the method followed in Example 7.4-2, develop a sensitivity curve depicting the sensitivity of the steady-state error in Examples 7.4-1 and 7.4-2 to process noise.

Problem 7-5

Using Eq. (7.3-3), show that the curve in Fig. 7.4-2 will asymptotically approach the line – ordinate = 0.247 × abscissa – as r grows without bound.

Problem 7-6

Utilizing the equation for propagation of the error in the estimate in a linear filter having the structure of a Kalman filter but without an explicitly defined gain matrix, show that, for such a filter, the error covariance equation across a measurement is

$$P_k(+) = (I-K_kH_k)\,P_k(-)\,(I-K_kH_k)^T + K_kR_kK_k^T$$

Problem 7-7

Refer to Table 4.3-1. The state estimate equation there contains the essence of the structure of the Kalman filter without explicitly defining the filter gain matrix K(t). The definition for K(t) given in the table is that which defines the optimal gain and thus the optimal filter. (a) Show that, for the case of a general K(t) the error differential equation is given by Eq. (4.3-13). (b) Show that the error covariance for the filter with a general gain matrix, in the case where $\underline{v}$ and $\underline{w}$ are not correlated, is [Eq. (7.2-3)],

$$\dot{P} = (F-KH)P + P(F-KH)^T + GQG^T + KRK^T$$

Problem 7-8

Consider the problem of estimating a random ramp function given by the 2-state set of equations:

$$\dot{x}_2 = x_1$$

$$\dot{x}_1 = 0$$

with the noisy measurement

$$z = x_2 + v$$

Show that the differential equations for the elements of the error covariance matrix P:

$$P \stackrel{\Delta}{=} \begin{bmatrix} p_{11} & p_{12} \\ p_{12} & p_{22} \end{bmatrix}$$

are

$$\dot{p}_{11} = -\frac{1}{r} p_{12}^2$$

$$\dot{p}_{12} = p_{11} - \frac{p_{12}p_{22}}{r}$$

$$\dot{p}_{22} = 2p_{12} - \frac{p_{22}^2}{r}$$

Observe that in the steady state ($\dot{p}_{11} = \dot{p}_{12} = \dot{p}_{22} = 0$), the elements of the error covariance matrix all vanish. While the process being observed is nonstationary and grows without bound, the estimate converges on the true values of the states. Notice also that, since the error covariance vanishes, the filter gain matrix vanishes. Show that if there is a small amount of white noise of spectral density q forcing the state – i.e.,

$$\dot{x}_2 = x_1$$

$$\dot{x}_1 = w$$

and if that forcing function is not accounted for in the filter design (i.e., the gains are allowed to go to zero), the error covariance grows according to the relations

$$\dot{p}_{11} = q$$

$$\dot{p}_{12} = p_{11}$$

$$\dot{p}_{22} = p_{12}$$

8. IMPLEMENTATION CONSIDERATIONS

In Chapter 7, the effects of either inadvertent or intended discrepancies between the true system model and the system model "assumed by the Kalman filter" were described. The rms estimation errors associated with the so-called *suboptimal* Kalman filter are always larger than what they would have been had the filter had an exact model of the true system, and they are often larger than the predicted rms errors associated with the filter gain matrix computation. In addition, the computations associated with calculating the true rms estimation errors are more complicated than those associated with the optimal filter. In this chapter, these same kinds of issues are addressed from the standpoint of real-time implementation of the Kalman filter equations. The point of view taken is that a simplified system model, upon which a Kalman filtering algorithm is to be based, has largely been decided upon; it remains to implement this algorithm in a manner that is both computationally efficient and that produces a state estimate that "tracks" the true system state in a meaningful manner.

In most practical situations, the Kalman filter equations are implemented on a digital computer. The digital computer used may be a small, relatively slow, or special purpose machine. If an attempt is made to implement the "theoretical" Kalman filter equations, often the result is that it is either impossible to do so due to the limited nature of the computer, or if it is possible, the resulting estimation simply does not correspond to that predicted by theory. These difficulties can usually be ascribed to the fact that the original model for the system was inaccurate, or that the computer really cannot *exactly* solve the

Kalman filter equations. These difficulties and the methods that may be used to overcome them are categorized in this chapter. The material includes a discussion of:

- modeling problems
- constraints imposed by the computer
- the inherently finite nature of the computer
- special purpose Kalman filter algorithms
- computer loading analysis

When reading this chapter, it is important to keep in mind that a unified body of theory and practice has not yet been developed in this area. Thus, the discussion encompasses a number of ideas which are not totally related. Also, it is to be emphasized that this chapter deals with *real-time* applications of Kalman filtering, and *not* covariance analyses or monte carlo simulations performed on large general-purpose digital computers. However, many of the concepts presented here are applicable to the latter situation.

8.1 MODELING PROBLEMS

Performance projections for data processing algorithms such as the Kalman filter are based on assumed models of the real world. Since these models are never exactly correct, the operation of the filter in a real-time computer and in a real-time environment is usually degraded from the theoretical projection. This discrepancy, commonly referred to as "divergence", can conveniently be separated into two categories: *apparent* divergence and *true* divergence (Refs. 1, 2, 3). In apparent divergence, the true estimation errors approach values that are larger, albeit bounded, than those predicted by theory. In true divergence, the true estimation errors eventually become "infinite." This is illustrated in Fig. 8.1-1.

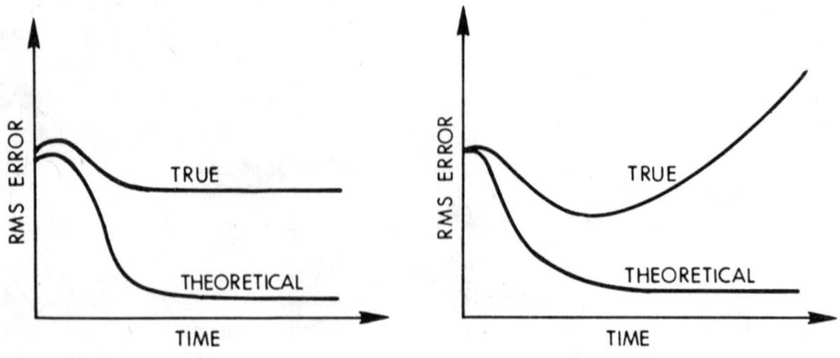

(a) Apparent Divergence (b) True Divergence

Figure 8.1-1 Two Kinds of Divergence

Apparent divergence can arise when there are modeling errors which cause the implemented filter to be suboptimal – i.e., it yields a larger rms error than that predicted by the filter covariance computations. This type of behavior was discussed in Chapter 7. True divergence occurs when the filter is not stable, in the sense discussed in Section 4.4 and Ref. 3, or when there are unmodeled, unbounded system states, which cause the true estimation to grow without bound. These concepts are illustrated in the following example.

Example 8.1-1

As shown in Fig. 8.1-2, the Kalman filter designer assumes that the state to be estimated, x_2, is simply a bias whereas, in fact, $x_2(t)$ is a bias, $x_2(0)$, plus the ramp, $x_1(0)t$. Recall from Example 4.3-1 that the filter is simply a first-order lag where the gain $k(t) \rightarrow 0$ as $t \rightarrow \infty$.

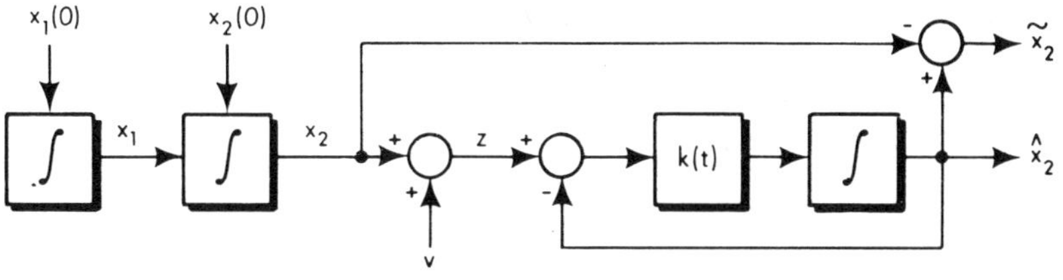

Figure 8.1-2 Example System Illustrating True Divergence

Thus eventually, $\tilde{x}_2(t)$ diverges as $x_1(0)t$. The analogous discrete filtering problem displays similar behavior. More complicated examples of divergence are given in Refs. 46 and 47.

Practical solutions to the true divergence problem due to modeling errors can generally be grouped into three categories: estimate unmodeled states, add process noise, or use finite memory filtering. In the first of these, if one suspects that there are unmodeled growing states in the real world, they are modeled in the filter for "insurance." Thus, in Example 8.1-1, a ramp state would have been included in the filter. However, this approach is generally considered unsatisfactory since it adds complexity to the filter and one can never be sure that all of the suspected unstable states are indeed modeled. The other two categories are treated below. More sophisticated approaches are also possible. For example, see Refs. 4 and 17.

FICTITIOUS PROCESS NOISE

An attractive solution for preventing divergence is the addition of fictitious process noise $\underline{w}(t)$ to the system model. This idea is easily explained in terms of Example 8.1-1. Suppose that the true system is thought to be a bias ($\dot{x}_2 = 0$), but that the designer purposely adds white noise ($\dot{x}_2 = w$), as shown in Fig. 8.1-3. The Kalman filter based on this model has the same form as that in Fig.

8.1-2. However, as shown in Fig. 8.1-3, the Kalman gain k(t) now approaches a constant *nonzero value* as $t \to \infty$. If, as in Example 8.1-1, the true system consists of a bias plus a *ramp*, the Kalman filter will now track the signal x_2 with a nongrowing error.

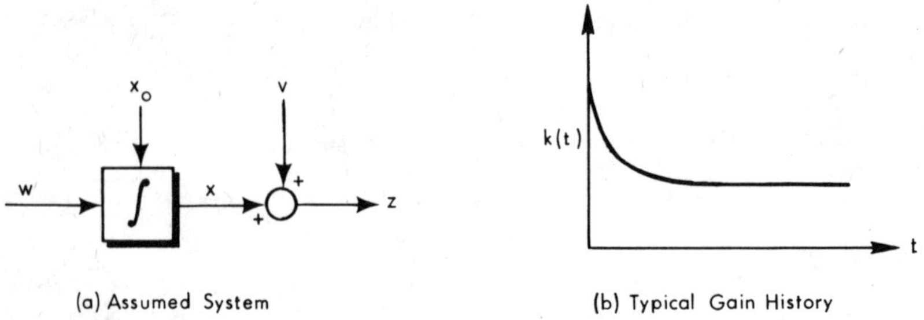

Figure 8.1-3 The Use of Fictitious Process Noise in Kalman Filter Design

The reason that this technique works can be seen by examining the Riccati equation for the error covariance matrix P:

$$\dot{P} = FP + PF^T - PH^TR^{-1}HP + GQG^T \tag{8.1-1}$$

In steady state, $\dot{P} = 0$ so that

$$FP + PF^T - PH^TR^{-1}HP = -GQG^T \tag{8.1-2}$$

If certain elements of Q are zero, corresponding elements in the steady-state value of P, and consequently K, are zero. Thus, with respect to the states assumed not driven by white noise, the filter disregards new measurements ("non-smoothable" states, see Section 5.2 and Ref. 5). However, if these elements of Q are assumed to be nonzero, then the corresponding elements of K will be nonzero, and the filter will always try to track the true system. Analogous remarks can be made about the discrete case.

The choice of the appropriate level of the elements of Q is largely heuristic, and depends to a great extent upon what is known about the unmodeled states. Some examples are given in Ref. 6.

FINITE MEMORY FILTERING

The basic idea of finite memory filtering is the elimination of old data, which are no longer thought to be meaningful (this is often called a "moving window.") This idea is conveniently explained in terms of an example.

Example 8.1-2

The true system is the same as that in Example 8.1-1. The designer wishes to estimate the assumed bias x_2, using a simple averager of the form

$$\hat{x}_2(t) = \frac{1}{T}\int_{t-T}^{t} z(\tau)\, d\tau$$

where

$$z(t) = x_2(t) + v(t)$$

Note that this averager only uses data T units of time into the past. If, in fact, $x_2(t)$ also contains a ramp, then

$$z(t) = x_2(0) + x_1(0)\, t + v(t)$$

It follows that

$$\hat{x}_2(t) = x_2(0) + x_1(0)\left(t - \frac{T}{2}\right) + \frac{1}{T}\int_{t-T}^{t} v(\tau)\, d\tau$$

The estimation error is

$$\hat{x}_2(t) - x_2(t) = \frac{1}{T}\int_{t-T}^{t} v(\tau)\, d\tau - \frac{x_1(0)\, T}{2}$$

It is easy to show that the rms value of this error is bounded for a fixed value of the data window T. Though an unknown ramp is present, the estimator is able to track the signal with a *bounded error*.

Although this example does serve to illustrate the concept of finite memory filtering, it does not represent a recursive filtering algorithm. In order to correspond to, and take advantage of, the discrete Kalman filter formulation, it is necessary to cast the finite memory filter in the form

$$\underline{\hat{x}}_{k+1}^{(N)} = \Psi_k \underline{\hat{x}}_k^{(N)} + K_k \underline{z}_k \tag{8.1-3}$$

where $\underline{\hat{x}}_k^{(N)}$ is an estimate of $\underline{x}$ at time t_k based on the last N data points, Ψ_k is the transition matrix of the filter, and K_k is the filter gain. Somehow, the finite memory property of the filter must be embodied in Ψ_k and K_k. Also, an additional term in Eq. (8.1-3), involving $\underline{z}_{k-N}$, may be needed to "subtract out" the effect of old measurements. The problem of casting the finite memory filter in the form of Eq. (8.1-3) has not yet been satisfactorily resolved, although some work based on least-squares ideas has been done (Ref. 7). However, three practically useful approximation techniques do exist; these are described in the following paragraphs.

Direct Limited Memory Filter Theory – If there is no process noise (Q=0), Jazwinski (Refs. 8, 9) has shown that the equations for the limited memory filter are $(j < k)$

$$\begin{aligned} \hat{\underline{x}}_k^{(N)} &= P_k^{(N)} \left(P_{k|k}^{-1}\hat{\underline{x}}_k - P_{k|j}^{-1}\hat{\underline{x}}_{k|j}\right) \\ \left(P_k^{(N)}\right)^{-1} &= P_{k|k}^{-1} - P_{k|j}^{-1} \end{aligned} \tag{8.1-4}$$

where

$\hat{\underline{x}}_k^{(N)}$ = best estimate of $\underline{x}_k$, given last $N = k-j$ measurements

$P_{k|k}$ = covariance of $\hat{\underline{x}}_k - \underline{x}_k$, that is, $P_k(+)$ in "normal" nomenclature

$P_{k|j}$ = covariance of $\hat{\underline{x}}_{k|j} - \underline{x}_k$ where $\hat{\underline{x}}_{k|j}$ is estimate of $\hat{\underline{x}}_k$ given j measurements – i.e., $\hat{\underline{x}}_{k|j} = \Phi(k,j)\,\hat{\underline{x}}_j(+)$

= $\Phi(k,j)\,P_j(+)\,\Phi^T(k,j)$

$\Phi(k,j)$ = transition matrix of the observed system

$P_k^{(N)}$ = covariance of $\hat{\underline{x}}_k^{(N)} - \underline{x}_k$

From Eq. (8.1-4), it is seen that the limited memory filter estimate is the weighted difference of two *infinite* memory filter estimates. This is unacceptable both because the weighting matrices involve inverses of the full dimension of $\underline{x}$ – a tedious computation – and because the computation involves two infinite memory estimates, the very calculation sought to be avoided.

An approximate method of solving Eq. (8.1-4) has been devised, that largely avoids the above difficulties. This method is illustrated in the following (see Fig. 8.1-4), for a hypothetical example where N = 8: (1) Run the regular

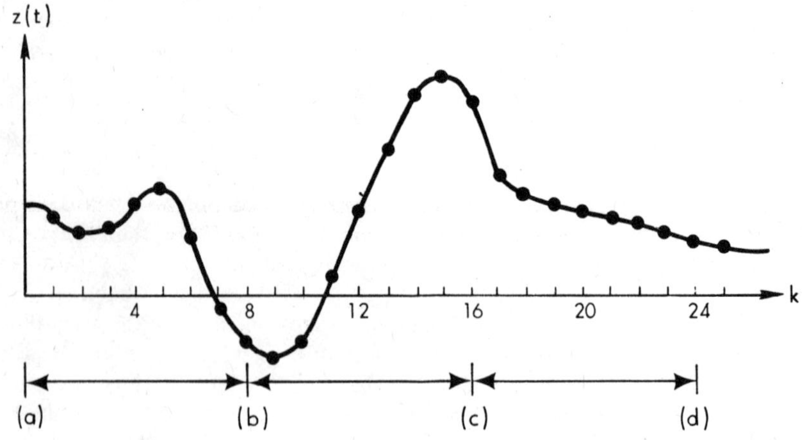

Figure 8.1-4 Measurement Schedule for the Limited Memory Example

Kalman filter from (a) to (b) in Fig. 8.1-4, obtaining $\hat{\underline{x}}_8$, $P_{8|8} = P_8(+)$; (2) Continue to run the regular Kalman filter from (b) to (c), obtaining $\hat{\underline{x}}_{16}$, $P_{16|16} = P_{16}(+)$; (3) Calculate $\hat{\underline{x}}_{16|8} = \Phi(16,8)\hat{\underline{x}}_8$, $P_{16|8} = \Phi(16,8)P_8(+) \cdot \Phi^T(16,8)$; (4) Calculate $\hat{\underline{x}}_{16}^{(8)}$ and $P_{16}^{(8)}$ from Eq. (8.1-4); and (5) Run the regular Kalman filter from (c) to (d) using $\hat{\underline{x}}_{16}^{(8)}$ and $P_{16}^{(8)}$ as initial conditions, etc. Note that for $k \geqslant 8$, $\hat{\underline{x}}_k$ is conditioned on the last 8 to 16 measurements.

Jazwinski (Refs. 8, 9) has applied this technique to a simulated orbit determination problem. The idea is to estimate the altitude and altitude rate of a satellite, given noisy measurements of the altitude. The modeling error in the problem is a 0.075% uncertainty in the universal gravitational constant times the mass of the earth, which manifests itself as an unknown bias. Scalar measurements are taken every 0.1 hr and the limited memory window is N = 10 (1.0 hr). The results are shown in Fig. 8.1-5. Note that the "regular" Kalman filter quickly diverges.

The ϵ Technique – As indicated earlier, filter divergence often occurs when the values of P and K, calculated by the filter, become unrealistically small and

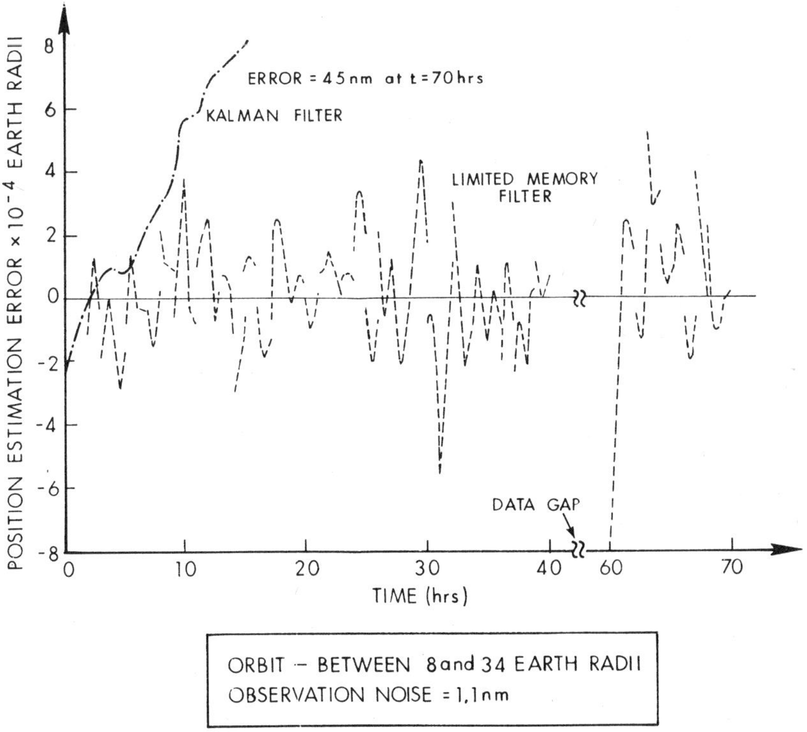

Figure 8.1-5 Position Estimation Error for the Regular and Limited Memory Kalman Filters (Refs. 8, 9)

the filter stops "paying attention" to new data. These ideas motivate the addition of a correction term to the regular Kalman filter update equation, which is based on the current measurement. This concept, which is due to Schmidt (Refs. 9, 10, 11), can be expressed mathematically as

$$\hat{\underline{x}}_{k+1}(+) - \Phi_k\hat{\underline{x}}_k(+) = \left[\hat{\underline{x}}_{k+1}(+) - \Phi_k\hat{\underline{x}}_k(+)\right]_{reg} + \epsilon'\Delta\hat{\underline{x}}_{k+1}(+) \tag{8.1-5}$$

where the subscript reg denotes "regular" – i.e.,

$$\left[\hat{\underline{x}}_{k+1}(+) - \Phi_k\hat{\underline{x}}_k(+)\right]_{reg} = K_{k+1}\left[z_{k+1} - H_{k+1}\Phi_k\hat{\underline{x}}_k(+)\right] \tag{8.1-6}$$

and where

$$\Delta\hat{\underline{x}}_{k+1}(+) = \left(H_{k+1}{}^T r_{k+1}{}^{-1} H_{k+1}\right)^{\#} H_{k+1}{}^T r_{k+1}{}^{-1}\left[z_{k+1} - H_{k+1}\Phi_k\hat{\underline{x}}_k(+)\right] \tag{8.1-7}$$

In Eq. (8.1-6), K_k is the "regular" Kalman gain based on the assumed values of F, H_k, Q_k, r_k, and P_0. In Eq. (8.1-7), # denotes the pseudoinverse, z_{k+1} *must be a scalar* (H_k is a row matrix, r_k is a scalar), and ϵ' is a scalar factor to be determined. Since H_k is a row vector, Eq. (8.1-7) reduces to

$$\Delta\hat{\underline{x}}_{k+1}(+) = H_{k+1}{}^T\left(H_{k+1}H_{k+1}{}^T\right)^{-1}\left[z_{k+1} - H_{k+1}\Phi_k\hat{\underline{x}}_k(+)\right] \tag{8.1-8}$$

Equations (8.1-7) and (8.1-8) represent the best estimate of $\underline{x}_{k+1} - \Phi_k\hat{\underline{x}}_k(+)$, *based only on the current measurement residual* (Ref. 12). Note that this estimate retains the desirable property that $\Delta\hat{\underline{x}}_{k+1} = \underline{0}$ if $z_{k+1} = H_k\,\Phi_k\,\hat{\underline{x}}_k(+)$. Since $\underline{x}_k$ generally has more components than the scalar measurement z_k, the problem is *underdetermined*, and use of the pseudo-inverse is required.

It is convenient to define a new scalar, ϵ, where

$$\epsilon' = \frac{\epsilon r_{k+1}}{H_{k+1}P_{k+1}(-)H_{k+1}{}^T + r_{k+1}} \tag{8.1-9}$$

Note that the numerator and denominator of Eq. (8.1-9) are scalars. If Eqs. (8.1-8) and (8.1-9) are substituted into Eq. (8.1-5), the update equations become

$$\hat{\underline{x}}_{k+1}(+) = \Phi_k\hat{\underline{x}}_k(+) + \frac{P_{k+1}(-)H_{k+1}{}^T + \epsilon\dfrac{r_{k+1}H_{k+1}{}^T}{H_{k+1}H_{k+1}{}^T}}{H_{k+1}P_{k+1}(-)H_{k+1}{}^T + r_{k+1}}\left[z_{k+1} - H_{k+1}\Phi_k\hat{\underline{x}}_k(+)\right] \tag{8.1-10}$$

From this equation, it is seen that the effective gain of this modified filter is

$$K = K_{reg} + K_{ow} \tag{8.1-11}$$

where K_{ow} is an "overweight" gain proportional to ϵ. Meaningful values of ϵ lie in the range of 0 to 1, and are usually chosen by experience or trial and error. Of course, $\epsilon = 0$ corresponds to the "regular" Kalman filter. When $\epsilon = 1$, $H_{k+1} \cdot \hat{\underline{x}}_{k+1}(+) = z_{k+1}$ – i.e., the estimate of the measurement equals the measurement.

Using the techniques in Chapter 7, it can be shown that the covariance matrix of the estimation error is

$$P_k(+) = [P_k(+)]_{reg} + \frac{\epsilon^2 r_k{}^2 H_k{}^T H_k}{[H_k P_k(-) H_k{}^T + r_k]\ [H_k H_k{}^T]^2} \tag{8.1-12}$$

assuming that the modeled values F, Q_k, H_k, r_k, and P_o are correct. Practically speaking then, this technique prevents $P_k(+)$ from becoming too small, and thus overweights the last measurement. This is clearly a form of a limited memory filter. A practical application of the "ϵ technique" is given in Ref. 10.

Fading Memory Filters and Age-Weighting (Refs. 9, 13-16) – Discarding old data can be accomplished by weighting them according to when they occurred, as illustrated in Fig. 8.1-6. This means that the covariance of the measurement noise must somehow be *increased* for past measurements. One manner of accomplishing this is to set

$$R_k{}^* = s^{j-k} R_k, \qquad k = j, j-1, j-2, \ldots \qquad s \geq 1 \tag{8.1-13}$$

where j is some number greater than or equal to one, R_k is the "regular" noise covariance, $R_k{}^*$ is the "new" noise covariance and j denotes the present time.

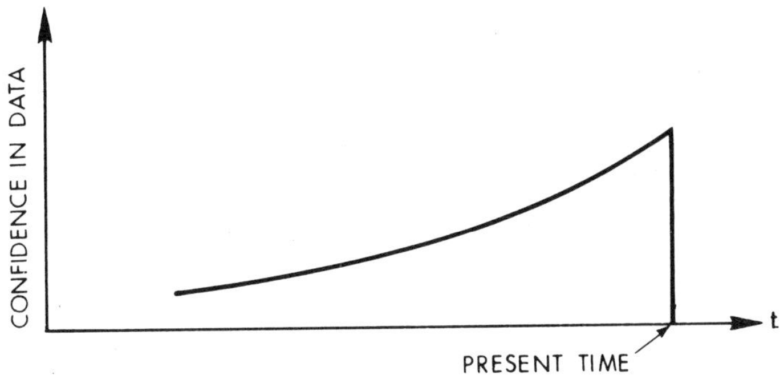

Figure 8.1-6 Age-Weighting Concept

For example, suppose R_k is the same for all k – i.e., $R_k = R$. A convenient way to think of the factor s is

$$s = e^{\Delta t/\tau} \tag{8.1-14}$$

where Δt is the measurement interval and τ is the age-weighting time constant. It then follows that

$$R^*_{k-m} = (e^{m\Delta t/\tau})R, \qquad m = 0, 1, 2, \ldots \tag{8.1-15}$$

This is illustrated in Fig. 8.1-7.

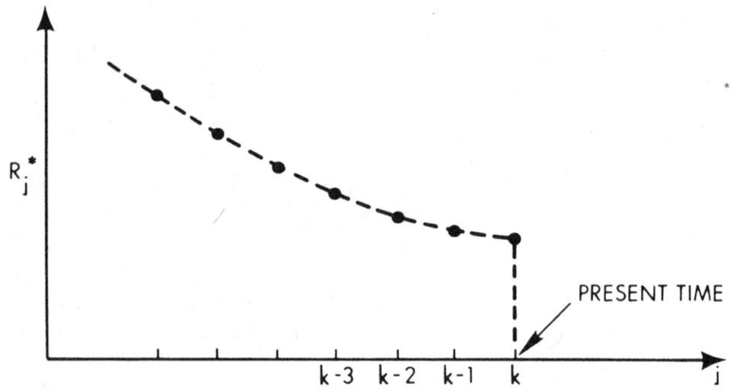

Figure 8.1-7 Conceptual Illustration of Age-Weighted Noise Covariance Matrix Behavior

A recursive Kalman filter can be constructed under these assumptions. The equations are:

$$\begin{aligned}
\hat{\underline{x}}_k(+) &= \Phi_{k-1}\hat{\underline{x}}_{k-1}(+) + K_k[\underline{z}_k - H_k\Phi_{k-1}\hat{\underline{x}}_{k-1}(+)] \\
K_k &= P'_k(-)H_k^T[H_kP'_k(-)H_k^T + R_k]^{-1} \\
P'_k(+) &= P'_k(-) - P'_k(-)H_k^T[H_kP'_k(-)H_k^T + R_k]^{-1}H_kP'_k(-) \\
P'_k(-) &= s\Phi_{k-1}P'_{k-1}(+)\Phi_{k-1}^T + Q_{k-1}
\end{aligned} \tag{8.1-16}$$

Comparison of this set of equations with the set for the Kalman filter will show that they are nearly identical. The only difference is the appearance of the age-weighting factor, s, in the equation for $P'_k(-)$. There is, however, a conceptual difference between these sets. In the case of the Kalman filter (s=1), P'_k (usually denoted by P_k) is the error covariance, $E[(\hat{\underline{x}}_k - \underline{x}_k)(\hat{\underline{x}}_k - \underline{x}_k)^T]$. If

$s \neq 1$, it can be shown that in general P'_k is *not* the error covariance. The true error covariance can easily be calculated using the techniques in Chapter 7.

Example 8.1-3

The age-weighted filter will now be applied to the simple example, used several times previously in this section and illustrated again in Fig. 8.1-8. For this case, it follows from Eq. (8.1-16) that the necessary equations are

$$p'_k(-) = s p'_{k-1}(+)$$

$$p'_k(+) = p'_k(-) - \frac{p'_k(-)^2}{\sigma^2 + p'_k(-)}$$

$$= s p'_{k-1}(+) - \frac{s^2 p'^2_{k-1}(+)}{\sigma^2 + s p'_{k-1}(+)}$$

$$= \frac{\sigma^2 s p'_{k-1}(+)}{\sigma^2 + s p'_{k-1}(+)}$$

where a lower case p'_k is used to emphasize the scalar nature of the example. For steady state, $p'_k(+) = p'_{k-1}(+) = p'_\infty$ and the solution is

$$p'_\infty = \sigma^2 \left(1 - \frac{1}{s}\right)$$

$$k_\infty = \left(1 - \frac{1}{s}\right)$$

Note that $k_\infty > 0$ for all $s > 1$. This is exactly the desired behavior; the gain does not "turn off" as it does in the "regular" Kalman filter.

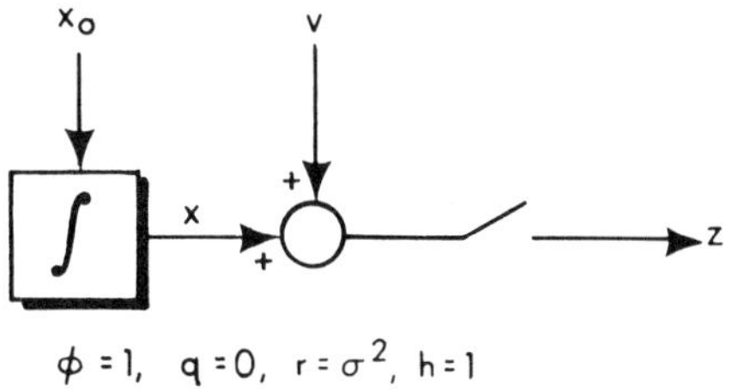

Figure 8.1-8 Bias System and Matrices (lower case φ, q, r, h denote scalars)

It can also be shown that for this problem as k becomes large the error covariance, $p_k(+) = E[(\hat{x}_k(+) - x_k)^2]$, approaches the limit

$$p_\infty = \lim_{k\to\infty} p_k(+) = \frac{s-1}{s+1}\,\sigma^2$$

when $s \geqslant 1$. This equation is summarized in Table 8.1-1 for s=1 and $s\to\infty$. Note that when $s\to\infty$, $\hat{x}_k$ equals the measurement. This implies that the filter also disregards the initial condition information, $E[x^2(0)]$; hence, the caption, "one-stage estimator."

TABLE 8.1-1 LIMITING CONDITIONS OF THE AGE-WEIGHTED FILTER EXAMPLE

	Regular Kalman Filter	One-Stage Estimator
s	1	∞
p_∞	0	σ^2
k_∞	0	1

8.2 CONSTRAINTS IMPOSED BY THE COMPUTER

Generally speaking, the class of digital computers available for a particular mission is largely constrained by considerations (e.g., weight and size restrictions) other than the complexity of the Kalman filter equations. This leads one to attempt to reduce the number of calculations performed in the implementation of the filter as much as possible. To accomplish this, the following techniques are often used: *reducing* the number of states, *decoupling* the equations (or otherwise simplifying the F matrix), and *prefiltering*.

DELETING STATES

Deleting states implies eliminating states in the system model upon which the filter is based, thus automatically reducing the number of states in the filter. Presently, state reduction is largely based upon engineering judgement and experience. However, some general guidelines (also see Section 7.1) are: (1) states with small rms values, or with a small effect on other state measurements of interest, can often be eliminated; (2) states that cannot be estimated accurately, or whose numerical values are of no practical interest, can often be eliminated; and (3) a large number of states describing the errors in a particular device can often be represented with fewer "equivalent" states. These ideas are illustrated below.

Example 8.2-1

The error dynamics of an externally-aided inertial navigation system can be represented in the form shown in Fig. 8.2-1 (Refs. 18-21). Typically, 25 to 75 states are required to describe such a system. However, long-term position error growth is eventually dominated

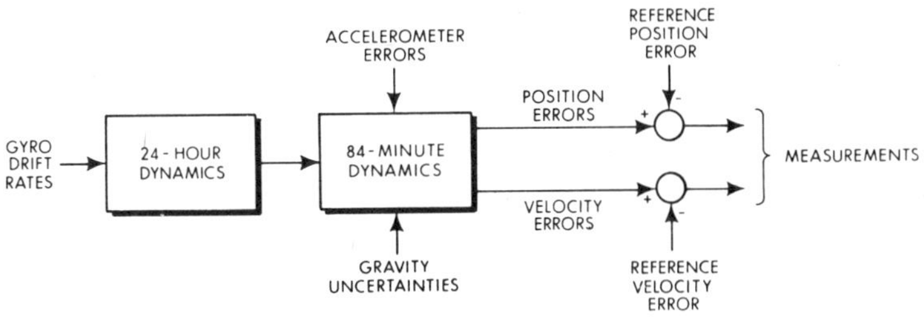

Figure 8.2-1 Error Flow in an Inertial Navigation System

by gyro drift rates and the associated 24-hr dynamics. Thus, if external position and velocity measurements are not available frequently, accelerometer errors, gravity uncertainties, and the 84-minute (Schuler) dynamics can generally be neglected in the filter. Of course, estimates of the high frequency components of velocity error are lost in this simplified filter. Even if measurements are available frequently, optimal covariance studies will often show that although the 84-minute dynamics must be modeled in the filter, error sources such as accelerometer errors, vertical deflections, and gravity anomalies cannot be estimated accurately relative to their *a priori* rms values. Thus, states representing these errors need not be modeled in the filter.

DECOUPLING STATES

One method of simplifying the Kalman filter equations is to decouple the equations. As mentioned in Section 7.1, the rationale here is that there are always fewer computations involved in solving two sets of equations of dimension n/2, as opposed to one set of dimension n. For example, suppose the model equations are

$$\begin{bmatrix} \dot{\underline{x}}_1 \\ \dot{\underline{x}}_2 \end{bmatrix} = \begin{bmatrix} F_{11} & F_{12} \\ F_{21} & F_{22} \end{bmatrix} \begin{bmatrix} \underline{x}_1 \\ \underline{x}_2 \end{bmatrix} + \begin{bmatrix} \underline{w}_1 \\ \underline{w}_2 \end{bmatrix} \tag{8.2-1}$$

If the elements of F_{12} and F_{21} are small, and if $\underline{w}_1$ and $\underline{w}_2$ are uncorrelated, it may be possible to uncouple $\underline{x}_1$ and $\underline{x}_2$ and work with the simplified set

$$\begin{aligned} \dot{\underline{x}}_1 &= F_{11}\underline{x}_1 + \underline{w}_1 \\ \dot{\underline{x}}_2 &= F_{22}\underline{x}_2 + \underline{w}_2 \end{aligned} \tag{8.2-2}$$

Of course, covariance studies would have to be performed to determine whether Eq. (8.2-2) was a reasonable approximation to Eq. (8.2-1).

Example 8.2-2

A more sophisticated approach to the decoupling problem was recently taken with regard to an Apollo application (Ref. 26). In general, the system equations can be partitioned to read

$$\underline{x} = \begin{bmatrix} \underline{x}_1 \\ \underline{x}_2 \end{bmatrix}, \quad P = \begin{bmatrix} P_{11} & P_{12} \\ P_{21} & P_{22} \end{bmatrix}, \quad H = [H_1 \quad H_2], K = \begin{bmatrix} K_1 \\ K_2 \end{bmatrix}$$

The goal is to eliminate the states $\underline{x}_2$ from the filter, thus reducing its complexity, but still somehow compensating for their loss. The true gain equation for K_1 can be written

$$K_1 = (P_{11}H_1^T + P_{12}H_2^T)\Big[H_1 (P_{11}H_1^T + P_{12}H_2^T) + H_2 (P_{21}H_1^T + P_{22}H_2^T) + R\Big]^{-1}$$

If P_{12} is small and P_{22} is dominated by its diagonal elements (and they only vary slightly about their initial conditions), then the following approximations can be made:

$$P_{12} \approx 0$$

$$P_{22} \approx P_{22}(0)$$

The gain equation reduces to

$$K_1 = P_1 H_1^T \Big[H_1 P_{11} H_1^T + R + H_2 P_{22}(0) H_2^T\Big]^{-1}$$

This is the "normal" equation for K_1, when $\underline{x}_2$ is neglected, plus the extra term $H_2P_{22}(0)H_2^T$.

For the Apollo application in question, $\underline{x}_1$ represented the position and velocity of the orbiting spacecraft, and $\underline{x}_2$ represented ten instrument biases. The measurements are range, range-rate, and pointing error. In Table 8.2-1, the performance of the optimal and reduced state filters are compared. Note that 80% of the performance loss, realized when $\underline{x}_2$ was deleted, is recovered by the inclusion of the term $H_2P_{22}(0)\,H_2^T$.

TABLE 8.2-1 COMPARISON OF OPTIMAL AND REDUCED-STATE FILTERS

	rms Filter Accuracy					
	x (ft)	y (ft)	z (ft)	$\dot{x}$ (fps)	$\dot{y}$ (fps)	$\dot{z}$ (fps)
Optimal Filter	1096	7417	919	0.410	8.21	1.51
No Compensation $H_2P_{22}(0)H_2^T = 0$	1271	8779	1768	0.637	9.63	2.67
Full Compensation	1112	9186	1015	0.442	9.46	1.63

Often, other kinds of simplifications can be made in the F matrix "assumed" by the filter. For example, it is common to approximate broadband error sources by white noise, thus eliminating certain terms in the F matrix. Even if this simplification does not actually decouple certain equations, it offers advantages in the computation of the transition matrix, particularly if use is made of the fact that certain elements in the F matrix are zero.

PREFILTERING

The concept of "prefiltering," (also referred to as "measurement averaging," or "data compression") is motivated by cases where measurements are available more frequently than it is necessary, desirable, or possible to process them. For example, suppose measurements of a state $\underline{x}$ are available every Δt. Suppose further that the time required by the computer to cycle through the Kalman filter equations (obtain $\hat{\underline{x}}_{k+1}$ and P_{k+1} from $\hat{\underline{x}}_k$, P_k, and $\underline{z}_k$) and to perform other required operations is ΔT, where $\Delta T = n\Delta t$ for $n > 1$. Clearly each individual measurement cannot be used. However, it is generally wasteful to use the measurement only every ΔT seconds. This problem is often resolved by using some form of *average* of the measurement (over ΔT) every ΔT seconds.

To illustrate what is involved, consider the scalar measurement

$$z_k = x_k + v_k \tag{8.2-3}$$

After every ΔT seconds, an averaged measurement z'_m is used, where

$$\begin{aligned} z'_m &= \frac{1}{n}\sum_{i=1}^{n} z_i \\ &= \frac{1}{n}\sum_{i=1}^{n} x_i + \frac{1}{n}\sum_{i=1}^{n} v_i \end{aligned} \tag{8.2-4}$$

where the index i runs over the measurements collected during the previous ΔT interval. In order to use the standard Kalman filtering equations, z'_m must be put in the form (state) + (noise). Therefore, by definition, the measurement noise v'_m must be (measurement) – (state), or

$$\begin{aligned} v'_m &= z'_m - x_m \\ &= \underbrace{\frac{1}{n}\sum_{i=1}^{n} v_i}_{\substack{\text{Reduced Noise} \\ \text{Due to} \\ \text{Smoothing the} \\ \text{Original Noise}}} + \underbrace{\left(\frac{1}{n}\sum_{i=1}^{n} x_i - x_m\right)}_{\substack{\text{Additional} \\ \text{Noise Due to} \\ \text{Smoothing} \\ \text{the Signal}}} \end{aligned} \tag{8.2-5}$$

As indicated in Eq. (8.2-5), the original noise v_k is indeed smoothed but there is an additional error due to smoothing the signal x_k.

To use the Kalman filter, the variance of v'_m must be specified. Since v_k and x_k are assumed independent, this calculation reduces to specifying the variance of the "reduced noise" and the "additional noise." Since v_k is an uncorrelated sequence, and if $E[v_k^2] = \sigma_v^2$, then

$$E\left[\left(\frac{1}{n}\sum_{k=1}^{n} v_k\right)^2\right] = \frac{\sigma_v^2}{n} \tag{8.2-6}$$

Properties of the "additional noise" depend on statistical properties of x_k. For example, suppose that n = 2, and that

$$E[x_i x_j] = \sigma_x^2 e^{-|i-j|\Delta t} \tag{8.2-7}$$

It then follows that

$$E\left[\left(\frac{1}{2}\sum_{k=1}^{2} x_k - x_2\right)^2\right] = \frac{\sigma_x^2}{2}(1 - e^{-\Delta t}) \tag{8.2-8}$$

It is very important to realize that the measurement noise v'_m, associated with the averaged measurement, is *not* white noise and – in addition – is *correlated* with the state x; both facts being due to the presence of the "additional noise." This must be accounted for in the Kalman filter mechanization or serious performance degradation may result. An approximate method for accomplishing this is indicated in Ref. 10. A more exact technique involves the so-called "delayed state" Kalman filter (Refs. 27, 28).

Recent work by Joglekar (Refs. 51 and 52) has attempted to quantify the concept of data compression and presents examples related to aircraft navigation. For example, Joglekar shows that the accuracy lost through data compression will be small if the plant noise $\underline{w}$ is small compared to the observation noise $\underline{v}$. Other examples of prefiltering and data compression are given in Ref. 24 (inertial navigation systems), Ref. 48 (communication systems), Ref. 49 (trajectory estimation) and Ref. 50 (usage in the extended Kalman filter).

8.3 THE INHERENTLY FINITE NATURE OF THE COMPUTER

A digital computer cannot exactly solve analytic equations because numerical algorithms must be used to approximate mathematical operations such as integration and differentiation (thus leading to *truncation* errors) and the word length of the computer is finite (thus leading to *roundoff* errors). The nature of the resultant errors must be carefully considered prior to computer implementation. Here, these errors are described quantitatively along with methods for reducing their impact.

ALGORITHMS AND INTEGRATION RULES

Kalman Filter Equations – In practical applications, the discrete form of the Kalman filter is always used. However, the discrete equations arise from sampling a continuous system. In particular, recall from Chapter 3 that the continuous system differential equation is "replaced" by

$$\underline{x}_{k+1} = \Phi_k \underline{x}_k + \underline{w}_k \tag{8.3-1}$$

Similarly, the covariance matrix Q_k of $\underline{w}_k$ is related to the spectral density matrix Q by the integral

$$\begin{aligned} Q_k &= E[\underline{w}_k \, \underline{w}_k^T] \\ &= \int_{t_k}^{t_{k+1}} \Phi(t_{k+1},\tau)\, Q(\tau)\, \Phi^T(t_{k+1},\tau)\, d\tau \end{aligned} \tag{8.3-2}$$

Further, recall that the discrete Kalman filter equations are

$$\begin{aligned} \hat{\underline{x}}_{k+1}(+) &= \Phi_k \hat{\underline{x}}_k(+) + K_k [\underline{z}_k - H_k \Phi_k \hat{\underline{x}}_k(+)] \\ P_{k+1}(-) &= \Phi_k P_k(+) \Phi_k^T + Q_k \end{aligned} \tag{8.3-3}$$

where

$$K_k = P_k(-) H_k^T \left[H_k P_k(-) H_k^T + R_k\right]^{-1} \tag{8.3-4}$$

Thus, exclusive of other matrix algebra, this implementation of the Kalman filter includes a matrix inverse operation and determination of Φ_k and Q_k. Algorithms designed to calculate matrix inverses are not peculiar to Kalman filtering and they are well-documented in the literature (see, for example, Ref. 30). Consequently, these calculations will not be discussed further here, although a method for avoiding the inverse is given in Section 8.4.

The state transition matrix Φ is obtained from

$$\frac{d\Phi(t,\tau)}{dt} = F(t)\,\Phi(t,\tau), \qquad \Phi(t,t) = I \tag{8.3-5}$$

and Q_k is given by Eq. (8.3-2). Differentiation of this equation using Leibniz' formula yields

$$\frac{dQ_k}{dt} = FQ_k + Q_k F^T + Q, \qquad Q_k(0) = 0 \tag{8.3-6}$$

Thus, solutions to matrix differential or integral equations are required. Excluding those cases where the solution is known in closed form, Eqs. (8.3-5) and (8.3-6) may be solved by either applying standard integration rules that will solve differential equations in general, or by developing algorithms that exploit the known properties of the equations in question. It is to be emphasized that the following discussion is slanted towards real-time applications. However, the results can generally be extrapolated to the off-line situation, where computer time and storage considerations are not as critical.

Integration Algorithms – Standard integration rules typically make use of Taylor series expansions (Refs. 31 and 32). To illustrate, the solution to the scalar differential equation

$$\dot{x} = f(x, t) \tag{8.3-7}$$

can be expressed as

$$x(t) = x(t_0) + \dot{x}(t_0)(t-t_0) + \frac{\ddot{x}(t_0)(t-t_0)^2}{2} + \frac{\dddot{x}(t_0)(t-t_0)^3}{6} + \ldots \tag{8.3-8}$$

where the initial condition $x(t_0)$ is known. The two-term approximation to Eq. (8.3-8) is

$$\begin{aligned} x_{k+1} &= x_k + \dot{x}_k \Delta t_k \\ &= x_k + f(x_k, t_k)\,\Delta t_k \end{aligned} \tag{8.3-9}$$

where the subscript k refers to time t_k, and where

$$\Delta t_k = t_{k+1} - t_k \tag{8.3-10}$$

The integration algorithm represented by Eq. (8.3-9) is called Euler's method. For many problems, this method is unsatisfactory – the approximate solution diverges quickly from the true solution.

A more accurate algorithm can be derived by retaining the first three terms of Eq. (8.3-8),

$$x_{k+1} = x_k + \dot{x}_k \Delta t_k + \frac{\ddot{x}_k {\Delta t_k}^2}{2} \tag{8.3-11}$$

The second derivative can be approximated according to

$$\ddot{x}_k = \frac{\dot{x}_{k+1} - \dot{x}_k}{\Delta t_k} \tag{8.3-12}$$

so that

$$x_{k+1} = x_k + \frac{\Delta t_k}{2}(\dot{x}_{k+1} + \dot{x}_k) \tag{8.3-13}$$

However, this is not satisfactory since $\dot{x}_{k+1}$ is not known. Recalling that $\dot{x}_k = f(x_k, t_k)$ exactly, and approximating x_{k+1} by

$$x_{k+1} = x_k + \dot{x}_k \Delta t_k \tag{8.3-14}$$

we see that $\dot{x}_{k+1}$ can be approximated as

$$\dot{x}_{k+1} = f(x_k + \dot{x}_k \Delta t_k, t_k) \tag{8.3-15}$$

Substitution of Eq. (8.3-15) into Eq. (8.3-13) yields the algorithm

$$x_{k+1} = x_k + \frac{\Delta t}{2}\,[f(x_k + \dot{x}_k \Delta t_k, t_k) + \dot{x}_k] \tag{8.3-16}$$

This algorithm is known as the "modified Euler method". All of the more sophisticated integration algorithms are extensions and modifications of the Taylor series idea. For example, the Runge-Kutta method uses terms through the fourth derivative.

The important point to be made is that theoretically, the accuracy of the algorithm increases as more terms of the Taylor series are used. However, computer storage requirements, execution time, and roundoff errors also increase. This tradeoff is illustrated in Fig. 8.3-1. In this figure, "truncation error" refers to the error in the algorithm due to the Taylor series approximation.

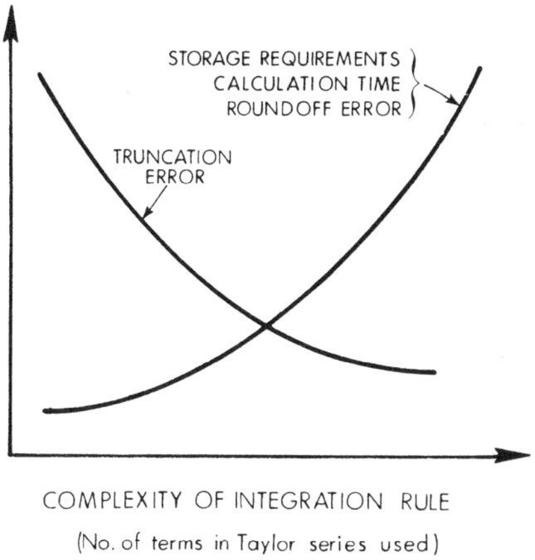

Figure 8.3-1 Integration Algorithm Tradeoffs

Special Algorithms for the Transition Matrix – As opposed to general integration algorithms, consider now the special case of the equation for the transition matrix

$$\frac{d\Phi}{dt} = F\Phi \tag{8.3-17}$$

If F is constant, it is well-known that

$$\begin{aligned}\Phi(t_2, t_1) &= \Phi(\Delta t)\\ &= e^{\Delta t F}\\ &= \sum_{n=0}^{\infty} \frac{\Delta t^n F^n}{n!}\end{aligned} \tag{8.3-18}$$

where $\Delta t = t_2 - t_1$. When Δt is less than the dominant time constants in the system, n on the order of 10 or 20 will furnish six to eight place decimal digit accuracy. For Δt much smaller, just two or three terms may suffice. For larger time intervals, Δt, the property

$$\Phi(t_3, t_1) = \Phi(t_3, t_2)\, \Phi(t_2, t_1) \tag{8.3-19}$$

is used so that

$$\Phi(\Delta T) = \Phi^n\left(\frac{\Delta T}{n}\right) \tag{8.3-20}$$

In most realistic cases, F is not time-invariant. However, the exponential time-series idea can still be used – i.e.,

$$\Phi(t_2, t_1) \cong \sum_{n=0}^{\infty} \frac{(t_2 - t_1)^n}{n!} F^n(t_1) \tag{8.3-21}$$

when $t_2 - t_1$ is much less than the time required for significant changes in F(t). Equation (8.3-19) can then be used to propagate Φ.

Example 8.3-1

Consider the case where F(t) is the 1 × 1 matrix sin Ωt, where Ω is 15 deg/hr (earth rate). Suppose the desired quantity is the transition matrix $\Phi(0.5\ \text{hr}, 0)$. Equations (8.3-21) and (8.3-19) are applied for an increasingly larger number of subdivisions of the 0.5 hr interval. The calculated value of $\Phi(0.5\ \text{hr}, 0)$ is plotted in Fig. 8.3-2, and it is seen that an acceptable interval over which the constant F matrix approximation can be used is less than 6 minutes.

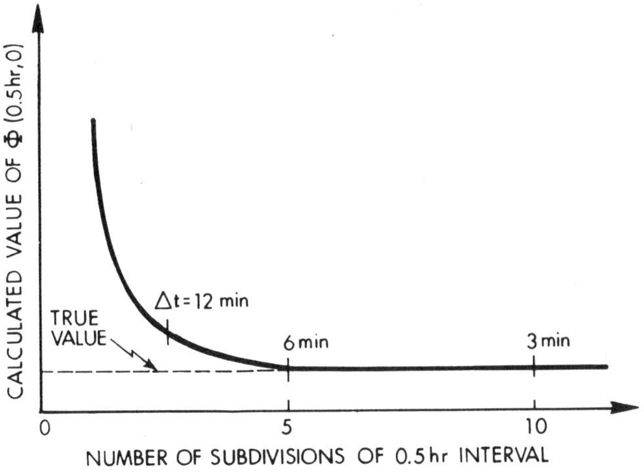

Figure 8.3-2 Calculated Value of Φ

Other examples of the "constant F approximation" are:

a) $\Phi_i \cong I + \Delta t\, F(t_i)$

$$\Phi(t + n\Delta t, t) = \prod_{i=1}^{n} \Phi_i$$

b) $\Phi_i \cong I + \Delta t\; F(t_i) + \dfrac{\Delta t^2}{2} F^2(t_i)$

$$\Phi(t + n\Delta t, t) = \prod_{i=1}^{n} \Phi_i$$

c) Multiplying out (a),

$$\Phi(t + n\Delta t, t) = I + \Delta t \sum_{i=1}^{n} F(t_i) + O(\Delta t^2)$$

$$\cong I + \Delta t \sum_{i=1}^{n} F(t_i)$$

where $O(\Delta t^2)$ indicates terms of order Δt^2 or higher. Of these algorithms, (b) is more accurate than (a) because more terms are used in the series, (a) and (b) are more accurate than (c), but (c) requires the least computer space and computation time.

Example 8.3-2

An algorithm considered for Φ in the C-5A aircraft navigation program (Ref. 10) is a modification of (b) and (c) just discussed. Multiplying (b) out,

$$\Phi(t + n\Delta t, t) = I + \Delta t \sum_{i=1}^{n} F(t_i) + \frac{1}{2} \Delta t^2 \sum_{i=1}^{n} F^2(t_i)$$

$$+ (\Delta t)^2 \sum_{i=1}^{n} F(t_i) \sum_{k=1}^{i-1} F(t_{k-1}) + O(\Delta t^3)$$

$$\cong I + \Delta t \sum_{i=1}^{n} F(t_i) + \frac{1}{2} \Delta t^2 \sum_{i=1}^{n} F_c^{\,2}$$

where F_c is the time-invariant portion of $F(t_i)$ – i.e.,

$$F(t_i) = F_c + F'(t_i)$$

and $F'(t_i)$ is time varying. Note that the cross-product term has been neglected. This term is both small and computationally costly.

Algorithms for Q_k – The design of algorithms that calculate Q_k is motivated by the same considerations just outlined for Φ. Consider first the case of applying standard integration rules to Eq. (8.3-6). Suppose, for example, it is desired to obtain Q_k by breaking the interval (t_k, t_{k+1}) into N equal steps of "length" Δt, and then applying Euler's method. Denote calculated values of Q_k after each integration step by $Q_k(\Delta t)$, $Q_k(2\Delta t)$, etc., where $Q_k(N\Delta t) = Q_k$, and let $F(t_k + i\Delta t) = F_i$. Then according to Eq. (8.3-6) and Eq. (8.3-9),

$$\begin{aligned} Q_k(\Delta t) &= Q_k(0) + [F_0 Q_k(0) + Q_k(0) F_0^T + Q]\,\Delta t \\ &= Q\Delta t \end{aligned} \tag{8.3-22}$$

and

$$\begin{aligned} Q_k(2\Delta t) &= Q_k(\Delta t) + [F_1 Q_k(\Delta t) + Q_k(\Delta t) F_1^T + Q]\,\Delta t \\ &= Q\Delta t + (F_1 Q\Delta t + Q\Delta t\, F_1^T + Q)\,\Delta t \\ &= 2Q\Delta t + (F_1 Q + Q F_1^T)\,\Delta t^2 \end{aligned} \tag{8.3-23}$$

where Q is assumed constant. If this process is continued, one obtains

$$Q_k = NQ\Delta t + \left(\sum_{i=1}^{N-1} iF_i\, Q + Q \sum_{i=1}^{N-1} iF_i^T \right) \Delta t^2 + O(\Delta t^3) \qquad (8.3\text{-}24)$$

In addition, like the equation for Φ, the equation for Q has certain known properties which can be exploited. In particular, note that Eq. (8.3-6) for Q is identical to the matrix Riccati equation with the nonlinear term absent. Therefore, it follows from Section 4.6 that if F and Q do not change appreciably over the interval $t_{k+1} - t_k$, then

$$Q_k = \Phi_{\lambda y}\,(t_{k+1} - t_k)\, \Phi_{\lambda\lambda}{}^T(t_{k+1} - t_k) \qquad (8.3\text{-}25)$$

where $\Phi_{\lambda y}$ and $\Phi_{\lambda\lambda}$ are submatrices of the transition matrix

$$\Phi = \left[\begin{array}{c|c} \Phi_{yy} & \Phi_{y\lambda} \\ \hline \Phi_{\lambda y} & \Phi_{\lambda\lambda} \end{array}\right] \qquad (8.3\text{-}26)$$

corresponding to the dynamics matrix

$$\left[\begin{array}{c|c} -F^T & 0 \\ \hline Q & F \end{array}\right] \qquad (8.3\text{-}27)$$

Thus, the problem reduces to determining algorithms to evaluate transition matrices. A more sophisticated application of the relationship between Q_k and the matrix Riccati equation is given in Ref. 33.

WORD LENGTH ERRORS

Computer errors associated with finite word length are best visualized by considering multiplication in a fixed-point machine. As illustrated in Fig. 8.3-3, the calculated value of the product a x b is

$$(ab)_{calc} = (ab)_{true} + r \qquad (8.3\text{-}28)$$

where r is a remainder that must be discarded. This discarding is generally done in two ways: *symmetric rounding* up or down to the nearest whole number or *chopping* down to the next smallest number (Ref. 34).

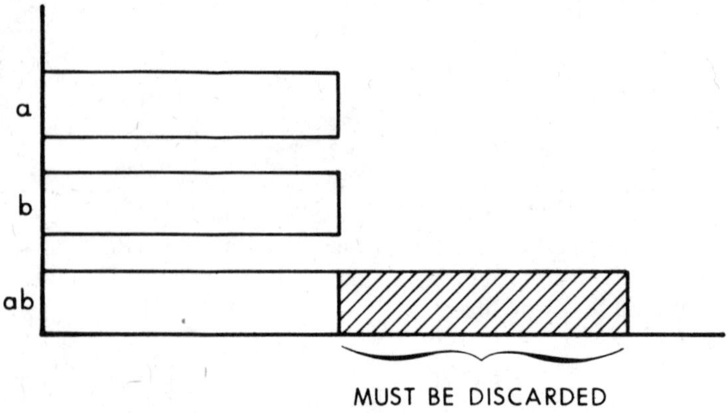

Figure 8.3-3 Multiplication in a Fixed-Point Machine

Symmetric rounding and chopping are illustrated in Fig. 8.3-4. Note that in symmetric rounding, the probability distribution p(r) of the remainder is symmetric about r = 0; whereas in chopping, the remainder r is either positive or negative depending upon the sign of ab. Thus if the products are in fact consistently positive or negative, truncating tends to produce systematic and larger errors than symmetric rounding. This is illustrated in Fig. 8.3-5.

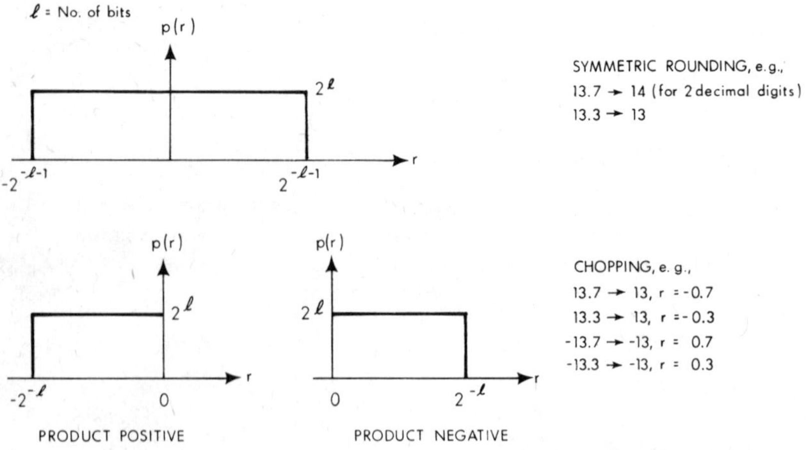

Figure 8.3-4 Distribution of the Remainder for Symmetric Rounding and Chopping (ℓ = number of bits)

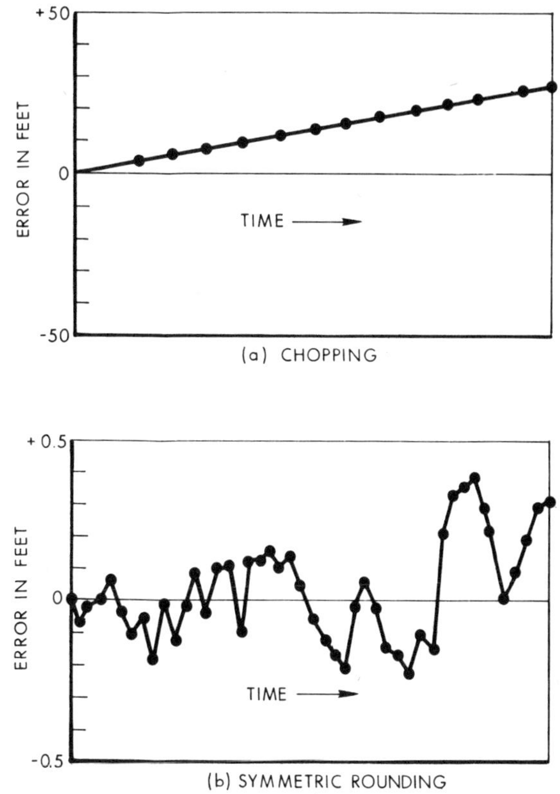

Figure 8.3-5 Errors Due to Finite Word Length (Ref. 34)

The errors caused by finite word length can be determined in two ways. One is to solve the Kalman filter equations by *direct simulation* on a machine with a very long word length, then solve the filter equations again using the shorter word length machine in question (or on a short word length machine simulation), and compare the results. The other is to use the probability distribution in Fig. 8.3-4 and theoretically compute the roundoff error. The choice of methods depends upon the availability of computers and the tradeoff between engineering time and computer time.

Analytic Methods (Ref. 34) – The basic idea of using probability distributions in Fig. 8.3-4 is the following. Suppose, for example, it is desired to calculate

$$\underline{y} = H\underline{x} \ , \qquad H\,(n\mathrm{x}n) \tag{8.3-29}$$

The value of $\underline{y}$ calculated by the computer is

$$(\underline{y})_{calc} = H\underline{x} + \underline{\epsilon}, \qquad \underline{x}, \underline{y}\ (n\times 1)$$
$$= (\underline{y})_{true} + \underline{\epsilon} \tag{8.3-30}$$

where

$$\epsilon = r_1 + r_2 + \ldots + r_n \text{ (per component)}$$

$$r = \text{word length error at each multiplication}$$

If it is assumed that all r's have the same distribution, symmetric rounding is used, and the r's are independent, then it follows from the *central limit theorem* (see Section 2.2) that ϵ is approximately gaussian and has the variance

$$\sigma_\epsilon^2 = n\sigma_r^2 = \frac{n}{12}\, 2^{-2\ell} \text{ (per component)} \tag{8.3-31}$$

From this equation, it is seen that roundoff error increases as the number of calculations (n) increase and decreases as the word length (ℓ) increases, an intuitively satisfying result.

With respect to Kalman filter applications, these concepts can be used to calculate the mean and covariance of the error in the calculated value of $\hat{\underline{x}}$, due to word length errors. These ideas can also be extended to floating-point machines, although the manipulations become more cumbersome.

EFFECT OF WORD LENGTH ERRORS

In Kalman filter implementations, word length errors tend to manifest themselves as errors in the calculation of the "filter-computed" covariance matrix $P_k(+)$ and in $\hat{\underline{x}}_k$. Since the equation for $P_k(+)$ does not account for word length errors, $P_k(+)$ tends to assume an unrealistically small value and perhaps even loses positive-definiteness and thus, numerical significance.* This causes inaccuracies in the calculation of the gain K_k which in turn may cause the estimate $\hat{\underline{x}}_k$ to diverge from the true value $\underline{x}_k$.

There are several ways in which sensitivity to word length errors can be reduced. These include minimizing the number of calculations, increasing word length (double precision), specialized Kalman filter algorithms, ϵ technique, choice of state variables, and good *a priori* estimates. The first of these involves techniques described elsewhere in the text (decoupling, reduced state filters, etc.), while the second is an obvious solution limited by computer size. Specialized Kalman filter algorithms are discussed in Section 8.4. The remaining alternatives are discussed separately in what follows.

*Positive-definiteness is generally only lost if the filter model is uncontrollable and thus the filter is not asymptotically stable.

The ϵ Technique – The ϵ technique was discussed in Section 8.1, in conjunction with finite memory filtering. Recall from that discussion that this technique consists of increasing the Kalman gain in such a manner so as to give extra weight to the most recent measurement. The resultant theoretical P matrix is increased according to Eq. (8.1-12). Therefore, the ϵ technique also produces the desired result for the problem at hand, preventing P_k from becoming unrealistically small. Often a modification of the technique is used, where the diagonal terms of $P_k(+)$ are not allowed to fall below a selected threshold.

The effectiveness of the ϵ technique is illustrated in Table 8.3-1. Note that without the ϵ technique, and contrary to the previous discussion, symmetric rounding is inferior to chopping. This is commented upon in Ref. 35 as being curious and points out that each problem may have unique characteristics which need to be examined closely.

TABLE 8.3-1 TYPICAL ERRORS FOR AN ORBITAL ESTIMATION PROBLEM FOR 20 ORBITS (REF. 35) – 150 NM CIRCULAR ORBIT, 28 BIT MACHINE

	Maximum Error	
	Without ϵ Technique	With ϵ Technique
Symmetric Rounding	32,000 ft	153 ft
Chopping	12,000 ft	550 ft

Choice of State Variables (Ref. 36) – Consider a spacecraft orbiting the moon and attempting to estimate altitude with an altimeter as the measurement source. There are three variables, two of which are independent and should be included as state variables: the radius of moon, R, the radius of orbit, r, and the altitude $h = r - R$. After a number of measurements, h can theoretically be determined with great accuracy. However, uncertainties in R and r remain large; thus, R and h or r and h should be used as state variables. Otherwise, $\hat{h} = \hat{r} - \hat{R}$ tends to have a large error due to roundoff errors in $\hat{r}$ and $\hat{R}$.

Good A Priori Estimates (Ref. 36) – For a fixed-point machine, roundoff causes errors in the last bit. Therefore, it as advantageous to fill the whole word with significant digits. For example, suppose the maximum allowable number is 1000 with a roundoff error of ± 1. Thus, if all numerical quantities can be kept near 1000, maximum error per operation $\cong 1/1000 = 0.1\%$. It follows that it is advantageous to obtain a good initial estimate, $\hat{\underline{x}}_0$. Then P_k and $\hat{\underline{x}}_k$ will not differ greatly from P_0 and $\hat{\underline{x}}_0$, and all of the quantities can be scaled to fill the entire word.

8.4 ALGORITHMS AND COMPUTER LOADING ANALYSIS

This chapter has emphasized the fact that in any real-time application of Kalman filtering, key issues are the accuracy of the implemented algorithm in the face of finite word length and other "noise" sources, and the associated computer burden in terms of memory and execution time. In this section, certain specialized algorithms are discussed from this perspective, including an example of "computer loading analysis."

PROCESSING MEASUREMENTS ONE AT A TIME

As early as 1962, it was known that the Kalman filter algorithm was in such a form as to avoid the requirement for taking the inverse of other than a scalar (Refs. 37 and 38). This is because the update equation for the covariance matrix,

$$P_k(+) = P_k(-) - P_k(-)\, H_k^T [H_k P_k(-)\, H_k^T + R_k]^{-1}\, H_k P_k(-) \qquad (8.4\text{-}1)$$

contains an inverse of the dimension of the measurement vector $\underline{z}$, and this dimension can always be taken to equal one by considering the simultaneous measurement components to occur *serially* over a zero (i.e., very short) time span.

Example 8.4-1

Suppose a system is defined by

$$P(-) = \begin{bmatrix} 1 & \frac{1}{2} \\ \frac{1}{2} & 1 \end{bmatrix}, \quad H = \begin{bmatrix} 1 & 0 \\ 0 & 1 \end{bmatrix}, \quad R = \begin{bmatrix} 1 & 0 \\ 0 & 1 \end{bmatrix}$$

where the subscripts k have been dropped for convenience. It follows from Eq. (8.4-1) that

$$P(+) = \begin{bmatrix} 1 & \frac{1}{2} \\ \frac{1}{2} & 1 \end{bmatrix} - \begin{bmatrix} 1 & \frac{1}{2} \\ \frac{1}{2} & 1 \end{bmatrix} \begin{bmatrix} 2 & \frac{1}{2} \\ \frac{1}{2} & 2 \end{bmatrix}^{-1} \begin{bmatrix} 1 & \frac{1}{2} \\ \frac{1}{2} & 1 \end{bmatrix}$$

$$= \begin{bmatrix} \frac{7}{15} & \frac{2}{15} \\ \frac{2}{15} & \frac{7}{15} \end{bmatrix} \qquad (8.4\text{-}2)$$

On the other hand, let us now assume that the two measurements are separated by an "instant" and process them individually. For this case, we first assume

$$H = [1 \;\; 0], \quad R = [1]$$

and obtain the *intermediate* result (denoted $P_i(+)$)

$$P_i(+) = \begin{bmatrix} 1 & \frac{1}{2} \\ \frac{1}{2} & 1 \end{bmatrix} - \begin{bmatrix} 1 & \frac{1}{2} \\ \frac{1}{2} & 1 \end{bmatrix} [2]^{-1} \begin{bmatrix} 1 & \frac{1}{2} \\ \frac{1}{2} & \frac{1}{2} \end{bmatrix}$$

$$= \begin{bmatrix} \frac{1}{2} & \frac{1}{4} \\ \frac{1}{4} & \frac{7}{8} \end{bmatrix}$$

Utilizing this result, the desired value of P(+) is calculated by next assuming

$$H = [0 \;\; 1], \quad R = [1]$$

It is easily verified that this yields Eq. (8.4-2). The advantage of this method is that only a 1×1 matrix need be inverted, thus avoiding the programming (storage) of a matrix inverse routine. However, some penalty may be incurred in execution time. Similar manipulations can be performed to obtain $\hat{\underline{x}}$.

Although this example considers a case where the elements of the measurement vector are uncorrelated (i.e., R is a diagonal matrix), the technique can be extended to the more general case (Ref. 39).

MATHEMATICAL FORM OF EQUATIONS

Recall that the covariance matrix update equations are, equivalently,

$$P(+) = (I-KH)\,P(-) \tag{8.4-3}$$

and

$$P(+) = (I-KH)\,P(-)\,(I-KH)^T + KRK^T \tag{8.4-4}$$

where

$$K = P(-)\,H^T\,[HP(-)\,H^T + R]^{-1}$$

and the subscripts have again been dropped for convenience. Bucy and Joseph (Ref. 36) have pointed out that Eq. (8.4-3), although simpler, is computationally inferior to Eq. (8.4-4). If $K \rightarrow K + \delta K$, it is easily shown that

$$\delta P(+) = -\delta K\;HP(-) \tag{8.4-5}$$

for Eq. (8.4-3). For Eq. (8.4-4), we instead find

$$\delta P(+) = \delta K[RK^T - HP(-)\,(I-KH)^T] + [KR - (I-KH)\,P(-)\,H^T]\;\delta K^T \tag{8.4-6}$$

Substituting for K, we obtain

$$\delta P(+) = 0 \tag{8.4-7}$$

to first order. Equation (8.4-4) is sometimes referred to as the "Joseph algorithm." Of course, despite its inherent accuracy, Eq. (8.4-4) consumes considerably more computation time than Eq. (8.4-3).

SQUARE-ROOT FORMULATIONS

In the square-root formulation, a matrix W is calculated instead of P, where P = WW^T. Thus, P(+) is always assured to be positive definite. The square-root formulation gives the same accuracy in single precision as does the conventional formulation in double precision. Unfortunately, the matrix W is not unique, and this leads to a proliferation of square-root algorithms (Refs. 11, 39-42, 53, 54). The square-root algorithm due to Andrews (Ref. 41) takes the form:

Update $$W(+) = W(-)\left[I - Z(U^T)^{-1}(U + V)^{-1} Z^T\right] \tag{8.4-8}$$

Extrapolation $$\frac{d}{dt} W(-) = FW(-) + \left(\frac{1}{2}\right) Q\left[W^T(-)\right]^{-1} \tag{8.4-9}$$

where

$$\begin{aligned} Z &= W^T(-)\, H^T \\ UU^T &= R + Z^T Z \\ VV^T &= R \end{aligned} \tag{8.4-10}$$

and where the subscripts have been dropped for convenience. Note that Eq. (8.4-9) can be verified by expanding the time derivative of P:

$$\begin{aligned} \dot{P} &= \dot{W}W^T + W\dot{W}^T \\ &= \left[FW + \frac{1}{2}Q(W^T)^{-1}\right]W^T + W\left[W^T F^T + \frac{1}{2}W^{-1}Q\right] \\ &= F(WW^T) + \frac{1}{2}Q + (WW^T)F^T + \frac{1}{2}Q \\ &= FP + PF^T + Q \end{aligned} \tag{8.4-11}$$

Schmidt (Ref. 11) has shown how to replace the differential equation in Eq. (8.4-9) with an equivalent difference equation.

Of course, a penalty is paid in that the "square root" of certain matrices such as R must be calculated; a somewhat tedious process involving eigenvalue-eigenvector routines. An indication of the number of extra calculations required to implement the square-root formulation can be seen from the sample case illustrated in Table 8.4-1, which is for a state vector of dimension 10 and a scalar measurement. This potential increase in computation time has motivated a search for efficient square root algorithms. Recently, Carlson (Ref. 53) has derived an algorithm which utilizes a lower triangle form for **W** to improve computation speed. Carlson demonstrates that this algorithm approaches or exceeds the speed of the conventional algorithm for low-order filters and reduces existing disadvantages of square-root filters for the high-order case.

TABLE 8.4-1 COMPARISON OF THE NUMBER OF CALCULATIONS INVOLVED IN THE CONVENTIONAL AND SQUARE-ROOT FORMULATIONS OF THE KALMAN FILTER (REF. 11)

Update:

	Square Roots	M&D	+	A&S	=	Equivalent M&D
Conventional	0	310		211		352
Square-Root	1	322		302		387

Extrapolation:

	Square Roots	M&D	+	A&S	=	Equivalent M&D
Conventional	0	2100		2250		2550
Square-Root	10	4830		4785		5837

Note: M&D = multiplications and divisions
A&S = additions and subtractions

Example 8.4-2 illustrates the manner in which square-root algorithms minimize roundoff error.

Example 8.4-2 (Ref. 39)

Suppose

$$P(-) = \begin{bmatrix} 1 & 0 \\ 0 & 1 \end{bmatrix}, \quad H = [1 \quad 0], \quad r = \epsilon^2$$

where $\epsilon \ll 1$ and to simulate computer word length roundoff, we assume $1 + \epsilon \neq 1$ but $1 + \epsilon^2 \cong 1$. It follows that the *exact* value for P(+) is

$$P(+) = \begin{bmatrix} \dfrac{\epsilon^2}{1+\epsilon^2} & 0 \\ 0 & 1 \end{bmatrix}$$

whereas the value calculated in the computer using the standard Kalman filter algorithm in Eq. (8.4-1) is

$$P(+) = \begin{bmatrix} 0 & 0 \\ 0 & 1 \end{bmatrix}$$

Using the square-root algorithm in Eq. (8.4-8), the result is

$$P(+) = \begin{bmatrix} \epsilon^2 & 0 \\ 0 & 1 \end{bmatrix}$$

Since

$$K = P(+)\,H^T R^{-1}$$

it follows that:

Exact $\quad K = \begin{bmatrix} \dfrac{1}{1+\epsilon^2} \\ 0 \end{bmatrix}$

Conventional $\quad K = \begin{bmatrix} 0 \\ 0 \end{bmatrix}$

Square-Root $\quad K = \begin{bmatrix} 1 \\ 0 \end{bmatrix}$

Clearly the conventional formulation may lead to divergence problems.

COMPUTER LOADING ANALYSIS

The burden that Kalman filtering algorithms place on real-time digital computers is considerable. Meaningful measures of this burden are storage and computation time. The first item impacts the memory requirements of the computer, whereas the second item helps to determine the rate at which measurements can be accepted. If a Kalman filter algorithm is to be programmed in real time, it is generally necessary to form some estimate of these storage and computation time requirements. These estimates can then be used to establish tradeoffs between computer size and speed and algorithm complexity. Of course, the constraints of the situation must be recognized; frequently the computer has been selected on the basis of other considerations, and the Kalman filter algorithm simply must be designed to "fit."

Computation time can be estimated by inspecting Kalman filter algorithm equations – i.e., counting the number of "adds," "multiplies," and "divides," multiplying by the individual computer execution times, and totalling the results. Additional time should also be added for logic (e.g., branching instructions), linkage between "executive" and subprograms, etc. An example of the results of such a calculation for several airborne computers is shown in Table 8.4-2 for a 20-state filter with a 9-state measurement. "Total cycle time"

TABLE 8.4-2 TYPICAL CYCLE TIME FOR KALMAN FILTER EQUATIONS (20 STATE FILTER, 9 STATE MEASUREMENT)

	1974 Computer (Fast) (μsec)	1968 Computer (Moderately Fast) (μsec)	1965 Computer (Fairly Slow) (μsec)	1963 Computer (Slow–No Floating Point Hardware) (μsec)
Load	1	2	8	120
Multiply	6	10	100	800
Divide	7	15	130	1000
Add	2	3	50	300
Store	2	2	8	120
Increment Index Register	1	2	8	80
Total Cycle Time (sec)	0.9	1.4	11.8	97.5

refers to the time required to compute the Kalman filter covariance and gain, update the estimate of the state, and extrapolate the covariance to the next measurement time. Considering the nature of the Kalman filter equations, one can see from the execution times in Table 8.4-2 that the computer multiply time tends to dominate the algebraic calculations. In Ref. 44, it is shown that logic time is comparable to multiply time.

Storage, or computer memory, can be divided into the space required to store the program instructions and the space required to store scalars, vectors, matrices, etc. If one does not have past experience to draw upon, the first step in determining the space required for the program instructions is to simply write the program out in the appropriate language. Then, using the knowledge of the particular machine in question, the number of storage locations can be determined. For example, the instruction A + B = C requires 1 1/2 28-bit words in the Litton LC-728 Computer.

When determining the storage required for matrices, etc., care must be taken to account for the fact that many matrices are symmetric, and that it may be possible to avoid storing zeros. For example, any $n \times n$ covariance matrix only requires $n(n + 1)/2$ storage locations. Similarly, transition matrices typically contain many zeros, and these need not be stored. Of course, overlay techniques should also be used when possible; this can be done for $P_k(+)$ and $P_k(-)$, for example.

For the example illustrated in Table 8.4-2, 3300 32-bit words of storage were required. Of these, 900 words were devoted to storage of the program instructions. A further example of storage requirements is illustrated in Table 8.4-3. The total storage requirement (exclusive of program linkage) is 793 32-bit words. The computer in question is the Raytheon RAC-230 and the speed of this machine is comparable with the "1968 computer" in Table 8.4-2 (Ref. 45). The Kalman filter equation cycle time is computed to be 9.2 msec.

TABLE 8.4-3 COMPUTER OPERATIONS REQUIRED FOR A 5-STATE FILTER WITH A 3-STATE MEASUREMENT*

Operation	Eraseable Storage	Permanent Storage	Program Instruction	Divide	Multiply	Add
1. Compute Φ	5	23	6	-	2	-
2. Compute $P_k(-) = \Phi P_{k-1}(+)\Phi^T$	-	29	58	-	250	300
3. Compute H	31	-	345	54	92	51
4. Compute $K_k = P_k(-)H^T \cdot [HP_k(-)H^T + R]^{-1}$	59	9	164	9	144	173
5. Compute $P_k(+) = [I - K_kH]\, P_k(-)$	-	2	22	-	75	125
6. Compute $\hat{\underline{x}}_k(+) = K_k\underline{z}_k$	3	2	35	-	18	21
Total	98	65	630	63	581	670

*There is no process noise in this example (Q = 0).

Note that for the 5-state filter in Table 8.4-3, program instructions comprise the major portion of computer storage, whereas the opposite is true for the 20-state filter in Table 8.4-2. The reason for this is that the matrix storage space tends to be proportional to the square of the state size. Conversely, program instruction storage is relatively independent of state size, so long as branch-type instructions are used. Branch instructions are those that set up looping operations. Thus, for matrix multiplication, it would require the same number of machine instructions to multiply a 2 X 2 matrix as a 20 X 20 matrix.

Recently, Gura and Bierman (Ref. 43) and Mendel (Ref. 44) have attempted to develop parametric curves of computer storage and time requirements, for several Kalman filter algorithms ("standard" Kalman, Joseph, square-root, etc.). A partial summary of their storage results for large n (n is the state vector dimension) is shown in Table 8.4-4, where m is the dimension of the measurement vectors. Mendel has considered only the "standard" Kalman filter algorithm, and has constructed plots of storage and time requirements vs. the dimension of the filter state, measurement, and noise.*

TABLE 8.4-4 KALMAN FILTER STORAGE REQUIREMENTS FOR LARGE N (PROGRAM INSTRUCTIONS NOT INCLUDED) (REF. 43)

	Storage Locations			
		$n \cong m$		
Algorithm	$n \gg m$	$n \cong 1$	$n \gg 1$	$m \gg n$
Standard Kalman*	$2.5\ n^2$	$3.5\ n(n + 1.3)$	$3.5\ n^2$	m^2
Joseph†	$2.5\ n^2$	$1.5\ n(n + 1)$	$1.5\ n^2$	m^2
Andrews Square-Root	$3\ n^2$	$5.5\ n(n + 0.8)$	$5.5\ n^2$	$2.5\ m^2$
Standard Kalman* (no symmetry)	$3\ n^2$	$5\ n(n + 0.6)$	$5\ n^2$	$2\ m^2$
Joseph† (no symmetry)	$3\ n^2$	$6\ n(n + 0.6)$	$6\ n^2$	$2\ m^2$

*Eq. (8.4-1)
†Eq. (8.4-4)
n is the state vector dimension
m is the measurement vector dimension

*Recall R may be singular – i.e., the dimension of $\underline{v}_k$ may be less than the dimension of $\underline{z}_k$.

REFERENCES

1. Price, C.F., "An Analysis of the Divergence Problem in the Kalman filter," *IEEE Transactions on Automatic Control*, Vol. AC-13, No. 6, December 1968, pp. 699-702.
2. Fitzgerald, R.J., "Error Divergence in Optimal Filtering Problem," Second IFAC Symposium on Automatic Control in Space, Vienna, Austria, September 1967.
3. Fitzgerald, R.J., "Divergence of the Kalman Filter," *IEEE Transactions on Automatic Control*, Vol. AC-16, No. 6, December 1971, pp. 736-747.
4. Nahi, N.E. and Schaefer, B.M., "Decision-Directed Adaptive Recursive Estimators: Divergence Prevention," *IEEE Transactions on Automatic Control*, Vol. AC-17, No. 1, February 1972, pp. 61-67.
5. Fraser, D.C., "A New Technique for the Optimal Smoothing of Data," Ph.D. Thesis, Massachusetts Institute of Technology, January 1967.
6. D'Appolito, J.A., "The Evaluation of Kalman Filter Designs for Multisensor Integrated Navigation Systems," The Analytic Sciences Corp., AFAL-TR-70-271. (AD 881206), January 1971.
7. Lee, R.C.K., "A Moving Window Approach to the Problems of Estimation and Identification," Aerospace Corp., El Segundo, California, Report No. TR-1001 (2307)-23, June 1967.
8. Jazwinski, A.H., "Limited Memory Optimal Filtering," *IEEE Transactions on Automatic Control*, Vol. AC-13, No. 5, October 1968, pp. 558-563.
9. Jazwinski, A.H., *Stochastic Processes and Filtering Theory*, Academic Press, New York, 1970.
10. Schmidt, S.F., Weinberg, J.D., and Lukesh, J.S., "Case Study of Kalman Filtering in the C-5 Aircraft Navigation System," *Case Studies in System Control*, University of Michigan, June 1968, pp. 57-109.
11. Schmidt, S.F., "Computational Techniques in Kalman Filtering," *Theory and Applications of Kalman Filtering*, Advisory Group for Aerospace Research and Development, AGARDograph 139, (AD 704 306), Feb. 1970.
12. Deutsch, R., *Estimation Theory*, Prentice-Hall, Inc., Englewood Cliffs, N.J., 1965.
13. Fagin, S.L., "Recursive Linear Regression Theory, Optimal Filter Theory, and Error Analysis of Optimal Systems," *IEEE International Convention Record*, March 1964, pp. 216-240.
14. Tarn, T.S. and Zaborsky, J., "A Practical Nondiverging Filter," *AIAA Journal*, Vol. 8, No. 6, June 1970, pp. 1127-1133.
15. Sacks, J.E. and Sorenson, H.W., "Comment on 'A Practical Nondiverging Filter,'" *AIAA Journal*, Vol. 9, No. 4, April 1971, pp. 767, 768.
16. Miller, R.W., "Asymptotic Behavior of the Kalman Filter with Exponential Aging," *AIAA Journal*, Vol. 9, No. 3, March 1971, pp. 537-539.
17. Jazwinski, A.H., "Adaptive Filtering," *Automatica*, Vol. 5, 1969, pp. 975-985.
18. Leondes, C.T., ed., *Guidance and Control of Aerospace Vehicles*, McGraw-Hill Book Co., Inc., New York, 1963.
19. Britting, K.R., *Inertial Navigation Systems Analysis*, John Wiley & Sons, New York, 1971.

20. Nash, R.A., Jr., Levine, S.A., and Roy, K.J., "Error Analysis of Space-Stable Inertial Navigation Systems," *IEEE Transactions on Aerospace and Electronic Systems*, Vol. AES-7, No. 4, July 1971, pp. 617-629.

21. Hutchinson, C.E. and Nash, R.A., Jr., "Comparison of Error Propagation in Local-Level and Space-Stable Inertial Systems," *IEEE Transactions on Aerospace and Electronic Systems*, Vol. AES-7, No. 6, November 1971, pp. 1138-1142.

22. Kayton, M. and Fried, W., eds., *Avionics Navigation Systems*, John Wiley & Sons, Inc., New York, 1969.

23. O'Halloran, W.F., Jr., "A Suboptimal Error Reduction Scheme for a Long-Term Self-Contained Inertial Navigation System," National Aerospace and Electronics Conference, Dayton, Ohio, May 1972.

24. Nash, R.A., Jr., D'Appolito, J.A., and Roy, K.J., "Error Analysis of Hybrid Inertial Navigation Systems," AIAA Guidance and Control Conference, Stanford, California, August 1972.

25. Bona, B.E. and Smay, R.J., "Optimum Reset of Ship's Inertial Navigation System," *IEEE Transactions on Aerospace and Electronic Systems*, Vol. AES-2, No. 4, July 1966, pp. 409-414.

26. Chin, P.P., "Real Time Kalman Filtering of APOLLO LM/AGS Rendezvous Radar Data," AIAA Guidance, Control and Flight Mechanics Conference, Santa Barbara, California, August 1970.

27. Brown, R.G., "Analysis of an Integrated Inertial-Doppler-Satellite Navigation System, Part 1, Theory and Mathematical Models," Engineering Research Institute, Iowa State University, Ames, Iowa, ERI-62600, November 1969.

28. Brown, R.G. and Hartmann, G.L., "Kalman Filter with Delayed States as Observables," *Proceedings of the National Electronics Conference*, Chicago, Illinois, Vol. 24, December 1966, pp. 67-72.

29. Klementis, K.A. and Standish, C.J., "Final Report – Phase I – Synergistic Navigation System Study," IBM Corp., Owego, N.Y., IBM No. 67-923-7, (AD 678 070), October 1966.

30. Faddeev, D.K. and Faddeeva, V.N., *Computational Methods of Linear Algebra*, W.H. Freeman and Co., San Francisco, Cal., 1963.

31. Kelly, L.G., *Handbook of Numerical Methods and Applications*, Addison-Wesley Publishing Co., Reading, Mass., 1967.

32. Grove, W.E., *Brief Numerical Methods*, Prentice-Hall, Inc., Englewood Cliffs, N.J., 1968.

33. D'Appolito, J.A., "A Simple Algorithm for Discretizing Linear Stationary Continuous Time Systems," *Proc. of the IEEE*, Vol. 54, No. 12, December 1966, pp. 2010, 2011.

34. Lee, J.S. and Jordan, J.W., "The Effect of Word Length in the Implementation of an Onboard Computer," ION National Meeting, Los Angeles, California, July 1967.

35. "Computer Precision Study II," IBM Corp., Rockville, Maryland, Report No. SSD-TDR-65-134, (AD 474 102), October 1965.

36. Bucy, R.S. and Joseph, P.D., *Filtering for Stochastic Processes With Applications to Guidance*, Interscience Publishers, N.Y., 1968.

37. Ho, Y.C., "The Method of Least Squares and Optimal Filtering Theory," Memo. RM-3329-PR, RAND Corp., Santa Monica, Cal., October 1962.

38. Ho, Y.C., "On the Stochastic Approximation Method and Optimal Filtering Theory, *J. Math. Analysis and Applications*, 6, 1963, pp. 152-154.

39. Kaminski, P.G., Bryson, A.E., Jr. and Schmidt, J.F., "Discrete Square Root Filtering: A Survey of Current Techniques," *IEEE Transactions on Automatic Control*, Vol. AC-16, No. 6, December 1971, pp. 727-736.

40. Bellantoni, J.F. and Dodge, K.W., "A Square Root Formulation of the Kalman-Schmidt Filter," *AIAA Journal*, Vol. 5, No. 7, July 1967, pp. 1309-1314.

41. Andrews, A., "A Square Root Formulation of the Kalman Covariance Equations," *AIAA Journal*, June 1968, pp. 1165, 1166.

42. Battin, R.H., *Astronautical Guidance*, McGraw-Hill Book Co., Inc., New York, 1964, pp. 338-339.

43. Gura, I.A. and Bierman, A.B., "On Computational Efficiency of Linear Filtering Algorithms," *Automatica*, Vol. 7, 1971, pp. 299-314.

44. Mendel, J.M., "Computational Requirements for a Discrete Kalman Filter," *IEEE Transactions on Automatic Control*, Vol. AC-16, No. 6, December 1971, pp. 748-758.

45. Dawson, W.F. and Parke, N.G., IV, "Development Tools for Strapdown Guidance Systems," AIAA Guidance, Control and Flight Dynamics Conference, Paper No. 68-826, August 1968.

46. Toda, N.F., Schlee, F.H. and Obsharsky, P., "The Region of Kalman Filter Convergence for Several Autonomous Modes," AIAA Guidance, Control and Flight Dynamics Conference, AIAA Paper No. 67-623, Huntsville, Alabama, August 1967.

47. Schlee, F.H., Standish, C.J. and Toda, N.F., "Divergence in the Kalman Filter," *AIAA Journal*, Vol. 5, No. 6, pp. 1114-1120, June 1967.

48. Davisson, L.D., "The Theoretical Analysis of Data Compression Systems," *Proceedings of the IEEE*, Vol. 56, pp. 176-186, February 1968.

49. Bar-Shalom, Y., "Redundancy and Data Compression in Recursive Estimation," *IEEE Trans. on Automatic Control*, Vol. AC-17, No. 5, October 1972, pp. 684-689.

50. Dressler, R.M. and Ross, D.W., "A Simplified Algorithm for Suboptimal Non-Linear State Estimation," *Automatica*, Vol. 6, May 1970, pp. 477-480.

51. Joglekar, A.N., "Data Compression in Recursive Estimation with Applications to Navigation Systems," Dept. of Aeronautics and Astronautics, Stanford University, Stanford, California, SUDAAR No. 458, July 1973.

52. Joglekar, A.N. and Powell, J.D., "Data Compression in Recursive Estimation with Applications to Navigation Systems," AIAA Guidance and Control Conference, Key Biscayne, Florida, Paper No. 73-901, August 1973.

53. Carlson, N.A., "Fast Triangular Formulation of the Square Root Filter," *AIAA Journal*, Vol. 11, No. 5, pp. 1259-1265, September 1973.

54. Dyer, P. and McReynolds, "Extension of Square-Root Filtering to Include Process Noise," *Journal of Optimization Theory and Applications*, Vol. 3, No. 6, pp. 444-458, 1969.

PROBLEMS

Problem 8-1

Verify that Eq. (8.1-12) does yield the correct covariance matrix when the ϵ technique is employed.

Problem 8-2

In Example 8.1-3, derive the equation for p_∞

Problem 8-3

Show that when the second measurement in Example 8.4-1 is processed, the correct covariance matrix is recovered.

Problem 8-4

Show formally why the filter in Example 8.1-1 is not asymptotically stable.

Problem 8-5

Apply the ϵ technique to the system in Examples 8.1-1 and 8.1-3. Compare the result with the other two techniques in the examples.

Problem 8-6

Show the relationship between the exponential series for the transition matrix [Eq. (8.3-18)] and the Euler and modified Euler integration algorithms in Section 8.3.

Problem 8-7

Using the definition $P = WW^T$, set up the equations for W, the square root of the matrix

$$P = \begin{bmatrix} 2 & 1 \\ 1 & 2 \end{bmatrix}$$

Note that the equations do not have a unique solution. What happens when W is constrained to be triangular?

9. ADDITIONAL TOPICS

This chapter presents brief treatments of several important topics which are closely related to the material presented thus far; each of the topics selected has practical value. For example, while it has been assumed until now that the optimal filter, once selected, is held fixed in any application, it is entirely reasonable to ask whether information acquired *during* system operation can be used to improve upon the *a priori* assumptions that were made at the outset. This leads us to the topic of *adaptive filtering*, treated in Section 9.1. One might also inquire as to the advantages of a filter chosen in a *form* similar to that of the Kalman filter, but in which the gain matrix, K(t), is specified on a basis other than to produce a statistically optimal estimate. This question is imbedded in the study of *observers*, Section 9.2, which originated in the treatment of state reconstruction for linear deterministic systems. Or, one might be interested in the class of estimation techniques which are not necessarily optimal in any statistical sense, but which yield recursive estimators possessing certain well-defined convergence properties. These *stochastic approximation* methods are examined in Section 9.3. The subject of *real-time parameter identification* can be viewed as an application of nonlinear estimation theory; it is addressed in Section 9.4. Finally, the very important subject of *optimal control* – whose mathematics, interestingly enough, closely parallels that encountered in optimal estimation – is treated in Section 9.5.

9.1 ADAPTIVE KALMAN FILTERING

We have seen that for a Kalman filter to yield optimal performance, it is necessary to provide the correct *a priori* descriptions of F, G, H, Q, R, and P(0). As a practical fact, this is usually impossible; *guesses* of these quantities must be advanced. Hopefully, the filter design will be such that the penalty for misguesses is small. But we may raise an interesting question – i.e., "Is it possible to deduce non-optimal behavior during operation and thus improve the quality of *a priori* information?" Within certain limits, the answer is *yes*. The particular viewpoint given here largely follows Mehra (Ref. 1); other approaches can be found in Refs. 2-4.

INNOVATIONS PROPERTY OF THE OPTIMAL FILTER

For a continuous system and measurement given by

$$\dot{\underline{x}} = F\underline{x} + G\underline{w} \qquad (9.1\text{-}1)$$

$$\underline{z} = H\underline{x} + \underline{v} \qquad (9.1\text{-}2)$$

and a filter given by

$$\dot{\hat{\underline{x}}} = F\hat{\underline{x}} + K\underline{\nu} \qquad (9.1\text{-}3)$$

$$\underline{\nu} = \underline{z} - H\hat{\underline{x}} \qquad (9.1\text{-}4)$$

the *innovations property* (Ref. 5) states that, if K is the optimal gain,

$$E[\underline{\nu}(t_1)\underline{\nu}(t_2)^T] = 0, \quad t_1 \neq t_2 \qquad (9.1\text{-}5)$$

In other words the *innovations process*, $\underline{\nu}$, is a white noise process. Heuristically, there is no "information" left in $\underline{\nu}$, if $\hat{\underline{x}}$ is an optimal estimate.

Equation (9.1-5) is readily proved. From Eqs. (9.1-2) and (9.1-4) we see that ($\tilde{\underline{x}} = \hat{\underline{x}} - \underline{x}$):

$$\underline{\nu} = H(\underline{x} - \hat{\underline{x}}) + \underline{v} = -H\tilde{\underline{x}} + \underline{v} \qquad (9.1\text{-}6)$$

Thus, for $t_2 > t_1$, we get

$$E[\underline{\nu}(t_2)\underline{\nu}(t_1)^T] = H(t_2)\,E[\tilde{\underline{x}}(t_2)\,\tilde{\underline{x}}^T(t_1)]\,H^T(t_1) - H(t_2)\,E[\tilde{\underline{x}}(t_2)\,\underline{v}^T(t_1)] + R(t_1)\,\delta(t_2 - t_1) \qquad (9.1\text{-}7)$$

From Eqs. (9.1-1, 2, 3, 4) it is seen that $\tilde{\underline{x}}$ satisfies the differential equation

$$\dot{\tilde{\underline{x}}} = (F - KH)\,\tilde{\underline{x}} - G\underline{w} + K\underline{v} \qquad (9.1\text{-}8)$$

The solution to this equation is

$$\underline{\tilde{x}}(t_2) = \Phi(t_2,t_1)\underline{\tilde{x}}(t_1) - \int_{t_1}^{t_2} \Phi(t_2,\tau)\,[G(\tau)\,\underline{w}(\tau) - K(\tau)\,\underline{v}(\tau)]\,d\tau \qquad (9.1\text{-}9)$$

where $\Phi(t_2, t_1)$ is the transition matrix corresponding to (F − KH). Using Eq. (9.1-9), we directly compute

$$E[\underline{\tilde{x}}(t_2)\,\underline{\tilde{x}}^T(t_1)] = \Phi(t_2, t_1)\,P(t_1) \qquad (9.1\text{-}10)$$

$$E[\underline{\tilde{x}}(t_2)\,\underline{v}^T(t_1)] = \Phi(t_2, t_1)\,K(t_1)\,R(t_1) \qquad (9.1\text{-}11)$$

Therefore, from Eqs. (9.1-7, 10, 11)

$$E[\underline{\nu}(t_2)\,\underline{\nu}^T(t_1)] = H(t_2)\,\Phi(t_2, t_1)\,[P(t_1)\,H^T(t_1) - K(t_1)\,R(t_1)] + R(t_1)\,\delta(t_1 - t_2) \qquad (9.1\text{-}12)$$

But for the optimal filter $K(t_1) = P(t_1)\,H^T(t_1)\,R^{-1}(t_1)$, therefore,

$$E[\underline{\nu}(t_2)\,\underline{\nu}^T(t_1)] = R(t_1)\,\delta(t_1 - t_2) \qquad (9.1\text{-}13)$$

which is the desired result. Note that Eq. (9.1-12) could have been employed in a *derivation* of the Kalman gain as that which "whitened" the process $\underline{\nu}(t)$.

ADAPTIVE KALMAN FILTER

At this point we restrict our attention to time-invariant systems for which Eqs. (9.1-1) and (9.1-3) are stable. Under this condition the autocorrelation function, $E[\underline{\nu}(t_2)\,\underline{\nu}^T(t_1)]$, is a function of $\tau = t_2 - t_1$ only, viz.:

$$E[\underline{\nu}(t_1 + \tau)\,\underline{\nu}^T(t_1)] = He^{(F-KH)|\tau|}\,(PH^T - KR) + R\delta(\tau) \qquad (9.1\text{-}14)$$

In the stationary, discrete-time case, corresponding results are

$$\begin{aligned} E[\underline{\nu}_k\,\underline{\nu}_{k-j}{}^T] &= HP(-)\,H^T + R \qquad \text{for } j = 0 \\ &= H[\Phi(I-KH)]^{j-1}\,\Phi\,(P(-)\,H^T - K\,[HP(-)\,H^T + R]) \\ &\qquad \text{for } j > 0 \end{aligned} \qquad (9.1\text{-}15)$$

which is independent of k. Here, Φ is the discrete system transition matrix. Note that the optimal choice, $K = P(-)\,H^T\,[HP(-)\,H^T + R]^{-1}$, makes the expression vanish for all $j \neq 0$.

In an adaptive Kalman filter, the innovations property is used as a criterion to test for optimality, see Fig. 9.1-1. Employing tests for whiteness, mean and covariance, the experimentally measured steady-state correlation function $E[\underline{\nu}_k \underline{\nu}_{k-j}^T]$ is processed to identify unknown Q and R, for known F, G and H. It can be shown that, for this case, the value of K which whitens the innovations process is the optimal gain. If F, G, and H are also unknown, the equations for identification are much more complicated. Care must be exercised in the identification of unknown system matrices from the innovations sequence; for example, it is known that whiteness of the innovations sequence is not a sufficient condition to identify an unknown system F matrix. Thus, non-unique solutions for the system F matrix can be obtained from an identification scheme based on the innovations sequence. The following simple example demonstrates use of the innovations sequence for adaptive Kalman filtering.

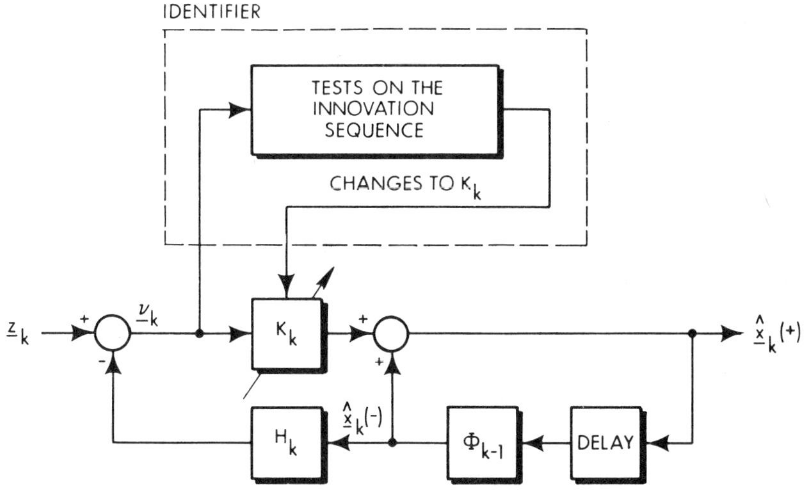

Figure 9.1-1 Adaptive Kalman Filter

Example 9.1-1

Suppose we have the continuous (scalar) system and measurement given by

$$\dot{x} = w, \qquad w \sim N(0,q)$$

$$z = x + v, \qquad v \sim N(0,r)$$

and utilize the measurement data to estimate x according to

$$\dot{\hat{x}} = k(z - \hat{x})$$

where k is based on a set of *incorrect* assumed values for q and r. The true values of q and r can be deduced as follows.

First, for steady-state system operation, the process $\nu(t) = z - \hat{x}$ is recorded and its autocorrelation function obtained. For a sufficient amount of information, this yields [f=0, h=1, in Eq. (9.1-12)] :

$$E[\nu(t)\,\nu(t-\tau)] = \varphi_{\nu\nu}(\tau) = e^{-k|\tau|}\,(p_\infty - kr) + r\,\delta(\tau)$$

Next, this experimental result is further processed to yield the numbers r and $(p_\infty - kr)$. Here, p_∞ is the steady-state value of p; it is the solution to the linear variance equation associated with Eq. (9.1-8),

$$\dot{p}_\infty = 0 = -2kp_\infty + q + k^2 r$$

or

$$p_\infty = \frac{q + k^2 r}{2k}$$

With the values of k, r, and p_∞ already known, this equation is solved for q. Thus, we have identified r and q based on analysis of data acquired during system operation, for this simple example.

In practice, the crux of the matter is processing $E[\underline{\nu}_k \underline{\nu}_{k-j}^T]$ to yield the quantities of interest. For the high-order systems of practical interest, the algorithms proposed in Refs. 1 and 3 may not work as well as theory would predict; other more heuristically motivated approaches may be both computationally simpler and more effective (e.g., Ref. 6). Nevertheless, the viewpoint presented herein is enlightening, and thus worthwhile. Extensions of the viewpoint to non-real-time adaptive smoothing may be accomplished through application to the forward filter. Here again, heuristically motivated and computationally simpler approaches may have a great deal to offer in practice.

9.2 OBSERVERS

In some estimation problems, it may be desired to reconstruct the state of a *deterministic*, linear dynamical system – based on *exact* observations of the system output. For deterministic problems of this nature, stochastic estimation concepts are not directly applicable. Luenberger (Refs. 7 and 8) formulated the notion of an *observer* for reconstructing the state vector of an *observable* deterministic linear system from exact measurements of the output.

Assume that m linearly independent, noise-free measurements are available from an nth-order system $(m < n)$. The initial system state, $\underline{x}_0$, is assumed to be a random vector. Then an observer of order (n–m) can be formulated which, by observing the system output, will reconstruct the current state of the system *exactly* in an asymptotic sense. Hence, an observer is a *reduced-order* estimator. A major application of observer concepts has been to deterministic feedback control problems, where the control law may depend on knowledge of *all* the system states, while only limited combinations of the states are measurable (Ref. 9).

As formulated by Luenberger, an observer is designed to be an *exponential estimator*, – i.e., for a time-invariant linear system, the estimation error will decay exponentially. Since there is no stochastic covariance equation which can be used to specify a unique optimal observer in the minimum mean square error sense, the eigenvalues of the observer can be chosen *arbitrarily* to achieve desired response characteristics. The observer response time chosen should be fast enough to provide convergence of the estimates within the time interval of interest. Observers can also be constructed to provide accurate state estimates for time-varying, deterministic systems – provided the observer response time is chosen to be short, relative to the system time variations.

In the sequel, the theory of reduced-order observers for continuous deterministic dynamic systems is presented. These results are then generalized to continuous stochastic estimation problems, containing both noisy and noise-free measurements. The *stochastic observer* unifies the concepts of deterministic Luenberger observer theory and stochastic Kalman filtering theory (Refs. 10 and 11). Only *continuous* linear systems are treated herein; however, analogous results have also been derived for discrete systems (Refs. 12, 13 and 14).

OBSERVERS FOR DETERMINISTIC SYSTEMS

In this section, Luenberger's theory of reduced-order observers is described in a form which facilitates extension to stochastic state estimation for time-varying systems. Consider a linear deterministic n^{th}-order system described by

$$\dot{\underline{x}}(t) = F(t)\,\underline{x}(t) + L(t)\,\underline{u}(t)\,; \qquad \underline{x}(t_0) = \underline{x}_0 \tag{9.2-1}$$

where $\underline{u}(t)$ is a deterministic (control) input. Observations of the state are available according to

$$\underline{z}(t) = H(t)\,\underline{x}(t) \tag{9.2-2}$$

H(t) is an m x n measurement matrix ($m < n$) which is assumed to be of full rank. Thus, $\underline{z}(t)$ represents m linearly independent combinations of the state vector, $\underline{x}(t)$. It is also assumed that the system described by Eqs. (9.2-1) and (9.2-2) is *completely observable* (the observability condition is defined in Section 3.5).

It is desired to provide an estimate of the state vector, $\hat{\underline{x}}(t)$, employing an $(n-m)^{th}$-order observer. To do this, introduce an $(n - m)$ dimensional vector $\underline{\xi}(t)$,

$$\underline{\xi}(t) = T(t)\,\underline{x}(t) \tag{9.2-3}$$

such that

$$\begin{bmatrix} T(t) \\ \hline H(t) \end{bmatrix} \tag{9.2-4}$$

is a nonsingular matrix. The vector $\underline{\xi}(t)$ represents (n – m) linear combinations of the system states which are independent of the measurements, $\underline{z}(t)$. It is therefore possible to obtain the inverse transformation

$$\underline{x}(t) = \begin{bmatrix} T(t) \\ \hline H(t) \end{bmatrix}^{-1} \begin{bmatrix} \underline{\xi}(t) \\ \hline \underline{z}(t) \end{bmatrix} \tag{9.2-5}$$

For convenience, define

$$\begin{bmatrix} T(t) \\ \hline H(t) \end{bmatrix}^{-1} \triangleq [A(t) \,|\, B(t)] \tag{9.2-6}$$

so that

$$\underline{x}(t) = A(t)\,\underline{\xi}(t) + B(t)\,\underline{z}(t) \tag{9.2-7}$$

The concept of observers is based on devising an $(n - m)^{th}$-order estimator for the transformed state vector $\underline{\xi}(t)$, which can then be used to reconstruct an estimate of the original state vector $\underline{x}(t)$, according to the relationship of Eq. (9.2-7). In the following development, the form of the dynamic observer is presented and the corresponding error equations derived.

At the outset, some constraint relationships can be established between the A, B, T and H matrices, viz*:

$$AT + BH = I \tag{9.2-8}$$

and

$$\begin{bmatrix} T \\ \hline H \end{bmatrix} [A \,|\, B] = I \tag{9.2-9}$$

These constraints, which are a direct consequence of the inverse relationship defined in Eq. (9.2-6), are useful in what follows.

A differential equation for $\underline{\xi}$ can be easily obtained by differentiating Eq. (9.2-3) and substituting from Eqs. (9.2-1) and (9.2-7). The result is

$$\dot{\underline{\xi}} = (TFA + \dot{T}A)\,\underline{\xi} + (TFB + \dot{T}B)\,\underline{z} + TL\underline{u} \tag{9.2-10}$$

By differentiating the appropriate partitions of Eq. (9.2-9), it is seen that the relationships $\dot{T}A = -T\dot{A}$ and $\dot{T}B = -T\dot{B}$ must hold. It is convenient to substitute these relationships into Eq. (9.2-10) to obtain an equivalent differential equation for $\underline{\xi}$, in the form

*Henceforth, the time arguments are dropped from variable quantities, except where necessary for clarification.

$$\dot{\underline{\xi}} = (\mathrm{TFA} - \mathrm{T\dot{A}})\,\underline{\xi} + (\mathrm{TFB} - \mathrm{T\dot{B}})\,\underline{z} + \mathrm{TL}\underline{u} \tag{9.2-11}$$

In order to reconstruct an estimate, $\hat{\underline{\xi}}$, of the vector $\underline{\xi}$, it is appropriate to design an observer which models the known dynamics of $\underline{\xi}$, given by Eq. (9.2-11), utilizing $\underline{z}$ and $\underline{u}$ as known inputs. We are therefore led to an observer of the form

$$\dot{\hat{\underline{\xi}}} = (\mathrm{TFA} - \mathrm{T\dot{A}})\,\hat{\underline{\xi}} + (\mathrm{TFB} - \mathrm{T\dot{B}})\,\underline{z} + \mathrm{TL}\underline{u} \tag{9.2-12a}$$

$$\hat{\underline{x}} = \mathrm{A}\,\hat{\underline{\xi}} + \mathrm{B}\underline{z} \tag{9.2-12b}$$

A block diagram of this observer is illustrated in Fig. 9.2-1.

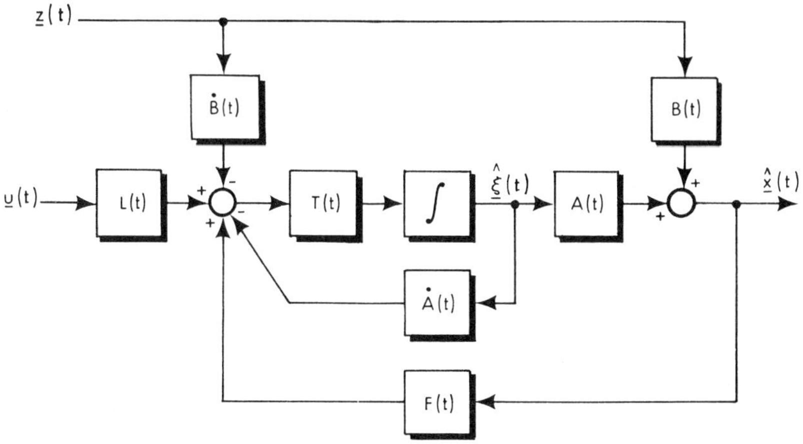

Figure 9.2-1 Deterministic Observer Block Diagram

It is important to note that for every initial state of the system, $\underline{x}(t_o)$, there exists an initial state $\hat{\underline{\xi}}(t_o)$ of the observer given by Eq. (9.2-12) such that $\hat{\underline{x}}(t) = \underline{x}(t)$ for any $\underline{u}(t)$, for all $t \geq t_o$. Thus, if properly initialized, the observer will track the true system state exactly. In practice, however, the proper initial condition is not known, so it is appropriate to consider the propagation of the observer error. As mentioned previously, observers exhibit the property that the observer error, defined by

$$\tilde{\underline{\xi}} = \hat{\underline{\xi}} - \underline{\xi} \tag{9.2-13}$$

decays exponentially to zero. This is easily demonstrated by subtracting Eq. (9.2-11) from Eq. (9.2-12a), to obtain the differential equation

$$\dot{\tilde{\underline{\xi}}} = (\mathrm{TFA} - \mathrm{T\dot{A}})\,\tilde{\underline{\xi}} \tag{9.2-14}$$

Note that if the observer is chosen to be asymptotically stable, then $\tilde{\underline{\xi}}(t)$ will tend uniformly and asymptotically to zero for arbitrary $\tilde{\underline{\xi}}(t_0)$. The stability of the observer and the behavior of $\tilde{\underline{\xi}}$ are both determined by the properties of the matrix $TFA - T\dot{A}$; the eigenvalues of this matrix can be chosen arbitrarily by appropriate specification of T, A and B, subject to the constraints of Eqs. (9.2-8) and (9.2-9).

Now consider the total state estimation error, $\tilde{\underline{x}}$, defined by

$$\tilde{\underline{x}} = \hat{\underline{x}} - \underline{x} \tag{9.2-15}$$

The differential equation for $\tilde{\underline{x}}$ is derived in the following manner. From Eqs. (9.2-7) and (9.2-12b), it is seen that $\tilde{\underline{x}}$ is related to $\tilde{\underline{\xi}}$ by

$$\tilde{\underline{x}} = A\,\tilde{\underline{\xi}} \tag{9.2-16}$$

It follows from Eq. (9.2-16) that $\tilde{\underline{x}}$ tends uniformly and asymptotically to zero, owing to the convergence properties of $\tilde{\underline{\xi}}$ previously discussed. The corresponding relationship

$$\tilde{\underline{\xi}} = T\tilde{\underline{x}} \tag{9.2-17}$$

can be obtained by premultiplying Eq. (9.2-16) by T and invoking the constraint TA=I from Eq. (9.2-9). Using these relationships, the differential equation for $\tilde{\underline{x}}$ can be derived as

$$\begin{aligned}\dot{\tilde{\underline{x}}} &= \dot{A}\,\tilde{\underline{\xi}} + A\,\dot{\tilde{\underline{\xi}}} \\ &= (\dot{A}T + ATF - AT\dot{A}T)\,\tilde{\underline{x}}\end{aligned} \tag{9.2-18}$$

Notice from Eq. (9.2-18) that the estimation error behavior depends on specifying the matrix products $\dot{A}T$ and AT. It would be more convenient, from a design standpoint, to specify the desired observer error characteristics in terms of fewer parameters. Fortunately, it is easy to demonstrate that Eq. (9.2-18) can be written in the equivalent form

$$\dot{\tilde{\underline{x}}} = (F - BHF - B\dot{H})\,\tilde{\underline{x}} \tag{9.2-19}$$

The transformation from Eq. (9.2-18) to Eq. (9.2-19) is left as an exercise for the reader. (Hint: show that $AT\,\tilde{\underline{x}} = \tilde{\underline{x}}$, and employ the constraints $AT + BH = I$ and $H\dot{A} = -\dot{H}A$).

From Eq. (9.2-19), it is apparent that for a given system described by F and H, *the estimation error depends only on the choice of B*. It can be shown that if the system is completely observable, the matrix B can be chosen to achieve any desired set of (n – m) eigenvalues for the error response. For the special case of a time-invariant system, the observer may be specified according to the following procedure:

- Choose an arbitrary set of eigenvalues, Λ (complex conjugate pairs)
- Pick B such that F − BHF has Λ as its nonzero set of eigenvalues
- Choose A and T consistent with AT + BH = I.

The choice of A and T which satisfies Eq. (9.2-8) is not unique. To illustrate this, suppose that an allowable pair of matrices (A*, T*) is chosen (a method for constructing such an allowable pair is given in Ref. 52). Then the pair (A,T) given by

$$A = A^*M^{-1}$$
$$T = MT^* \qquad (9.2\text{-}20)$$

also satisfies Eq. (9.2-8), where M is *any* nonsingular matrix. The set of all allowable pairs (A, T) defines an *equivalent* class of observers which exhibit the same error behavior.

Example 9.2-1

Consider the second-order example illustrated in Fig. 9.2-2. This system model might be representative of a simplified one-dimensional tracking problem, where x_1 and x_2 are position and velocity of the tracked object, respectively. The system dynamics are

$$\dot{\underline{x}} = \begin{bmatrix} \dot{x}_1 \\ \dot{x}_2 \end{bmatrix} = \begin{bmatrix} 0 & 1 \\ 0 & -\beta \end{bmatrix} \begin{bmatrix} x_1 \\ x_2 \end{bmatrix} + \begin{bmatrix} 0 \\ \ell \end{bmatrix} u$$

with measurements of position available according to

$$z = [1 \quad 0] \begin{bmatrix} x_1 \\ x_2 \end{bmatrix}$$

It is desired to construct an observer for this system.

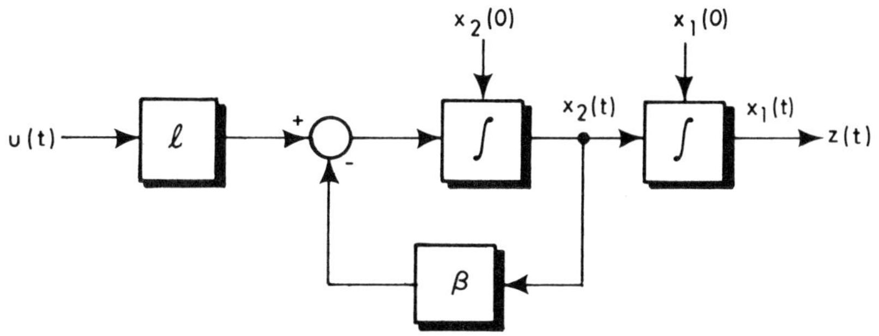

Figure 9.2-2 Second-Order Tracking System Example

From Eq. (9.2-9), HB = I, so that B is constrained to be of the form

$$B = \begin{bmatrix} 1 \\ b_2 \end{bmatrix}$$

The error dynamics can be written as [Eq. (9.2-19)]

$$\dot{\tilde{\underline{x}}} = (F - BHF)\,\tilde{\underline{x}}$$

where

$$F - BHF = \begin{bmatrix} 0 & 0 \\ 0 & -(\beta + b_2) \end{bmatrix}$$

Since n − m = 1 for this problem, the observer is of first order and is specified by the single eigenvalue, $\lambda = -(\beta + b_2)$. It is desirable to choose the observer time constant to be significantly smaller than the system constant. Arbitrarily choose $\lambda = -5\beta$, which implies $b_2 = 4\beta$. A possible choice of A and T, satisfying Eqs. (9.2-8) and (9.2-9), is

$$A = \begin{bmatrix} 0 \\ 1 \end{bmatrix}, \quad T = [-4\beta \;\; 1]$$

This completes the specification of the first-order observer. The corresponding block diagram is illustrated in Fig. 9.2-3. Although the choice of A and T is not unique, it is easily demonstrated that *any allowable* choice of A and T leads to the equivalent observer configuration of Fig. 9.2-3. This is left as an exercise for the reader.

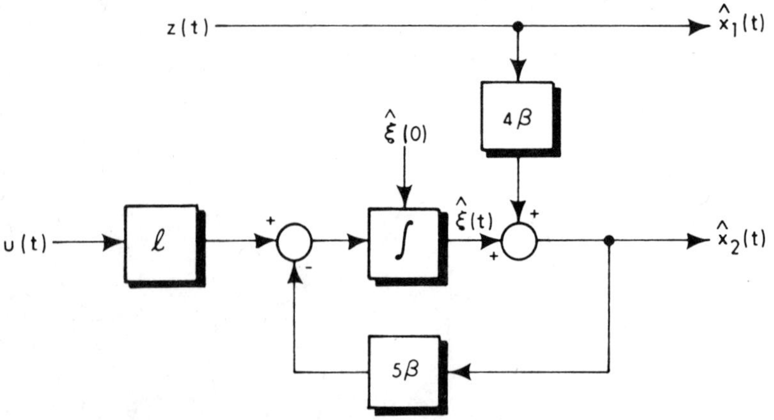

Figure 9.2-3 Observer Configuration for Simplified Tracking Example

The observer performance may be illustrated by a numerical example. Assume that the system parameters are

$$\beta = 1.0\ \text{sec}^{-1},\ \ell = 1.0,\ u(t) = -1.0\ \text{ft/sec}^2$$

$$x_1(0) = 1.0\ \text{ft},\ x_2(0) = 1.0\ \text{ft/sec}$$

The observer is initialized by choosing

$$\hat{x}_1(0) = x_1(0) = 1.0 \text{ ft}, \ \hat{x}_2(0) = 0$$

so that $\hat{\xi}(0) = -4.0$ ft/sec. The resulting system and observer outputs are plotted in Fig. 9.2-4. Note that x_1 is estimated without error since it is directly measured, while the error in the estimate of x_2 decays exponentially to zero with a time constant of 0.2 sec.

OBSERVERS FOR STOCHASTIC SYSTEMS

The formulation for deterministic observers can be extended to encompass stochastic systems. Stochastic observer theory forges the link between reduced-state deterministic observers and optimal Kalman filtering; in some applications, the stochastic observer offers a convenient approach for the design of reduced-state filtering algorithms. In the sequel, a *heuristic* approach is taken to the design of stochastic observers.

Consider the system described by

$$\dot{\underline{x}} = F\underline{x} + G\underline{w}, \qquad \underline{w} \sim N(\underline{0}, Q) \tag{9.2-21}$$

For simplicity, deterministic inputs are not included in the system model; any deterministic inputs are assumed to be compensated, and can be removed from the formulation without loss of generality. The measurements are assumed to be partly deterministic and partly stochastic, viz:

$$\underline{z} = \begin{bmatrix} \underline{z}_1 \\ \hline \underline{z}_2 \end{bmatrix} = \begin{bmatrix} H_1 \\ \hline H_2 \end{bmatrix} \underline{x} + \begin{bmatrix} \underline{v}_1 \\ \hline \underline{0} \end{bmatrix}, \qquad \underline{v}_1 \sim N(\underline{0}, R_1) \tag{9.2-22}$$

Of the m measurements, m_1 are noisy with measurement noise spectral density R_1, and m_2 are noise-free ($m_2 = m - m_1$). As before, we assume that H_2 is of full rank.

A configuration is sought for the stochastic observer that has the form of the $(n - m)^{th}$-order deterministic observer when all the measurements are noise-free ($m_1 = 0$), and becomes the nth-order Kalman filter when all the measurements are noisy ($m_2 = 0$). For combined noisy and noise-free measurements, the conventional continuous Kalman filter cannot be implemented, due to singularity of the measurement noise covariance matrix. In Section 4.5, an approach is given for modifying the Kalman filter to circumvent the singularity of R for certain special cases. *The concept of stochastic observers presented herein is applicable to the more general case and reduces exactly to the modified Kalman filter of Section 4.5, under equivalent conditions.*

It is immediately apparent from the deterministic observer block diagram, Fig. 9.2-1, that the noisy measurements must be processed differently than the noise-free measurements to avoid the appearance of white noise on the output. It is necessary to provide additional filtering of the noisy measurements to avoid this situation.

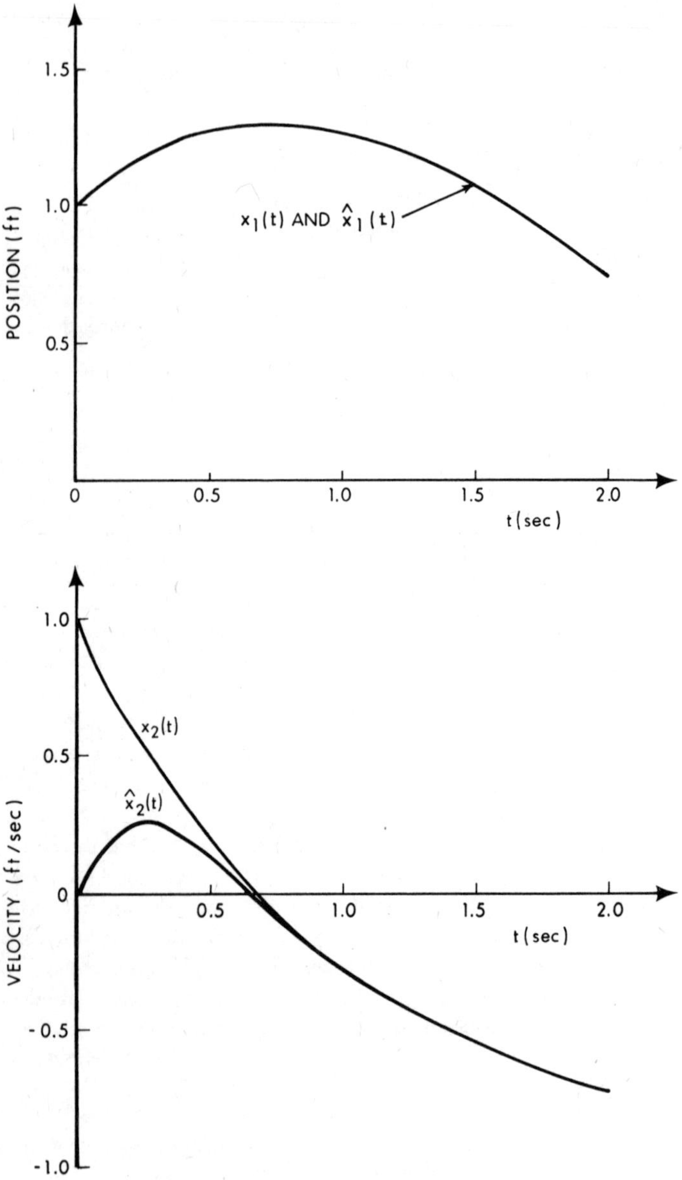

Figure 9.2-4 Actual and Estimated States vs. Time for Simplified Tracking Example

Define the $(n - m_2) \times n$ transformation

$$\underline{\xi} = T\underline{x} \tag{9.2-23}$$

Such that

$$\begin{bmatrix} T \\ \hline H_2 \end{bmatrix} \tag{9.2-24}$$

is a nonsingular matrix. It then follows that

$$\underline{x} = A\,\underline{\xi} + B_2\,\underline{z}_2 \tag{9.2-25}$$

where

$$AT + B_2H_2 = I \tag{9.2-26}$$

By analogy to the deterministic problem, $\underline{\xi}$ can be shown to satisfy the differential equation

$$\dot{\underline{\xi}} = (TFA - T\dot{A})\,\underline{\xi} + (TFB_2 - T\dot{B}_2)\,\underline{z}_2 + TGw \tag{9.2-27}$$

This is the same expression previously derived in Eq. (9.2-11), except that the deterministic input, $L\underline{u}$, has been replaced by the stochastic input, $G\underline{w}$.

It is reasonable to postulate that the stochastic observer should be designed to process the noise-free measurements, $\underline{z}_2$, according to the deterministic observer formulation, while processing the noisy measurements, $\underline{z}_1$, according to the stochastic Kalman filter formulation. Accordingly, the stochastic observer can be postulated to have the form

$$\hat{\underline{x}} = A\,\hat{\underline{\xi}} + B_2\,\underline{z}_2 \tag{9.2-28a}$$

$$\dot{\hat{\underline{\xi}}} = (TFA - T\dot{A})\,\hat{\underline{\xi}} + (TFB_2 - T\dot{B}_2)\,\underline{z}_2 + TB_1\,(\underline{z}_1 - H_1\hat{\underline{x}}) \tag{9.2-28b}$$

An additional $n \times m_1$ free gain matrix, B_1, has been incorporated for processing the noisy measurements, $\underline{z}_1$. Notice that the noisy measurements appear only as *inputs* to the observer dynamics in a manner analogous to the Kalman filter formulation, and are *not* fed forward directly into the state estimates, as are the noise-free measurements. A block diagram for this stochastic observer configuration is illustrated in Fig. 9.2-5. The choice of the observer specified by Eqs. (9.2-28) has been constrained so that in the absence of the $\underline{z}_2$ measurements, the observer structure will be identical to the nth-order Kalman filter. It is easily shown that, for $m_2 = 0$,

$$T = A = I$$

$$\dot{A} = 0 \tag{9.2-29}$$

and Fig. 9.2-5 reduces to the continuous Kalman filter shown in Fig. 4.3-1. In this case, the optimal choice of the gain B_1 corresponds to the Kalman gain. For the other extreme, where all m measurements are noise-free ($m_1 = 0$), Fig. 9.2-5 immediately reduces to the deterministic $(n - m)^{th}$-order observer depicted in Fig. 9.2-1.

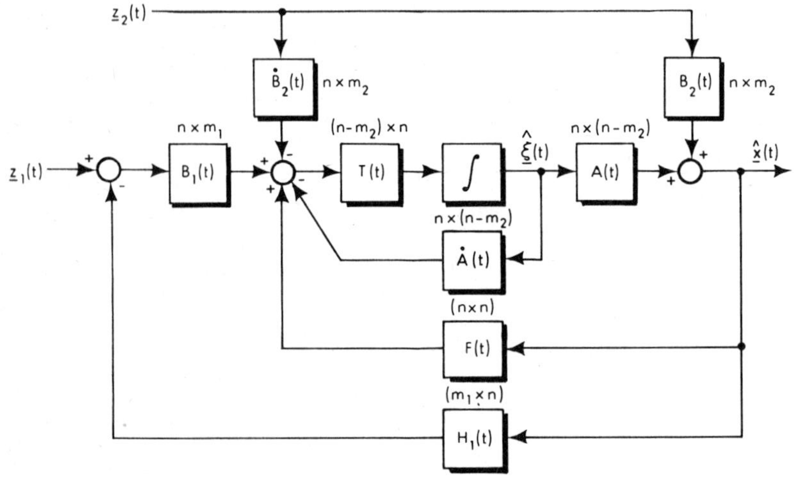

Figure 9.2-5 Stochastic Observer Block Diagram

The estimation error dynamics for the stochastic observer may be determined in a manner analogous to that used to derive the deterministic observer dynamics. From Eqs. (9.2-27) and (9.2-28b), the expression for $\dot{\tilde{\underline{\xi}}}$ is

$$\dot{\tilde{\underline{\xi}}} = (TFA - T\dot{A})\,\tilde{\underline{\xi}} + TB_1\,(\underline{z}_1 - H_1\hat{\underline{x}}) - TG\underline{w} \tag{9.2-30}$$

Noting that

$$\tilde{\underline{\xi}} = T\tilde{\underline{x}} \tag{9.2-31a}$$

$$\underline{z}_1 - H_1\hat{\underline{x}} = -H_1\tilde{\underline{x}} + \underline{v}_1 \tag{9.2-31b}$$

the differential equation for the estimation error, $\tilde{\underline{x}}$, is then obtained as

$$\dot{\tilde{\underline{x}}} = \dot{A}\,\tilde{\underline{\xi}} + A\,\dot{\tilde{\underline{\xi}}}$$

$$= (\dot{A}T + ATF - AT\dot{A}T - ATB_1H_1)\,\tilde{\underline{x}} + ATB_1\underline{v}_1 - ATG\underline{w} \tag{9.2-32}$$

Following the deterministic observer analysis, an equivalent differential equation for $\tilde{\underline{x}}$ can be expressed in terms of the gain B_2 as

$$\dot{\tilde{\underline{x}}} = (F - B_2 H_2 F - B_2 \dot{H}_2 - ATB_1 H_1)\tilde{\underline{x}} + ATB_1 \underline{v}_1 - (I - B_2 H_2) G \underline{w} \quad (9.2\text{-}33)$$

Consider replacing B_1 by ATB_1 in Eq. (9.2-33). Using the identity $TA = I$, it can be seen that the error dynamics are unaffected by the substitution. Hence, Eq. (9.2-33) may be simplified by replacing the term ATB_1 by B_1, yielding the following equivalent expression for the estimation error dynamics:

$$\dot{\tilde{\underline{x}}} = (F - B_2 H_2 F - B_2 \dot{H}_2 - B_1 H_1)\tilde{\underline{x}} + B_1 \underline{v}_1 - (I - B_2 H_2) G \underline{w} \quad (9.2\text{-}34)$$

Notice that, in the case of no measurement or process noise, Eq. (9.2-34) reduces to the deterministic observer error dynamics of Eq. (9.2-19). In the absence of noise-free measurements, B_2 and H_2 are zero and Eq. (9.2-34) reduces to the standard Kalman filter error dynamics of Eq. (4.3-13), where B_1 is identified as the Kalman gain matrix.

OPTIMAL CHOICE OF B_1 AND B_2

The stochastic observer design may now be optimized by choosing B_1 and B_2 to minimize the mean square estimation error. The error covariance equation, determined from Eq. (9.2-34), is

$$\dot{P} = (F - B_2 H_2 F - B_2 \dot{H}_2 - B_1 H_1) P + P(F - B_2 H_2 F - B_2 \dot{H}_2 - B_1 H_1)^T$$
$$+ B_1 R_1 B_1^T + (I - B_2 H_2) GQG^T (I - B_2 H_2)^T \quad (9.2\text{-}35)$$

Both B_1 and B_2 may now be chosen to minimize the trace of $\dot{P}$ (see Problem 4-8). Minimizing first with respect to B_1, yields the optimal gain

$$B_1^{opt} = PH_1^T R_1^{-1} \quad (9.2\text{-}36)$$

Substituting B_1^{opt} into the covariance equation gives

$$\dot{P} = (F - B_2 H_2 F - B_2 \dot{H}_2) P + P(F - B_2 H_2 F - B_2 \dot{H}_2)^T$$
$$- PH_1^T R_1^{-1} H_1 P + (I - B_2 H_2) GQG^T (I - B_2 H_2)^T \quad (9.2\text{-}37)$$

The optimum choice of B_2 can be determined by minimizing the trace of $\dot{P}$ in Eq. (9.2-37), with respect to B_2. This computation leads readily to

$$B_2^{opt} = [PF^T H_2^T + GQG^T H_2^T + P\dot{H}_2^T]\,[H_2 GQG^T H_2^T]^{-1} \quad (9.2\text{-}38)$$

To complete the specification of the optimal stochastic observer, it is necessary to properly initialize the filter estimates and the covariance matrix. Due to the exact measurements, *discontinuities* occur at $t = 0^+$ Single stage estimation theory can be used to determine the initial conditions $\hat{\underline{x}}(0^+)$ and $P(0^+)$ from $\underline{z}_2(0)$ and the prior estimates, $\hat{\underline{x}}(0)$ and $P(0)$:

$$\hat{\underline{x}}(0^+) = \hat{\underline{x}}(0) + P(0)H_2^T(0)\,[H_2(0)\,P(0)\,H_2^T(0)]^{-1}\,[\underline{z}_2(0) - H_2(0)\hat{\underline{x}}(0)] \tag{9.2-39}$$

$$P(0^+) = P(0) - P(0)\,H_2^T(0)\,[H_2(0)\,P(0)\,H_2^T(0)]^{-1}\,H_2(0)\,P(0) \tag{9.2-40}$$

The initial estimate depends on the initial measurement, $\underline{z}_2(0)$, and cannot be determined *a priori*. The initial condition for the observer, $\hat{\underline{\xi}}(0^+)$, is related to $\hat{\underline{x}}(0^+)$ by

$$\hat{\underline{\xi}}(0^+) = T(0)\hat{\underline{x}}(0^+) \tag{9.2-41}$$

Note that the optimal gain B_2^{opt}, specified according Eq. (9.2-38), depends on the presence of process noise. In order to compute B_2^{opt}, it is necessary to invert the matrix $H_2GQG^TH_2^T$. This matrix will be nonsingular, yielding a unique solution for B_2^{opt}, if the first derivatives of the noise-free measurements contain white process noise. If some of the derivatives of the noise-free measurements are *also* free of white noise, then the choice of B_2 is not completely specified. Under these conditions, B_2 may be chosen to give desirable error convergence properties, as in the case of completely deterministic observers.

SPECIALIZATION TO CORRELATED MEASUREMENT ERRORS

In many filtering problems of practical interest, the measurements may be modeled as containing correlated measurement errors, where the measurement errors are described by *first-order* differential equations driven by white noise. Consider the nth-order system, with m measurements, described by

$$\begin{aligned} \dot{\underline{x}} &= F\underline{x} + G\underline{w} \\ \underline{z} &= H\underline{x} + \underline{v} \end{aligned} \tag{9.2-42}$$

where the measurement noise, $\underline{v}$, satisfies the differential equation

$$\dot{\underline{v}} = E\underline{v} + \underline{w}_1 \tag{9.2-43}$$

The $(n + m)^{th}$-order augmented system, with $\underline{x}'^T = [\underline{x}^T \,|\, \underline{v}^T]$ is described by

$$\begin{aligned} \dot{\underline{x}}' &= F'\underline{x}' + G'\underline{w}' \\ \underline{z}_2 &= H'_2\underline{x}' \end{aligned} \tag{9.2-44}$$

where

$$F' = \left[\begin{array}{c|c} F & 0 \\ \hline 0 & E \end{array}\right], G' = \left[\begin{array}{c|c} G & 0 \\ \hline 0 & I \end{array}\right], H'_2 = [H \quad I], \; Q' = \left[\begin{array}{c|c} Q & 0 \\ \hline 0 & Q_1 \end{array}\right] \tag{9.2-45}$$

The nth-order optimal stochastic observer is now derived for this problem.

A useful property of observers is that the estimation error is orthogonal to the noise-free measurements, so that

$$H'_2 \underline{\tilde{x}}' = H\underline{\tilde{x}} + \underline{\tilde{v}} = \underline{0} \tag{9.2-46}$$

The proof of this result is left as an exercise for the reader. [Hint: Premultiply the relation $\underline{\tilde{x}}' = A\,\underline{\tilde{\xi}}$ by H'_2, and use the constraint of Eq. (9.2-9)]. Hence, the covariance matrix, $P' = E\,[\underline{\tilde{x}}'\,\underline{\tilde{x}}'^T]$, can be expressed as

$$P' = \left[\begin{array}{c|c} P & -PH^T \\ \hline -HP & HPH^T \end{array}\right] \tag{9.2-47}$$

where P is the covariance matrix of the error $\underline{\tilde{x}}$. The pertinent observer matrices may be partitioned according to

$$B_2 = \left[\begin{array}{c} B_{21} \\ \hline B_{22} \end{array}\right], \; T = [T_1 \,|\, T_2], \; A = \left[\begin{array}{c} A_1 \\ \hline A_2 \end{array}\right] \tag{9.2-48}$$

where B_{21} is n x m, T_1 is n x n and A_1 is n x n. Substituting Eqs. (9.2-45) and (9.2-47) into the optimal gain expression of Eq. (9.2-38) gives

$$B_{21}^{opt} = [P(\dot{H} + HF - EH)^T + GQG^TH^T]\;[HGQG^TH^T + Q_1]^{-1} \tag{9.2-49}$$

$$\begin{aligned} B_{22}^{opt} &= [-HP(\dot{H} + HF - EH)^T + Q_1]\;[HGQG^TH^T + Q_1]^{-1} \\ &= I - HB_{21}^{opt} \end{aligned} \tag{9.2-50}$$

A possible selection of A and T that satisfies the constraint $AT + B_2H'_2 = I$ is given by

$$A = \left[\begin{array}{c} I \\ \hline -H \end{array}\right]$$

$$T = [I - B_{21}H \,|\, -B_{21}] \tag{9.2-51}$$

The observer dynamics are described by

$$\hat{\underline{x}}' = A\,\hat{\underline{\xi}} + B_2\underline{z}$$

$$\dot{\hat{\underline{\xi}}} = (TF'A - T\dot{A})\,\hat{\underline{\xi}} + (TF'B_2 - T\dot{B}_2)\underline{z} \tag{9.2-52}$$

Substituting for T, F′, $\dot{A}$ and B_2 leads to the result

$$TF'A - T\dot{A} = F - B_{21}H_1$$

$$TF'B_2 - T\dot{B}_2 = FB_{21} - B_{21}E - \dot{B}_{21} - B_{21}H_1B_{21} \tag{9.2-53}$$

where H_1 has been defined as

$$H_1 = (\dot{H} + HF - EH) \tag{9.2-54}$$

A block diagram of the optimal observer is illustrated in Fig. 9.2-6, for the observer defined by Eqs. (9.2-52) and (9.2-53). Since we are not interested in estimating the noise, $\underline{v}$, only the estimates $\hat{\underline{x}}$ are shown in the figure. Assume that the prior estimate of P′ is given by

$$P'(0) = \left[\begin{array}{c|c} P(0) & 0 \\ \hline 0 & R(0) \end{array}\right] \tag{9.2-55}$$

For an initial estimate $\hat{\underline{x}}(0)$, the discontinuities in $\hat{\underline{x}}(0^+)$ and $P(0^+)$ are determined by appropriately partitioning Eqs. (9.2-39) and (9.2-40), resulting in

$$\hat{\underline{x}}(0^+) = \hat{\underline{x}}(0) + P(0)H^T(0)\,[H(0)\,P(0)\,H^T(0) + R(0)]^{-1}\,[\underline{z}(0) - H(0)\hat{\underline{x}}(0)]$$

$$P(0^+) = P(0) - P(0)H^T(0)\,[H(0)\,P(0)\,H^T(0) + R(0)]^{-1}\,H(0)\,P(0) \tag{9.2-56}$$

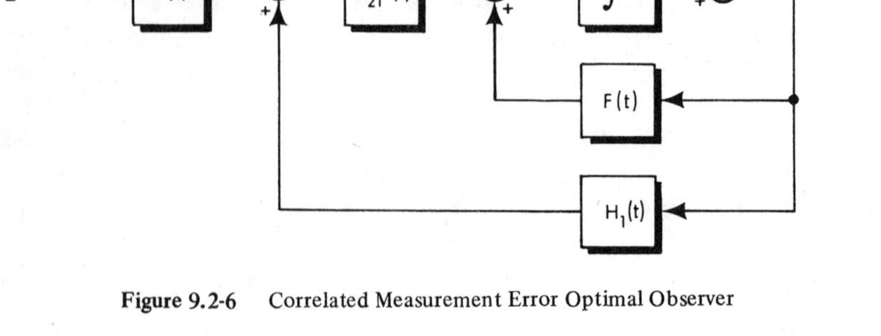

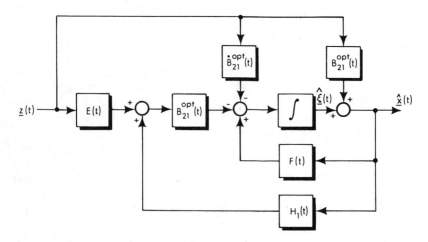

Figure 9.2-6 Correlated Measurement Error Optimal Observer

The optimal reduced-order observer for the special case of correlated measurement errors is identical to the modified Kalman filter derived in Section 4.5, using a completely different approach. The Kalman filter approach to the problem requires that special allowances be made to circumvent the singularity of the R matrix. Using the structure of stochastic observer theory, however, the solution to the correlated measurement error problem follows directly.

9.3 STOCHASTIC APPROXIMATION

Most of the material contained in this book is concerned with obtaining *optimal* estimates of a random vector $\underline{x}$, or a vector random process $\underline{x}(t)$, from noise-corrupted measurement data. Recall that an optimal estimate is one which minimizes an appropriate functional of the estimation error; examples of such criteria – maximum likelihood, least squares, etc. – are discussed in the introduction to Chapter 4. This section considers a class of estimation techniques, called *stochastic approximation methods*, that are not necessarily optimal in any statistical sense, but which yield recursive estimates with certain well-defined convergence properties.

The motivation for stochastic approximation methods is that optimal estimation criteria often depend upon assumptions about the statistical characteristics of $\underline{x}(t)$, and its associated measurement data, which may not hold true in practice. For instance, the Kalman filter yields a minimum variance estimate of $\underline{x}(t)$, provided the latter satisfies a linear stochastic differential equation driven by gaussian noise and measurements are linearly related to $\underline{x}(t)$ with additive gaussian noise. If the dynamics of $\underline{x}(t)$ and its observations are dominated by nonlinear effects that cannot be accurately approximated by linearization, or if the noise processes are nongaussian, the corresponding optimal estimation algorithm is often too complex to mechanize. More generally, if the noise statistics are unknown or undefined, the optimal estimate may be indeterminate. In such circumstances it is sometimes possible, through use of stochastic approximation methods, to obtain a sequence of estimates for $\underline{x}(t)$ that either asymptotically approaches the true value (when $\underline{x}(t)$ is constant), or possesses a statistically bounded error (in the time-varying case). The mathematical assumptions needed to prove these convergence properties are generally much weaker than those required to determine optimal estimators. Furthermore, most stochastic approximation algorithms are recursive linear functions of the measurement data that can be readily mechanized in a computer. Consequently, they offer attractive alternatives to optimal estimation techniques in some applications.

THE SCALAR CONSTANT-PARAMETER CASE

Stochastic approximation methods were first developed as iterative procedures for determining a solution, x_o, to the scalar nonlinear algebraic equation,

$$g(x) = 0 \tag{9.3-1}$$

where $g(\hat{x}_k)$ cannot be evaluated exactly for trial values, or "estimates," $\hat{x}_1, \ldots \hat{x}_k, \ldots$, of the argument, x. This is analogous to the classical *deterministic* problem of obtaining a solution to Eq. (9.3-1) for a *known* function g(x), which can be treated by any of several classical numerical techniques – Newton's method, successive approximations, etc. Most such numerical techniques have the common property that an approximate solution for x_0 is obtained by iteratively performing the calculation

$$\hat{x}_{k+1} = \hat{x}_k + k_k\, g(\hat{x}_k) \tag{9.3-2}$$

where k_k, k = 1, 2, . . . , is an appropriately chosen sequence of "gains" (denoted by $\{k_k\}$). In the method of successive approximations $k_k = -\mathrm{Sgn}\,[g'(\hat{x}_k)]$ for all values of k; in Newton's method*

$$k_k = -\,1/g'(\hat{x}_k) \tag{9.3-3}$$

The objective at each stage is to apply a correction to the most recent estimate of x_0, which yields a better estimate. Conditions under which the sequence $\{\hat{x}_k\}$, generated by Eq. (9.3-2), converges to a solution of Eq. (9.3-1) can be stated in terms of restrictions on the sequence of gains, on the function g(x) in the vicinity of the solution and on the initial "guess", x_1 (see Ref. 15).

The *stochastic* case refers to situations where $g(\hat{x}_k)$ cannot be evaluated exactly; instead, for each trial value of x a noise-corrupted observation

$$m_k = g(\hat{x}_k) + v_k \tag{9.3-4}$$

is generated. In the sequel it is convenient to assume that g(x) is a monotonically increasing or decreasing function having a unique solution to Eq. (9.3-1), and $\{v_k\}$ is a sequence of zero mean independent random variables having bounded variances. Furthermore it is assumed that g(x) has finite, nonzero slope as illustrated in Fig. 9.3-1 – i.e.,

$$0 < a \leqslant |g'(x)| \leqslant b < \infty$$

Somewhat less restrictive conditions could be imposed; however, those given above suffice for many applications of interest.

A practical example of the type of problem described above might be the nonlinear control system illustrated in Fig. 9.3-2, where x represents the value of a control gain and g(x) represents the steady state error between the system output and a constant input, as an *unknown* function of x. If the objective is to

---

$$^*g'(\hat{x}_k) \stackrel{\Delta}{=} dg(x)/dx|_{x=\hat{x}_k}, \quad \mathrm{Sgn}[\alpha] \stackrel{\Delta}{=} \begin{cases} 1; & \alpha > 0 \\ 0; & \alpha = 0 \\ -1; & \alpha < 0 \end{cases}$$

determine the proper gain setting to achieve zero steady state error, an experiment can be devised whereby successive gain settings are tried and the resulting steady-state error is measured with an error v_k.

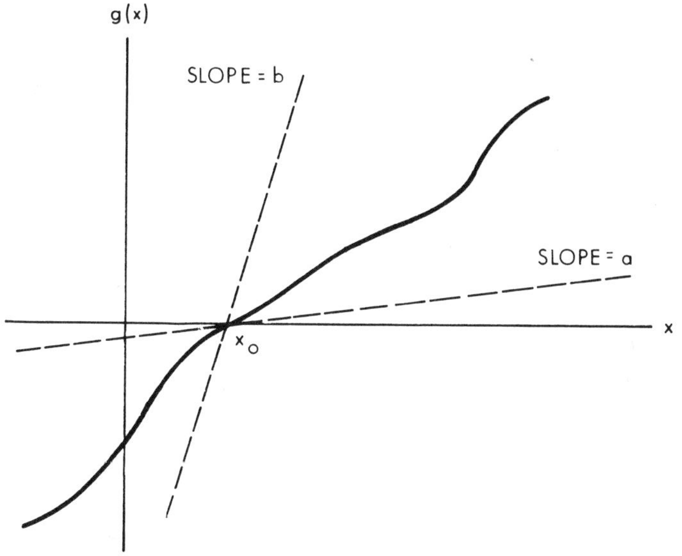

Figure 9.3-1 Graphical Illustration of the Class of Functions, g(x)

The background of iterative methods available for determining the solution to Eq. (9.3-1) in cases where g(x) is a known function led investigators, beginning in the early 1950's, to inquire whether a recursion relation of the form

$$\hat{x}_{k+1} = \hat{x}_k + k_k m_k \tag{9.3-5}$$

can also generate a sequence that in some sense converges to x_0 when $g(\hat{x}_k)$ is observable only through Eq. (9.3-4). Equation (9.3-5) has the same form as its deterministic counterpart, Eq. (9.3-2), except that $g(\hat{x}_k)$ is replaced by the measurement m_k. Because m_k has a random component, Eq. (9.3-5) is called a *stochastic approximation algorithm*.

A common definition of convergence applied to the random sequence generated by Eq. (9.3-5) is *mean square convergence*, defined by

$$\lim_{k\to\infty} E[(x_0-\hat{x}_k)^2] = 0 \tag{9.3-6}$$

The pioneering work of Robbins and Munroe (Ref. 16) demonstrates that the solution to Eq. (9.3-5) converges in mean square, if the gains k_k satisfy the conditions

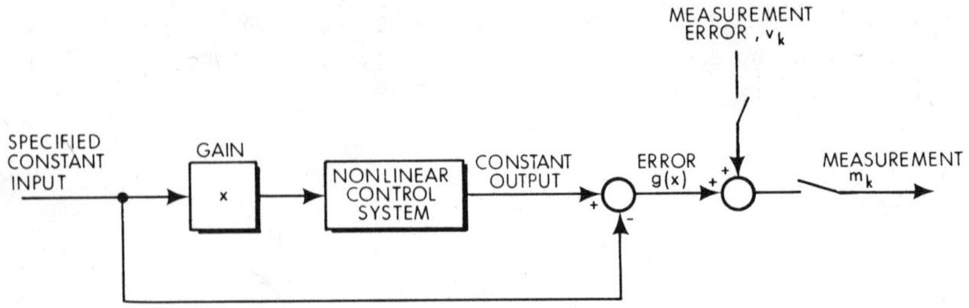

Figure 9.3-2 A Control System Application Where Stochastic Approximation Methods can be Applied to Determine the Proper DC Gain

$$\mathrm{Sgn}(k_k) = -\,\mathrm{Sgn}\left(g'(x_k)\right) \tag{9.3-7a}$$

$$\lim_{k\to\infty} (k_k) = 0 \tag{9.3-7b}$$

$$\sum_{k=1}^{\infty} |k_k| = \infty \tag{9.3-7c}$$

$$\sum_{k=1}^{\infty} {k_k}^2 < \infty \tag{9.3-7d}$$

We shall not supply convergence proofs here; however, the reader should qualitatively understand why the conditions in Eq. (9.3-7) are generally necessary for convergence. For instance, Eq. (9.3-7a) insures that the correction to each estimate $\hat{x}_k$ is in the proper direction, by analogy with the classical techniques for determining zeros of known functions. Hence, although g(x) is unknown, the sign of its derivative must be known in order to choose a proper sequence of gains. The condition that the gain sequence approaches zero [Eq. (9.3-7b)] is needed to insure that

$$\lim_{k\to\infty} \hat{x}_{k+1} = \hat{x}_k \tag{9.3-8}$$

in Eq. (9.3-5), otherwise $\left\{ \hat{x}_k \right\}$ cannot converge in a mean square sense to x_o. [Note that condition (9.3-7b) is implied by condition (9.3-7d).] The conditions in Eqs. (9.3-7c) and (9.3-7d) are needed to insure that $|k_k|$ approaches zero at the proper rate – not too fast, Eq. (9.3-7c), but not too slowly either, Eq. (9.3-7d); an example that specifically demonstrates the necessity of these conditions is given in the problem section.

Following Robbins and Munroe, others derived alternative conditions for convergence of stochastic approximation algorithms and investigated ways of choosing the gain sequence $\{k_k\}$ in Eq. (9.3-5) to achieve a high rate of convergence (Refs. 17, 18 and 19). Algorithms have been obtained for problems in estimation (Refs. 20 through 25), optimization (Refs. 26 and 27), and pattern classification (Refs. 28, 29, and 30). The subsequent discussion in this section pursues the viewpoint of our main interest, viz., estimation theory.

Suppose that an unknown constant x_0 is to be estimated from observations of the form

$$z_k = h(x_0) + v_k \tag{9.3-9}$$

where h(x) is a known function and $\{v_k\}$ is a sequence of independent zero mean random variables, having otherwise unknown statistical properties. If a new function g(x) is defined by

$$g(x) \triangleq h(x_0) - h(x) \tag{9.3-10}$$

then the problem of determining x_0 is equivalent to finding the solution to

$$g(x) = 0 \tag{9.3-11}$$

Subtracting $h(\hat{x}_k)$ from both sides of Eq. (9.3-9), using Eq. (9.3-10), and defining

$$m_k \triangleq z_k - h(\hat{x}_k) \tag{9.3-12}$$

we obtain equivalent measurements

$$m_k = g(\hat{x}_k) + v_k \tag{9.3-13}$$

Thus, estimation of x_0 is recast as the problem of approximating the solution to Eq. (9.3-11) from the measurement data $\{m_k\}$ in Eq. (9.3-13). Applying the algorithm in Eq. (9.3-5), we can estimate x_0, recursively, from the relation

$$\begin{aligned} \hat{x}_{k+1} &= \hat{x}_k + k_k m_k \\ &= \hat{x}_k + k_k\,[z_k - h(\hat{x}_k)] \end{aligned} \tag{9.3-14}$$

where the gain sequence $\{k_k\}$ is chosen to satisfy the conditions in Eq. (9.3-7). Equation (9.3-14) is similar in form to the discrete Kalman filtering algorithm described in Section 4.2. In particular, the new estimate at each stage is a linear function of the difference between the new measurement and the most recent estimate of $h(x_0)$. However, notable differences are that h(x) can be a nonlinear function and the gains $\{k_k\}$ are determined *without making any assumptions about the statistical properties of* x_o *and* $\{v_k\}$.

As an example, suppose x_o is observed through linear measurements of the form

$$z_k = x_o + v_k \tag{9.3-15}$$

Then Eq. (9.3-14) becomes

$$\hat{x}_{k+1} = \hat{x}_k + k_k (z_k - \hat{x}_k) \tag{9.3-16}$$

Using the fact that $g(x) = x_o - x$, the condition in Eq. (9.3-7a) reduces to $k_k > 0$; hence Eq. (9.3-14) becomes a linear filter with positive gains. Some gain sequences which satisfy the other convergence conditions in Eq. (9.3-7) are

$$\left.\begin{aligned} k_k &= \frac{\alpha}{k} \\ k_k &= \frac{\alpha}{\beta + k} \\ k_k &= \frac{\alpha + k}{\beta + k^2} \end{aligned}\right\} \quad \alpha,\ \beta > 0;\ k = 1,2,\ldots \tag{9.3-17}$$

The class of estimation algorithms discussed above can be extended to situations where the observation function, $h(x_o)$, varies with time – i.e.,

$$z_k = h_k(x_o) + v_k \tag{9.3-18}$$

This category includes the important problem of estimating the coefficient of a known time-varying function – e.g.,

$$z_k = x_o \sin \omega t_k + v_k \tag{9.3-19}$$

Defining

$$g_k(x) \triangleq h_k(x_o) - h_k(x) \tag{9.3-20}$$

and corresponding observations,

$$m_k \triangleq g_k(\hat{x}_k) + v_k \tag{9.3-21}$$

we seek the value of x that satisfies

$$g_k(x) = 0 \tag{9.3-22}$$

for all values of k. To this end, an algorithm having the same form as Eq. (9.3-14) is employed, viz.:

$$\hat{x}_{k+1} = \hat{x}_k + k_k\,[z_k - h_k(\hat{x}_k)] \tag{9.3-23}$$

In order for the sequence $\{\hat{x}_k\}$ to converge to x_o in the case of a time-varying observation function, conditions must be imposed which take into account the variations in $g_k(x)$ with the index k, as well as the properties of the gain sequence $\{k_k\}$. Henceforth, attention will be restricted to linear problems – i.e.,

$$h_k(x_o) = h_k x_o \tag{9.3-24}$$

where h_k may be a function of k. For this case, the following conditions, in addition to those in Eq. (9.3-7), are imposed (Ref. 24):

$$|g'_k(x)| = |h_k| < d < \infty \tag{9.3-25a}$$

$$\sum_{k=1}^{\infty} |k_k|\,|h_k| = \infty \tag{9.3-25b}$$

The condition in Eq. (9.3-25a) is analogous to that in Fig. 9.3-1, which bounds g(x) by a linear function with a finite slope. The condition in Eq. (9.3-25b) is a generalization of Eq. (9.3-7c) (the latter is implied by the former) needed to insure that the measurements contain sufficient information about x_o. For example, if the measurements in Eq. (9.3-19) are taken at uniform intervals of length $2\pi/\omega$ sec, beginning at t = 0, then $h_k = \sin(2\pi k)$ is zero for all values of k, causing Eq. (9.3-25b) to be violated – i.e., $\Sigma|k_k||h_k| = 0 \neq \infty$. In this case, the data would contain no information about x_o.

The condition in Eq. (9.3-25b) is physically similar to the concept of stochastic observability (see Section 4.4), which describes the information content of measurement data associated with linear dynamical systems. To demonstrate this, we begin by recalling that a discrete-time dynamic system is uniformly stochastically observable if

$$\alpha_1 I \leqslant \sum_{i=k-N}^{k} \Phi(i,k)^T H_i^T R_i^{-1} H_i \Phi(i,k) \leqslant \alpha_2 I \tag{9.3-26}$$

for some value of N, with $\alpha_1 > 0$ and $\alpha_2 > 0$, where $\Phi(i,k)$ is the transition matrix from stage k to stage i, H_i is the measurement matrix and R_i is the measurement noise covariance matrix. It follows from Eq. (9.3-26) that

$$\sum_{i=1}^{\infty} \Phi(i,k)^T H_i^T R_i^{-1} H_i \Phi(i,k) = \infty \tag{9.3-27}$$

Applied to the problem of estimating x_0 in Eq. (9.3-24) for which

$$\left.\begin{aligned} \Phi(i,k) &= 1 \\ H_i &= h_i \end{aligned}\right\} \text{ for all i,k}$$

with the assumption that the additive measurement noise has constant variance,* σ^2, Eq. (9.3-27) reduces to

$$\frac{1}{\sigma^2} \sum_{k=1}^{\infty} h_k^2 = \infty \tag{9.3-28}$$

Now, the *optimal least-squares estimate* for x_0 in Eq. (9.3-24), obtained from a set of k−1 measurements of the form

$$z_j = h_j x_0 + v_j$$

is given by (see the discussions of least-squares estimation in the introduction to Chapter 4),

$$\hat{x}_k = \frac{\sum\limits_{j=1}^{k-1} z_j h_j}{\sum\limits_{j=1}^{k-1} h_j^2} \tag{9.3-29}$$

That is, $\hat{x}_k$ in Eq. (9.3-29) is the value of x_k which minimizes the function

$$\sum_{j=1}^{k-1} (z_j - h_j x_k)^2$$

*This assumption is for convenience of exposition only; the analogy we are making here will hold under much weaker conditions.

If the additive measurement errors have bounded variance, it is easy to prove* that $\hat{x}_k$ will converge to x_o, provided the denominator in Eq. (9.3-29) grows without bound as k approaches infinity – i.e., provided x_o is *observable* according to Eq. (9.3-28). *Thus, observability implies convergence of the least-squares estimate.*

On the other hand Eq. (9.3-29) can be written in the recursive form

$$\hat{x}_{k+1} = \hat{x}_k + \frac{h_k}{\sum_{j=1}^{k} h_j^2} (z_k - h_k \hat{x}_k) \tag{9.3-30}$$

Comparing Eqs. (9.3-23) and (9.3-30), we can regard the latter as a specific *stochastic approximation algorithm* having a gain given by

$$k_k = \frac{h_k}{\sum_{j=1}^{k} h_j^2} \tag{9.3-31}$$

In order that the condition in Eq. (9.3-25b) be satisfied, it must be true that

$$\sum_{k=1}^{\infty} |k_k|\, |h_k| = \sum_{k=1}^{\infty} \frac{h_k^2}{\sum_{j=1}^{k} h_j^2} = \infty \tag{9.3-32}$$

It can be shown that Eq. (9.3-32) will hold provided (see Ref. 31 on the Abel-Dini Theorem)

$$\sum_{j=1}^{\infty} h_j^2 = \infty \tag{9.3-33}$$

which is equivalent to Eq. (9.3-28). *Consequently, for least-squares estimation, stochastic observability and Eq. (9.3-25b) imply the same conditions on h_k.*

---

*

$$\lim_{k\to\infty} \hat{x}_k = \lim_{k\to\infty} \frac{\sum_{j=1}^{k-1} x_o h_j^2 + v_k}{\sum_{j=1}^{k-1} h_j^2} = \lim_{k\to\infty} \left\{ x_o + \frac{v_k}{\sum_{j=1}^{k-1} h_j^2} \right\} = x_o \text{ in the mean square sense}$$

provided $E[v_k^2]$ is bounded.

However, Eq. (9.3-25b) has more general applicability, in the sense that it provides convergence conditions for gain sequences that are not derived from a particular optimization criterion.

Example 9.3-1

To illustrate the performance of stochastic approximation methods, consider the estimation of a gaussian random variable x_0 with a mean μ_0 and variance σ_0^2. Suppose that x_0 is observed from linear measurements defined by

$$z_k = x_0 + v_k \tag{9.3-34}$$

where $\{v_k\}$ is a sequence of zero mean, independent gaussian random variables having variance σ^2. A commonly used stochastic approximation algorithm is one having gains of the form

$$k_k = \frac{1}{k} \tag{9.3-35}$$

This choice is motivated by least-squares estimators of the type given in Eq. (9.3-30), whose gains are the same as in Eq. (9.3-35) in the special case, $h_k = 1$. Applying this sequence of gains, the algorithm for estimating x_0 becomes

$$\hat{x}_{k+1} = \hat{x}_k + \frac{1}{k}(z_k - \hat{x}_k); \qquad k = 1, 2, \ldots \tag{9.3-36}$$

If the variances of x_0 and $\{v_k\}$ are known, the mean square estimation error can be computed. In particular, for this example

$$E\left[(\hat{x}_k - x_k)^2\right] = \frac{\sigma^2}{k} \tag{9.3-37}$$

It is instructive to compare the result in Eq. (9.3-37) with the error that would be achieved if an optimal Kalman filter, based upon knowledge of μ_0, σ_0^2, and σ^2, had been used. The filter equation has the form

$$\hat{x}_{k+1} = \hat{x}_k + \frac{1}{k + \dfrac{\sigma^2}{\sigma_0^2}}(z_k - \hat{x}_k) \tag{9.3-38}$$

and the corresponding mean square estimation error is given by

$$E\left[(\hat{x}_k - x_0)^2\right] = \frac{\sigma^2}{k + \dfrac{\sigma^2}{\sigma_0^2}} \tag{9.3-39}$$

The mean square estimation errors, computed from Eqs. (9.3-37) and (9.3-39), are shown in Fig. 9.3-3 for the case $\sigma_0^2 = 0.5\,\sigma^2$ – i.e., the error variance in the *a priori* estimate of x_0 is smaller than the measurement error. The Kalman filter takes advantage of this information by using gains that give less weight to the first few measurements than does the algorithm in Eq. (9.3-36), thereby providing a lower estimation error. This comparison simply emphasizes the fact that it is wise to take advantage of any available knowledge about the statistics of x_0 and $\{v_k\}$.

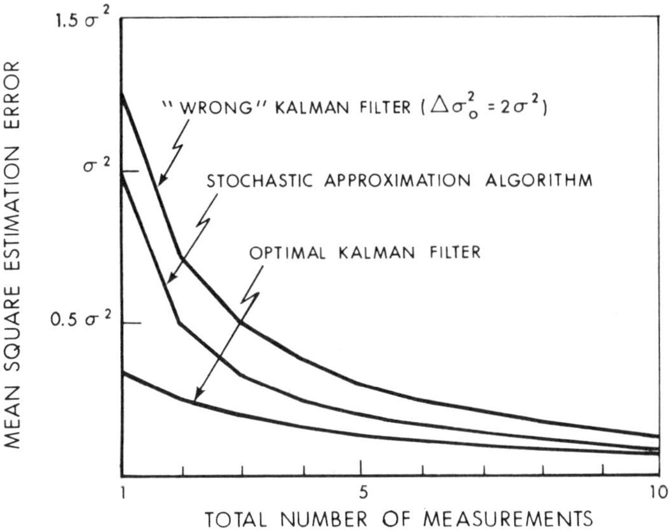

Figure 9.3-3 Comparison of rms Estimation Error for Various Filtering Algorithms

On the other hand, if a Kalman filter design is based upon assumed values for the statistical parameters which are incorrect, the algorithm in Eq. (9.3-36) may yield a *better* estimate. To illustrate, suppose x_0 has zero mean and variance

$$E\left[x_0^2\right] = \sigma_0^2 + \Delta\sigma_0^2$$

where σ_0^2 is the assumed value and $\Delta\sigma_0^2$ represents an error. Then the Kalman filter in Eq. (9.3-28) produces an actual mean square estimation error given by

$$E\left[(\hat{x}_k - x_0)^2\right] = \frac{\sigma^2}{k + \frac{\sigma^2}{\sigma_0^2}} + \frac{\Delta\sigma_0^2}{\left(1 + k\frac{\sigma_0^2}{\sigma^2}\right)^2} \tag{9.3-40}$$

By contrast, the mean square estimation error associated with Eq. (9.3-36) remains the same as in Eq. (9.3-37), which is independent of σ_0^2. The mean square error computed from Eq. (9.3-40) is also shown in Fig. 9.3-3, for the case $\Delta\sigma_0^2 = 2\sigma^2$. Evidently, the improperly designed Kalman filter now performs consistently worse than the stochastic approximation algorithm, reflecting the fact that the latter generally tends to be less sensitive to changes in *a priori* statistics.

We conclude, from the above discussion and example, that a stochastic approximation algorithm is a reasonable alternative to an optimal estimation technique when the statistical parameters on which the latter is based, are not well known. In addition, in many nonlinear estimation problems the linear stochastic approximation algorithm is generally more easily mechanized. Finally, in real-time filtering applications where measurements must be processed rapidly due to high data rates, easily computed gain sequences such as that in Eq. (9.3-35) may be preferable to the generally more complicated optimal gains. This advantage is particularly applicable to the vector case described in the next section. Thus this class of estimation techniques should be included within the data processing repertoire of any engineer concerned with filtering and estimation problems. For additional details the reader is referred to the cited references, particularly Ref. 24.

GENERALIZATIONS

The Vector Constant-Parameter Case – To extend the preceding discussion to the problem of estimating a set of n constant parameters arranged in a vector $\underline{x}_o$, we again consider a scalar linear measurement of the form

$$z_k = \underline{h}_k^T \underline{x}_o + v_k \tag{9.3-41}$$

where $\underline{h}_k$ is a time-varying vector and v_k is a sequence of zero mean independent random variables having bounded variance. The case of vector measurements can be included within this category by considering individual measurements one at a time. By analogy with Eq. (9.3-23), the corresponding vector stochastic approximation algorithm takes the form

$$\hat{\underline{x}}_{k+1} = \hat{\underline{x}}_k + \underline{k}_k (z_k - \underline{h}_k^T \hat{\underline{x}}_k) \tag{9.3-42}$$

Conditions for convergence of the sequence of vectors $\{\underline{x}_k\}$ to $\underline{x}_o$ can be found in Ref. 24. They represent a nontrivial multidimensional generalization of the conditions required in the scalar parameter case; we do not include them here in keeping with our purpose of presenting a principally qualitative description of stochastic approximation methods. However, it is useful to mention a particular gain sequence that often yields convergent estimates. In particular, if*

$$\underline{k}_k = \frac{\underline{h}_k}{\sum_{j=1}^{k} \|\underline{h}_j\|^2} \tag{9.3-43}$$

---

*$\|\underline{h}_j\|^2 \triangleq \underline{h}_j^T \underline{h}_j$

subject to the conditions

$$0 < \gamma_1 \leqslant \|\underline{h}_j\| \leqslant \gamma_2 < \infty$$

$$\sum_{j=k}^{k+N} \underline{h}_j \underline{h}_j^T > \alpha I; \qquad \text{for some } \alpha, N > 0 \tag{9.3-44}$$

then $\{\underline{x}_k\}$ converges to $\underline{x}_o$. The form of Eq. (9.3-43) is suggested by the least-squares gain sequence in Eq. (9.3-31) for the scalar parameter case. It has the advantage of being easily computed and the conditions for convergence can be readily checked.

Time-Varying Parameters – Stochastic approximation methods were first developed for estimating constant parameters. However, analogous methods can be applied for estimating variables that are time-varying. To illustrate, suppose it is desired to estimate x_k from the measurements

$$z_k = h_k(x_k) + v_k \tag{9.3-45}$$

where x_k varies from sample to sample. Specifically the variation in x_k is assumed to be governed by the stochastic difference equation

$$x_{k+1} = \varphi_k(x_k) + w_k \tag{9.3-46}$$

where $\varphi_k(x)$ is a sequence of known functions, and $\{w_k\}$ is a sequence of zero mean random variables.

By analogy with the constant parameter case, an algorithm of the form

$$\hat{x}_{k+1} = \varphi(\hat{x}_k) + k_k\,[z_k - h_k(\hat{x}_k)] \tag{9.3-47}$$

is suggested for estimating x_k. Equation (9.3-47) differs from Eq. (9.3-23) in that $\varphi(\hat{x}_k)$ replaces $\hat{x}_k$ to account for the predicted change in x_k at each stage. In general, we cannot expect the sequence $\{\hat{x}_k\}$ to converge to x_k; however, it is often possible to achieve an estimation error that is statistically bounded according to the conditions

$$E\left\{(\hat{x}_k - x_k)^2\right\} \leqslant b_k$$

$$\lim_{k\to\infty} b_k = b_o < \infty \tag{9.3-48}$$

Criteria for achieving statistically bounded estimates, for a scalar nonlinear estimation problem, are provided in Ref. 24. In the case of linear dynamic systems, any Kalman filter whose gains may be incorrect due to errors in the assumed noise covariance matrices, or because of imperfect knowledge of the system dynamics, can be viewed as a time-varying stochastic approximation algorithm. Conditions for the boundedness of the resulting estimation error are given in Refs. 32 and 33. Aside from these special cases, development of simple algorithms that yield bounded estimation errors for time-varying parameters – particularly when the dynamics and/or the measurements are nonlinear – is, in many respects, still an open problem requiring further research.

9.4 REAL-TIME PARAMETER IDENTIFICATION

The problem of identifying constant parameters in a system can be regarded as a special case of the general state estimation problem discussed throughout this book, where the parameters are a set of random variables, $\underline{a}$, satisfying the differential equation

$$\dot{\underline{a}}(t) = \underline{0} \tag{9.4-1}$$

The need to estimate parameters can arise in a number of ways. For example, in linear filtering theory it is assumed that the elements of the matrices F, H, . . ., etc., that describe the dynamics and measurement data for the linear system, are known. In practice, this condition is not always satisfied, and the task of estimating certain unknown parameters may be included as part of the overall state estimation problem. Similarly, in control system design, it is necessary that we know the dynamics of the plant to be controlled, so that an appropriate control law can be derived. If some plant parameters are unknown, a parameter estimation algorithm may be a necessary part of the system control loop.

In this section, we present an example of real-time parameter identification for a specific control system – viz, an airframe autopilot – where several dynamic quantities are unknown. The equations of motion are nominally linear, in the form

$$\dot{\underline{x}}_s(t) = F_s(\underline{a})\, \underline{x}_s(t) + \underline{\ell}_s(\underline{a})\, u(t) \tag{9.4-2}$$

where u(t) is a known control input. In the more general case, process noise can be included in Eq. (9.4-2); however it is omitted from this particular illustration. For this application, real-time estimates of $\underline{x}_s(t)$ and $\underline{a}$ are needed so that satisfactory control of the system can be maintained. That is, u(t) is a feedback control depending upon the current best estimate of $\underline{x}(t)$; the nature of this dependence is discussed in the sequel. Since u(t) is assumed to be *known*, it is not necessary to know *how* it is generated in order to design a filter for $\underline{x}(t)$; it can simply be regarded as a time-varying known input analogous to $\underline{u}(t)$ in Table 4.4-1.

Both $\underline{a}$ and $\underline{x}_s(t)$ are to be estimated in real-time from the linear noisy measurement data

$$\underline{z}_k = H_s \underline{x}_s(t_k) + \underline{v}_k \tag{9.4-3}$$

If $\underline{a}$ and $\underline{x}_s(t)$ are combined into a composite state vector $\underline{x}(t)$, the combined equations of motion for the system become

$$\dot{\underline{x}}(t) = \begin{bmatrix} \dot{\underline{a}}(t) \\ \hline \dot{\underline{x}}_s(t) \end{bmatrix} = \begin{bmatrix} \underline{0} \\ \hline F_s(\underline{a})\,\underline{x}_s(t) + \underline{\ell}_s(\underline{a})\,u(t) \end{bmatrix} \triangleq \underline{f}(\underline{x}(t), t) \tag{9.4-4}$$

In this form it is evident that the composite estimation problem is nonlinear because the product $F_s(\underline{a})\,\underline{x}_s(t)$ is a nonlinear function of $\underline{a}$ and $\underline{x}_s$.

MODELING UNKNOWN PARAMETERS

Applications where the need for parameter identification often arises, are in systems where the dynamics are nominally time-invariant, but where it is expected that certain parameters will change with time, while the system is operating. The exact nature of this time variation is frequently unknown; however, the range of variation may be sufficiently large to require that it be included in modeling the equations of motion. To illustrate, suppose that the nominal equations of motion are

$$\dot{\underline{x}}_s(t) = F_s(\underline{a}_0, t)\,\underline{x}_s(t) \tag{9.4-5}$$

with parameters, $\underline{a}_0$. One approach to allowing for time variations in the parameters is to model $\underline{a}_0$ as a truncated power series,

$$\underline{a}_0 = \underline{a}_1 + \underline{a}_2 t + \ldots + \underline{a}_{n+1} t^n \tag{9.4-6}$$

If Eq. (9.4-6) is substituted into Eq. (9.4-5), the equations of motion take the form of Eq. (9.4-4) with the vector $\underline{a}$ being composed of the *constant* sets of coefficients $\underline{a}_1, \ldots, \underline{a}_{n+1}$.

One apparent difficulty with the above modeling technique is that the dimension of $\underline{a}$ – and hence, also, the dimension of the composite state vector in Eq. (9.4-4) – increases with every term added in Eq. (9.4-6). Because the complexity of the filter algorithm required to estimate $\underline{a}$ increases with the number of parameters, it is desirable that Eq. (9.4-6) have as few terms as possible. However, a practical, finite number of terms may not adequately describe the change in $\underline{a}_0$, over the time interval of interest.

An alternative method useful in accounting for time-varying parameters, that avoids introducing a large number of state variables, is to assume that the time-variation in $\underline{a}_0$ is partially random in nature. In particular, we replace the constant parameter model ($\dot{\underline{a}} = 0$) in Eq. (9.4-1) with the expression

$$\dot{\underline{a}}(t) = \underline{w}_1(t) \tag{9.4-7}$$

where $\underline{w}_1(t) \sim N(\underline{0}, Q_1)$. The strength of the noise should correspond roughly to the possible range of parameter variation. For example, if it is known that the

i^{th} element of $\underline{a}$ is likely to change by an amount Δa_i over the interval of interest, Δt, then require

$$i^{th} \text{ diagonal element of } Q_1 \triangleq \frac{\Delta a_i{}^2}{\Delta t} \tag{9.4-8}$$

In practice, it is frequently observed that Eq. (9.4-7) is a good model for the purpose of filter design, even though the parameter changes may actually be deterministic in nature. In fact, the random model often permits dropping all but the first term in Eq. (9.4-6), thus keeping the number of state variables at a minimum. Combining Eq. (9.4-7) with Eqs. (9.4-2) and (9.4-3), we obtain the model

$$\underline{\dot{x}}(t) = \begin{bmatrix} \underline{\dot{a}}(t) \\ \hline \underline{\dot{x}}_s(t) \end{bmatrix} = \begin{bmatrix} \underline{0} \\ \hline F_s(\underline{a}(t))\,\underline{x}_s(t) + \underline{\ell}_s(\underline{a}(t))\,u(t) \end{bmatrix} + \begin{bmatrix} \underline{w}_1(t) \\ \hline \underline{0} \end{bmatrix}$$

$$\underline{z}_k = [0 \,|\, H_s(t_k)] \begin{bmatrix} \underline{a}(t_k) \\ \hline \underline{x}_s(t_k) \end{bmatrix} + \underline{v}_k \tag{9.4-9}$$

The equations of motion and observation given in Eq. (9.4-9) have the same form as the nonlinear system dynamics in Eq. (6.0-1) with linear discrete measurements,

$$\underline{z}_k = H_k\,\underline{x}(t_k) + \underline{v}_k \tag{9.4-10}$$

if we make the identifications

$$\underline{w}(t) \triangleq \begin{bmatrix} \underline{w}_1(t) \\ \hline \underline{0} \end{bmatrix}$$

$$\underline{f}(\underline{x}(t), t) \triangleq \begin{bmatrix} \underline{0} \\ \hline F_s(\underline{a}(t))\,\underline{x}_s(t) + \underline{\ell}_s(\underline{a}(t))\,u(t) \end{bmatrix}$$

$$H_k \triangleq [0 \,|\, H_s(t_k)] \tag{9.4-11}$$

In addition, if statistical models are assumed for $\underline{a}(t_o)$, $\underline{x}_s(t_o)$ and $\underline{v}_k$, the Bayesian nonlinear filtering methods discussed in Chapter 6 can be applied. The extended Kalman filter, described in Table 6.1-1, is used for the application discussed in this section. However, the reader should be aware that there are many different methods that can be used for parameter identification – e.g., maximum likelihood (Refs. 34 and 35), least-squares (Ref. 36), equation error (Refs. 37 and 38), stochastic approximation (see Section 9.3 and Ref. 24), and correlation (Ref. 39). Some discussion of the relative applicability of these techniques is in order.

Some of the alternative identification methods mentioned above are advocated for situations where the unknown parameters are assumed to be constant with unknown statistics, and where an algorithm is desired that yields perfect (unbiased, consistent) estimates in the limit, as an infinite number of measurements is taken. The equation error, stochastic approximation, and correlation methods are all in this category. However, when the parameters are time-varying, as in Eq. (9.4-7), the convergence criteria do not apply. Whether or not these methods will operate satisfactorily in the time-varying case can only be ascertained by simulation.

Other methods are based upon various optimization criteria – e.g., maximum likelihood and least-squares. The maximum likelihood estimate is one that maximizes the joint probability density function for the set of unknown parameters; however, it is typically calculated by a non-real-time algorithm that is not well suited for control system applications. In addition the probability density function can have several peaks, in which case the optimal estimate may require extensive searching. The least-squares criterion seeks an estimate that "best" fits the observed measurement data, in the sense that it minimizes a quadratic penalty function; in general, the estimate is different for each quadratic penalty function. In contrast with the minimum variance estimate, the least-squares estimate does not require knowledge of noise statistics.

The autopilot design application treated here demonstrates the parameter identification capability of the minimum variance estimation algorithms discussed in Chapter 6. The latter provide a logical choice of identification techniques when the random process statistics are known and the requirement exists for real-time filtered estimates; both of these conditions exist in this example.

Example 9.4-1

The framework of the parameter identification problem described here is the design of an adaptive pitch-plane autopilot for an aerodynamic vehicle having the functional block diagram shown in Fig. 9.4-1. This example treats motion in a single plane only – referred to as the pitch-plane. The airframe dynamics are assumed linear, with a number of unknown time-varying parameters. The objective is to obtain a feedback control signal u(t), that causes the airframe lateral acceleration to closely follow the input steering command c(t).

If the airframe dynamics were completely known, feedback compensation could be selected to obtain the desired autopilot response characteristics using any appropriate system design technique. Because certain parameters are unknown, an autopilot design is sought that identifies (estimates) the parameters in real time and adjusts the feedback compensation accordingly to maintain the desired response characteristics. In this application, the method used to determine the control input is the so-called *pole assignment technique,* whereby u(t) is specified as a function of $\hat{\underline{a}}$, $\hat{\underline{x}}_s$, and c(t) from the relation

$$u(t) = h_0(\hat{\underline{a}})\, c(t) - \underline{h}^T(\hat{\underline{a}})\, \hat{\underline{x}}_s(t) \tag{9.4-12}$$

Assuming the parameter estimates are perfect, the feedback gains, $\underline{h}$, are chosen such that the autopilot "instantaneous closed-loop poles" have desired values; h_0 is an input gain chosen to provide a specified level of input-output dc gain. If the airframe parameters are

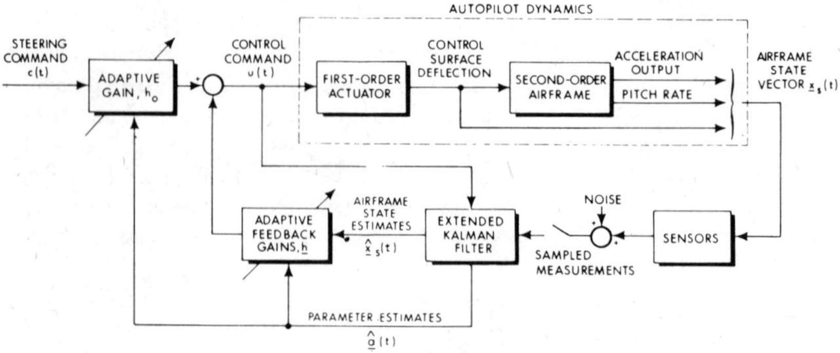

Figure 9.4-1 Adaptive Autopilot Including Parameter Identification

known and constant, h_0 and $\underline{h}$ will be constant; the mathematical details for deriving the functional expressions for the gains can be found in Ref. 40. However, because $\underline{a}$ is initially unknown and time-varying, the estimates $\hat{\underline{a}}$ will vary and the control gains must be periodically recalculated; this is the adaptive control feature of the autopilot. For the purpose of developing the parameter identification algorithm, we need say no more about u(t); the fact that it is known to the designer as a function of the real time estimates of $\underline{x}_s$ and $\underline{a}$ is sufficient.

The airframe configuration in this application is for the tail-controlled vehicle illustrated in Fig. 9.4-2. The equations of motion for the airframe can be expressed in the vector-matrix form

$$\begin{bmatrix} \dot{q}(t) \\ \dot{a}(t) \\ \dot{\delta}(t) \end{bmatrix} = \begin{bmatrix} M_q & \dfrac{1}{V}\dfrac{M_\alpha}{L_\alpha} & M_\delta - \dfrac{M_\alpha L_\delta}{L_\alpha} \\ VL_\alpha & -L_\alpha & -\lambda VL_\delta \\ 0 & 0 & -\lambda \end{bmatrix} \begin{bmatrix} q(t) \\ a(t) \\ \delta(t) \end{bmatrix} + \begin{bmatrix} 0 \\ \lambda VL_\delta \\ \lambda \end{bmatrix} u(t) \tag{9.4-13}$$

or, in abbreviated notation,

$$\dot{\underline{x}}_s(t) = F_s \underline{x}_s(t) + \underline{\ell}_s u(t)$$

$$F_s \triangleq \begin{bmatrix} f_{11} & f_{12} & f_{13} \\ f_{21} & f_{22} & f_{23} \\ 0 & 0 & f_{33} \end{bmatrix}; \quad \underline{\ell}_s \triangleq \begin{bmatrix} 0 \\ \ell_2 \\ \ell_3 \end{bmatrix} = \begin{bmatrix} 0 \\ -f_{23} \\ -f_{33} \end{bmatrix} \tag{9.4-14}$$

The quantity λ represents known actuator dynamics, V is unknown airspeed, and M_q, M_δ, etc., are unknown stability derivatives. In the context of Eq. (9.4-9), the set of unknown parameters is defined to be

$$\underline{a} = \begin{bmatrix} f_{11} \\ f_{12} \\ f_{13} \\ f_{21} \\ f_{22} \\ f_{23} \end{bmatrix} \tag{9.4-15}$$

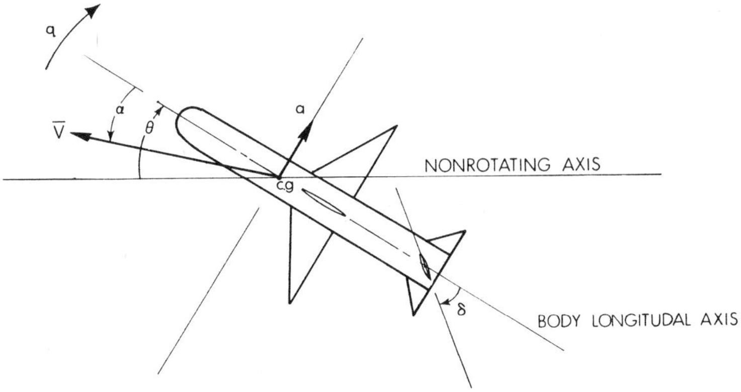

Figure 9.4-2 Airframe Configuration

For this demonstration, the vehicle is assumed to be thrusting longitudinally at a level of 25-g's, causing rapid variations in the parameters in Eq. (9.4-14) through their dependence upon airspeed. The trajectory duration is three seconds. The simulation truth model uses piecewise-linear time functions to represent the parameters; the filter model assumes that the parameters vary randomly according to Eq. (9.4-7). Measurements of the three airframe state variables are assumed to be available in the form

$$\underline{z}_k = \underline{x}_s(t_k) + \underline{v}_k \tag{9.4-16}$$

where $\{\underline{v}_k\}$ is a gaussian white sequence. With these assumptions, Eqs. (9.4-14) and (9.4-16) fit the format of Eq. (9.4-9) and the extended Kalman filter algorithm can be directly applied. A digital computer monte carlo simulation of this model was performed under the following conditions:

Measurement noise rms level:	pitch rate gyro, 4.5×10^{-4} rad/sec accelerometer, 0.16 ft/sec^2 deflection angle resolver, 6.3×10^{-3} rad
Input command, c(t):	square wave – amplitude = 10 ft/sec^2 and frequency = 1.67 Hz
Measurement interval:	$t_{k+1} - t_k = 0.02$ sec

The parameter tracking accuracy achieved for the system in Fig. 9.4-1 is illustrated by the curves in Fig. 9.4-3, which display both the truth model and the estimated parameter values as functions of time.

It is observed from Fig. 9.4-3 that some parameters are identified less accurately than others; in particular, f_{11} and f_{22} are relatively poorly estimated. This is physically explained by the fact that these two parameters describe the open loop damping characteristics of the missile airframe. Since the damping is very light, it has little influence on the airframe state variables $\underline{x}_s(t)$, and hence, on the measurement data. Thus, f_{11} and f_{22} are *relatively* unobservable. However, this behavior does not adversely affect autopilot design because parameters that have little influence on airframe behavior tend also to have little effect on the airframe control law.

Another phenomenon demonstrated by this application is that the character of the input signal u(t) in Eq. (9.4-13) influences the parameter identification accuracy – an effect that

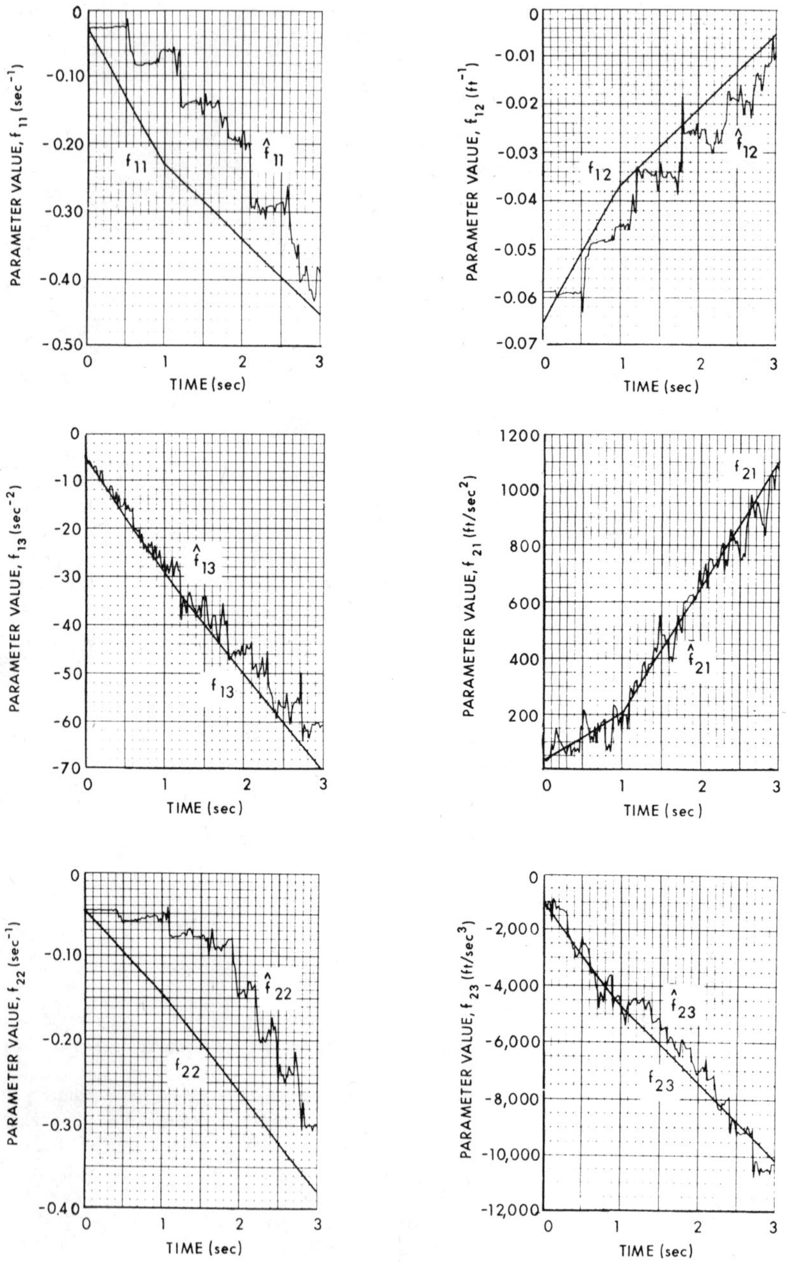

Figure 9.4-3 Comparison of Parameter Estimates and Truth Model Parameter Values

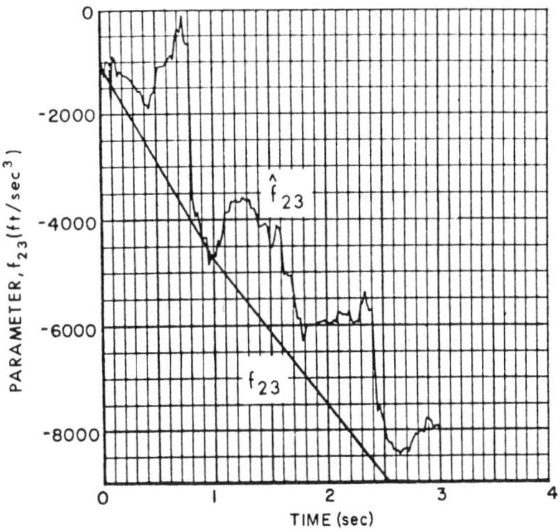

(a) Input Square Wave Period 1.6 sec

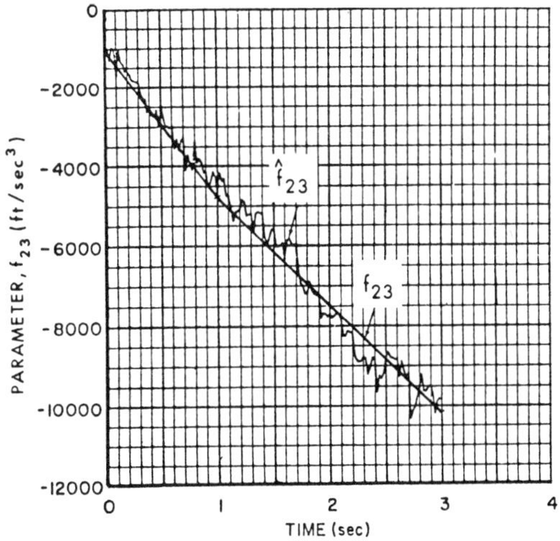

(b) Input Square Wave Period 0.2 sec

Figure 9.4-4 Demonstration of Identification Accuracy Sensitivity to Control Input Signal Frequency is not observed in truly linear filtering problems.* This occurs because the extended Kalman filter gains are dependent upon $\underline{x}(t)$ which, in turn, is dependent upon u(t). To illustrate this behavior, two simulations were performed under identical conditions except that different values of the square wave signal frequency were used for c(t). One parameter in Eq. (9.4-13), f_{23}, exhibited marked sensitivity to the frequency changes, as demonstrated in Fig. 9.4-4.

The above example illustrates the use of the extended Kalman filter for obtaining estimates of state variables in problems that are inherently nonlinear. Practical problems in parameter identification are currently receiving considerable attention in many fields – e.g., high-speed aircraft, ship control and nuclear power plant control. Work on the design of *optimal inputs* for system parameter identification is also a subject of current research (e.g., Ref. 41).

9.5 OPTIMAL CONTROL OF LINEAR SYSTEMS

An important application for the estimation techniques discussed in this book is the field of control system design. The concept of control arises when the equations of motion of a system contain a set of "free" variables $\underline{u}(t)$, whose functional form can be prescribed by a designer to alter the system dynamic properties. A linear stochastic system having this capability is represented by the equation

$$\dot{\underline{x}}(t) = F(t)\,\underline{x}(t) + G(t)\,\underline{w}(t) + L(t)\,\underline{u}(t) \tag{9.5-1}$$

where L(t) is a gain matrix describing the influence of $\underline{u}(t)$ on the state vector. If $\underline{u}(t)$ is explicitly a function of time only, it is referred to as an *open loop control*. If $\underline{u}(t)$ is explicitly a function of $\underline{x}(t)$ also, it is called a *closed loop* or *feedback* control.

We note that Eq. (9.5-1) has the same form as the equations for the continuous system in Table 4.4-1, where $\underline{u}(t)$ is referred to as a deterministic input. The linear open loop control system falls into the same category, because the control variables are specified by the designer; therefore, they can be considered as known (deterministic).

To mechanize a feedback control, $\underline{u}[\underline{x}(t), t]$, the state variables $\underline{x}(t)$ must be accessible. Typically $\underline{x}(t)$ can be observed only through the available measurements $\underline{z}(t)$. In designing linear control systems, the latter are assumed to be given by the familiar linear expression

$$\underline{z}(t) = H(t)\,\underline{x}(t) + \underline{v}(t) \tag{9.5-2}$$

where $\underline{v}(t)$ is gaussian measurement noise. As we shall subsequently demonstrate, the role of state estimation is to process the measurement data so that an

*Note from Table 4.4-1 that the covariance matrix for a linear system is unaffected by the presence of $\underline{u}(t)$.

appropriate approximation to the desired feedback control law – $\underline{u}[\hat{\underline{x}}(t),t]$ – can be achieved. An example of this type was treated in Section 9.4; however, in this section we are explicitly concerned with the *design* of the control law.

Just as various optimization criteria have been employed to derive optimal estimation algorithms, the concept of *optimal control* arises when $\underline{u}(t)$ is chosen to minimize a performance index, or figure of merit, for the controlled system. For linear systems with certain specific types of performance indices, there are significant similarities in the solutions to the control and estimation problems. This section briefly discusses the control problem, compares it with the estimation problem, and indicates how both subjects are combined in the design of an optimal stochastic control system.

DETERMINISTIC OPTIMAL LINEAR SYSTEMS – DUALITY

First, we discuss the control problem for deterministic linear systems where the noise processes, $\underline{w}(t)$ and $\underline{v}(t)$, in Eqs. (9.5-1) and (9.5-2) are absent. To determine a control law for the system described by

$$\dot{\underline{x}}(t) = F(t)\,\underline{x}(t) + L(t)\,\underline{u}(t) \tag{9.5-3}$$

it is desirable to impose a performance criterion that leads to a unique choice of $\underline{u}(t)$. An important class of problems, successfully treated from this point of view, is the so-called *regulator* problem, wherein $\underline{x}(t)$ is assumed to have an initial value $\underline{x}_0$ at time t_0 and the control is chosen to drive the state toward zero. This objective is stated more precisely by requiring that $\underline{u}(t)$ minimize a performance index, J, which provides a measure of the size of $\underline{x}(t)$. In addition, the index should include a weighting on the magnitude of $\underline{u}(t)$ to limit the amount of control effort expended in nulling the state vector. A form of J that is found to be convenient and useful for linear systems is the *quadratic performance index*, defined by

$$J = \underline{x}(t_f)^T\,V_f\,\underline{x}(t_f) + \int_{t_0}^{t_f} [\underline{x}^T(t)\,V(t)\,\underline{x}(t) + \underline{u}^T(t)\,U(t)\,\underline{u}(t)]\,dt \tag{9.5-4}$$

where V_f, V(t) and U(t) are specified weighting matrices and t_f is a specified final time. The matrices V_f and V(t) are usually required to be symmetric and positive semidefinite; U(t) is symmetric and positive definite.

The positive quadratic terms in $\underline{x}$ in Eq. (9.5-4) provide performance measures that tend to achieve the desired reduction in the state when $\underline{u}(t)$ is chosen to minimize J. However, the amount of reduction achieved is compromised with the control effort expended by the inclusion of a quadratic term in $\underline{u}(t)$. The smaller the elements of U(t), relative to those of V_f and V(t), the larger will be the magnitude of the optimal control and the greater will be the reduction in $\underline{x}(t)$.

The procedure outlined above for designing a feedback control system is referred to as an *optimal control problem* formulation, where a control law is sought that minimizes a scalar index of performance, J. Such problems are extensively treated in the control systems literature (see Refs. 36, 42-44). Solutions for the optimal controls are obtained by various mathematical techniques – the principal ones being the *calculus of variations*, *Pontryagin's maximum principle*, and *dynamic programming*. It is not our intention here to provide a detailed exposition of control theory; instead we discuss the analogy between the control and estimation problems for linear systems, and demonstrate the role of estimation theory in the control of linear stochastic systems. For these purposes, a relatively simple "special-purpose" derivation (Ref. 45) of the optimal feedback control law, which minimizes the index in Eq. (9.5-4), will suffice.

Let us assume that a time-varying symmetric matrix S(t) exists, defined on the interval $t_o \leqslant t \leqslant t_f$, such that

$$S(t_f) = V_f \tag{9.5-5}$$

where V_f is the weighting matrix for the terminal value of the state appearing in Eq. (9.5-4). For the present, no additional conditions are imposed on S(t); hence, its values at other times can be quite arbitrary. Observe that Eq. (9.5-5) implies

$$\underline{x}^T(t_f)\, S(t_f)\, \underline{x}(t_f) = \underline{x}^T(t_f)\, V_f \underline{x}(t_f) \tag{9.5-6}$$

Now we form

$$\frac{d}{dt}(\underline{x}^T S \underline{x}) = \dot{\underline{x}}^T S\underline{x} + \underline{x}^T \dot{S}\underline{x} + \underline{x}^T S\dot{\underline{x}} \tag{9.5-7}$$

where it is tacitly assumed that S(t) is differentiable. For notational convenience the explicit dependence of $\underline{x}$, S, and other time-varying quantities upon t is frequently omitted in the sequel. If $\dot{\underline{x}}$ is substituted from Eq. (9.5-3) into Eq. (9.5-7), the result can be written as

$$\begin{aligned}\frac{d}{dt}(\underline{x}^T S\underline{x}) &= \underline{x}^T(F^T S + SF + \dot{S})\underline{x} + \underline{u}^T L^T S\underline{x} + \underline{x}^T S L\underline{u} \\ &= \underline{x}^T(F^T S + SF + \dot{S} + V)\underline{x} + \underline{u}^T L^T S\underline{x} + \underline{x}^T S L\underline{u} + \underline{u}^T U\underline{u} - \underline{x}^T V\underline{x} - \underline{u}^T U\underline{u}\end{aligned} \tag{9.5-8}$$

The final expression of Eq. (9.5-8) is obtained simply by adding and subtracting the terms in the integrand of the quadratic performance index to the right side of Eq. (9.5-7).

Examination of Eq. (9.5-8) reveals that it can be rewritten in the form

$$\frac{d}{dt}(\underline{x}^T S\underline{x}) = (\underline{x}^T SL + \underline{u}^T U)\, U^{-1}\, (L^T S\underline{x} + U\underline{u}) - \underline{x}^T V\underline{x} - \underline{u}^T U\underline{u} \tag{9.5-9}$$

if we impose the following condition on S(t):

$$F^T S + SF + \dot{S} + V = SLU^{-1}\, L^T S \tag{9.5-10}$$

Equation (9.5-9) is readily verified by inspection of Eqs. (9.5-8) and (9.5-10). Until now, S(t) has been an arbitrary symmetric matrix, except for its value at t_f specified by Eq. (9.5-5); hence, we have the freedom to require that S(t) satisfy Eq. (9.5-10), recognized as the matrix Riccati equation, on the interval $t_o \leqslant t < t_f$.

Finally we observe that

$$\int_{t_o}^{t_f} \frac{d}{dt}(\underline{x}^T S\underline{x})\, dt = \underline{x}(t_f)^T\, S(t_f)\, \underline{x}(t_f) - \underline{x}_o{}^T\, S(t_o)\underline{x}_o \tag{9.5-11}$$

Combining Eqs. (9.5-4), (9.5.5), (9.5-9), and (9.5-11), we obtain the following equivalent expression for the performance index:

$$J = \underline{x}_o{}^T\, S(t_o)\, \underline{x}_o + \int_{t_o}^{t_f} (\underline{x}^T\, SL + \underline{u}^T U)U^{-1}\, (L^T S\underline{x} + U\underline{u})\, dt \tag{9.5-12}$$

Both $S(t_o)$ and $\underline{x}_o$ are independent of the control $\underline{u}(t)$; consequently, J can be minimized by minimizing the integral alone in Eq. (9.5-12). Furthermore, it is evident that the integrand in Eq. (9.5-12) is nonnegative and can be made identically zero by requiring

$$L^T S\underline{x} + U\underline{u} = \underline{0} \tag{9.5-13}$$

Solving Eq. (9.5-13) for $\underline{u}(t)$ and Eq. (9.5-10) for S(t), subject to the condition imposed in Eq. (9.5-5), provides the complete solution to the optimal control problem, which is summarized as follows:

$$\begin{aligned} &\underline{u}(t) = -U^{-1}(t)\, L^T(t)\, S(t)\, \underline{x}(t) \\ &\dot{S}(t) = -F^T(t)\, S(t) - S(t)\, F(t) + S(t)\, L(t)\, U^{-1}(t)\, L^T(t)\, S(t) - V(t) \\ &S(t_f) = V_f \end{aligned} \tag{9.5-14}$$

Observe that the optimal control law, derived above, has the linear form

$$\underline{u}(t) = -C(t)\, \underline{x}(t) \tag{9.5-15}$$

with the feedback gains C(t) given by

$$C(t) = U^{-1}(t)\, L^T(t)\, S(t) \tag{9.5-16}$$

The latter are evaluated by integrating the Riccati differential equation for S(t) backward in time from the terminal condition, $S(t_f) = V_f$. If all of the system state variables are measured directly [H = I in Eq. (9.5-2)] with negligible measurement error, the control law can be mechanized as illustrated in Fig. 9.5-1. If the matrix H in Eq. (9.5-2) is singular, an approximation to $\underline{x}(t)$ can be generated with an "observer," as demonstrated in Fig. 9.5-2 (see the discussion in Section 9.2). The case where appreciable measurement errors may be present is treated in the next section.

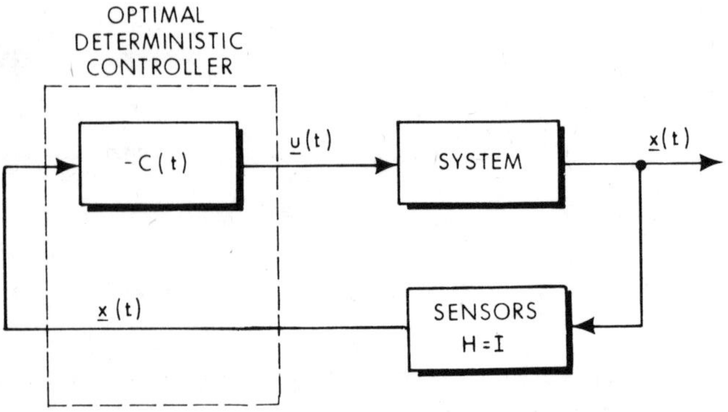

Figure 9.5-1 Optimal Control System Configuration, H = I

There are important structural similarities between the linear control law, derived above, and the Kalman filtering algorithm for continuous systems, discussed in Section 4.3. In particular, the control gains C(t) are determined by solution of a matrix Riccati equation in Eq. (9.5-14), similar in form to that in Table 4.3-1 which specifies the filter gains, K(t). Therefore, all the conditions under which solutions to the filtering problem exist, have their counterparts in the control problem. For this reason the control and estimation problems are said to be the *duals* of each other.

One consequence of duality is the fact that the concepts of observability and controllability, discussed in Section 4.4, are also defined for control systems. The control system is said to be *controllable* if the integral

$$\int_{t_o}^{t} \Phi(t,\tau)\, L(\tau)\, U^{-1}(\tau) L^T(\tau)\, \Phi^T(t,\tau)\, d\tau$$

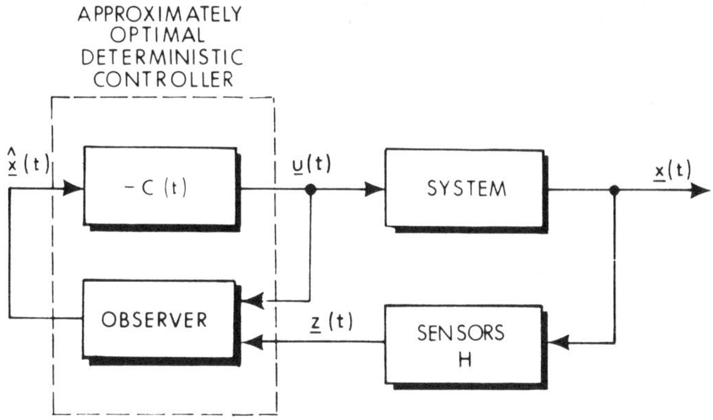

Figure 9.5-2 Approximately Optimal Control System Configuration; H is Singular

is positive definite for some $t > 0$, and *observable* if the integral

$$\int_{t_o}^{t} \Phi^T(\tau, t)\, V(\tau)\, \Phi(\tau, t) d\tau$$

is positive definite for some $t > 0$. The quantity Φ is the transition matrix associated with F in Eq. (9.5-3). Comparing the above definitions with those given in Section 4.4, we find that controllability in the control system, defined in terms of F, L, and U^{-1}, corresponds to observability in the Kalman filter, defined in terms of F^T, H^T and R^{-1}. Similarly, observability for the control system, defined in terms of F^T and V, corresponds to controllability of the Kalman filter, defined in terms of F and GQG^T. Thus, *for each controllable dynamic system in a filtering problem, there exists a corresponding observable dynamic system for a control problem, and vice versa.*

A list of duality transformations for the control and filtering problems is given in Table 9.5-1. This comparison is useful because only one body of theory is needed to treat both situations.

OPTIMAL LINEAR STOCHASTIC CONTROL SYSTEMS – SEPARATION PRINCIPLES

When the noise processes, $\underline{w}(t)$ and $\underline{v}(t)$, are present in Eqs. (9.5-1) and (9.5-2), the optimal control problem must be posed differently than for the deterministic case treated in the preceding section. Because the exact value of the state vector is unknown, due to the noisy measurement data, it is

TABLE 9.5-1 DUALITY RELATIONSHIPS

Filtering		Control
P	$\longleftrightarrow$	S
F	$\longleftrightarrow$	F^T
GQG^T	$\longleftrightarrow$	V
H	$\longleftrightarrow$	L^T
R	$\longleftrightarrow$	U
K	$\longleftrightarrow$	C^T
t_o	$\longleftrightarrow$	t_f
Observability	$\longleftrightarrow$	Controllability
Controllability	$\longleftrightarrow$	Observability

meaningless to talk about minimizing the index J in Eq. (9.5-4), which depends upon $\underline{x}(t)$. Instead, a statistical measure of performance is needed; this can be obtained by defining a new performance index $\overline{J}$, which is the average value of J, written as

$$\overline{J} = E\left[\underline{x}(t_f)^T V_f \underline{x}(t_f) + \int_{t_o}^{t_f} [\underline{x}^T(t) V(t) \underline{x}(t) + \underline{u}^T(t) U(t) \underline{u}(t)]\, dt\right] \tag{9.5-17}$$

The *optimal stochastic control problem* is to choose $\underline{u}(t)$ so that $\overline{J}$ is minimized, subject to the equations of motion given in Eq. (9.5-1). If a feedback control is sought, $\underline{u}(t)$ will depend upon the measurement data in Eq. (9.5-2).

To describe the solution to the control problem defined above, we first note that regardless of what method is used to generate $\underline{u}(t)$, the conditional mean (optimal estimate) of the state vector $\hat{\underline{x}}(t)$ can always be determined by applying a Kalman filter to the measurement data. This is true because $\underline{u}(t)$, in Eq. (9.5-1), is effectively a known time-varying input to the linear system when the control law is specified. Consequently, $\hat{\underline{x}}(t)$ can be determined from the algorithm in Table 4.4-1. With this fact in mind, we can state the optimal control law which minimizes $\overline{J}$; it consists of two separate cascaded functions. First, a conventional Kalman filter is employed in the manner outlined above to obtain an optimal estimate of the state vector. Then the control command, $\underline{u}(t)$, is generated according to the relation

$$\underline{u}(t) = -C(t)\, \hat{\underline{x}}(t) \tag{9.5-18}$$

where C(t) is the set of control gains derived in the preceding section. A functional diagram of the control law is shown in Fig. 9.5-3. The realization of the optimal linear stochastic control law, in terms of distinct filtering and control functions, is referred to as the *separation principle*.

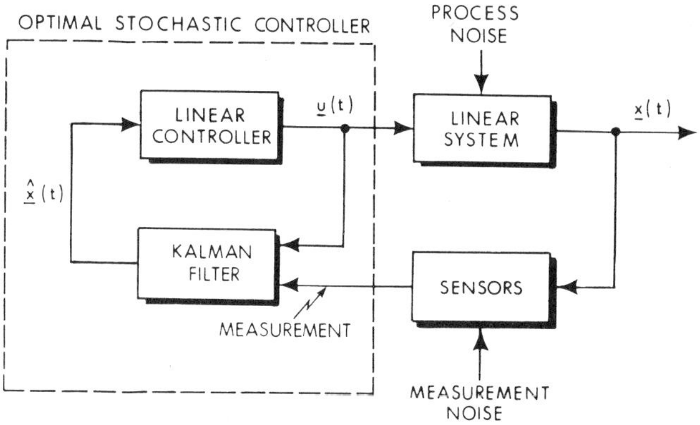

Figure 9.5-3 Optimal Stochastic Control System Configuration

The separation principle is usually derived by applying the theory of dynamic programming to the minimization of $\bar{J}$ in Eq. (9.5-17) (Ref. 44). To provide insight as to why the principle is true, we offer a relatively simple plausibility argument, without the mathematical rigor of a complete proof. Using a procedure similar to that described in the preceding section for deriving the optimal deterministic control law, it can be shown that the performance index can be written in the form

$$\begin{aligned}\bar{J} = E[\underline{x}_o^T S(t_o)\underline{x}_o] &+ \int_{t_o}^{t_f} \text{trace}\,[SGQG^T]\,dt \\ &+ E\left[\int_{t_o}^{t_f} (\underline{x}^T SL + \underline{u}^T U)U^{-1}(L^T S\underline{x} + U\underline{u})\,dt\right]\end{aligned} \tag{9.5-19}$$

where S(t) is the solution to the matrix Riccati equation in Eq. (9.5-14). The derivation of Eq. (9.5-19) is left as an exercise. The principal difference between Eqs. (9.5-12) and (9.5-19) is that the latter contains the term $SGQG^T$, which arises from the process noise in Eq. (9.5-1). The first and second terms in Eq. (9.5-19) are independent of $\underline{u}(t)$ and can be disregarded in the minimization. Interchanging the order of expectation and integration, the third term in Eq. (9.5-19) becomes

$$\overline{J}_u \triangleq \int_{t_o}^{t_f} E[(\underline{x}^T SL + \underline{u}^T U)U^{-1}(L^T S\underline{x} + U\underline{u})] = 0 \tag{9.5-20}$$

At any given time t, having collected all the measurement data up to time t, the integrand in Eq. (9.5-20) is instantaneously minimized with respect to the control by solving the expression

$$\frac{\partial}{\partial \underline{u}} E[(\underline{x}^T SL + \underline{u}^T U)U^{-1}(L^T S\underline{x} + U\underline{u})] = \underline{0} \tag{9.5-21}$$

for $\underline{u}(t)$, where the expectation is conditioned on the available measurement data. The resulting value of the control is given by

$$\underline{u}(t) = -U^{-1}(t)\, L(t)\, S(t)\, \hat{\underline{x}}(t) \tag{9.5-22}$$

Observe that it is expressed in terms of the optimal state estimate, $\hat{\underline{x}}(t)$. Substituting Eq. (9.5-22) into Eq. (9.5-20) and carrying out the indicated expectation operation produces

$$\overline{J}_u = \text{trace}\left\{\int_{t_o}^{t_f} SL^T U^{-1} LSP\, dt\right\} \tag{9.5-23}$$

which is independent of $\underline{x}(t)$; $\overline{J}_u$ depends only upon the trajectory-independent quantities S, L, U and P. Therefore, if $\underline{u}(t)$ is specified by Eq. (9.5-22), the integrand in Eq. (9.5-20) is minimized at each point in time. Consequently, the integral itself, and hence $\overline{J}$, is minimized by the choice of control given in Eq. (9.5-22).

The separation principle cited above is most frequently associated with linear gaussian stochastic systems having quadratic performance indices. However, it has been demonstrated (Refs. 46-49) that optimal control laws for linear stochastic systems with more general types of performance indices, including the possibility of an explicit constraint on the control magnitude as well as nongaussian measurement noise, also satisfy a form of separation principle. In the latter case, the control law consists of a Kalman filter cascaded with a controller that computes $\underline{u}(t)$ in terms of $\hat{\underline{x}}(t)$. However, for nonquadratic performance indices, the controller portion of the control law can be a nonlinear function of $\hat{\underline{x}}(t)$; furthermore, it generally depends upon the statistics of the process and measurement noises. In many cases the resulting controller computation cannot be carried out in closed form. The structure of this more general separation principle is similar to that shown in Fig. 9.5-3; however, the controller can now be nonlinear. The reader is referred to the works cited above for further details.

The purpose of this section has been to demonstrate the role that optimal estimation theory plays in developing optimal control laws for linear stochastic systems. For nonlinear systems, the optimal control laws generally do not separate into an optimal filter cascaded with a control command computation (Refs. 50 and 51). However, in these cases it may be possible to linearize the nonlinearities and design an approximately optimal linearized system using the separation principle. Thus, from this point of view, estimation theory is seen to be exceptionally useful for a wide variety of feedback control system design problems.

REFERENCES

1. Mehra, R.K., "On the Identification of Variances and Adaptive Kalman Filtering," *IEEE Trans. on Automatic Control*, Vol. AC-15, No. 2, April 1970, pp. 175-184.
2. Weiss, I.M., "A Survey of Discrete Kalman-Bucy Filtering with Unknown Noise Covariances," AIAA Guidance, Control and Flight Mechanics Conference, Paper No. 70-955, Santa Barbara, California, August 1970.
3. Mehra, R.K., "On-Line Identification of Linear Dynamic Systems with Applications to Kalman Filtering," *IEEE Trans. on Automatic Control*, Vol. AC-16, No. 1, February 1971, pp. 12-21.
4. Abramson, P.D., Jr., "Simultaneous Estimation of the State and Noise Statistics in Linear Dynamic Systems," Ph.D. Thesis, Massachusetts Institute of Technology, TE-25, May 1968.
5. Kailath, T., "An Innovations Approach to Least Squares Estimation – Part I: Linear Filtering in Additive White Noise," *IEEE Trans. on Automatic Control*, Vol. AC-13, No. 6, December 1968.
6. Gelb, A., Dushman, A., and Sandberg, H.J., "A Means for Optimum Signal Identification," *NEREM Record*, Boston, November 1963.
7. Luenberger, D.G., "Observing the State of a Linear System," *IEEE Transactions on Military Electronics*, Vol. MIL 8, April 1964.
8. Luenberger, D.G., "Observers for Multivariable Systems," *IEEE Transactions on Automatic Control*, Vol. AC-11, No. 2, April 1966.
9. Uttam, B.J., "On the Stability of Time-Varying Systems Using an Observer for Feedback Control," *International Journal of Control*, December 1971.
10. Uttam, B.J. and O'Halloran, W.F., Jr., "On Observers and Reduced Order Optimal Filters for Linear Stochastic Systems," *Proc. of the Joint Automatic Control Conference*, August 1972.
11. O'Halloran, W.F., Jr. and Uttam, B.J., "Derivation of Observers for Continuous Linear Stochastic Systems From the Discrete Formulation," Sixth Asilomar Conference, Pacific Grove, California, November 1972.
12. Aoki, M. and Huddle, J.R., "Estimation of a State Vector of a Linear Stochastic System with a Constrained Estimator," *IEEE Transactions on Automatic Control*, Vol. AC-12, No. 4, August 1967.
13. Tse, E. and Athans, M., "Optimal Minimal-Order Observer Estimators for Discrete Linear Time-Varying Systems," *IEEE Transactions on Automatic Control*, Vol. AC-15, No. 4, August 1970.
14. Novak, L.M., "Optimal Minimal-Order Observers for Discrete-Time Systems – A Unified Theory," *Automatica*, Vol. 8, July 1972.

15. Todd, J., *Survey of Numerical Analysis*, McGraw-Hill Book Co., Inc., New York, 1962.

16. Robbins, H. and Munroe, S., "A Stochastic Approximation Method," *Ann. Math. Statistics*, Vol. 22, No. 1, 1951, pp. 400-407.

17. Wolfowitz, J., "On the Stochastic Approximation Method of Robbins and Munroe," *Ann. Math. Statistics*, Vol. 23, No. 3, 1952, pp. 457-466.

18. Schmetterer, L., "Stochastic Approximation," *Proc. 4th Berkeley Symp. on Mathematical Statistics and Probability*, Vol. 1, Los Angeles: University of California Press, 1958, pp. 587-609.

19. Derman, C., "Stochastic Approximation," *Ann. Math. Statistics*, Vol. 27, 1956, pp. 879-886.

20. Ho, Y.C., "On the Stochastic Approximation Method and Optimal Filtering Theory," *Journal of Mathematical Analysis and Applications*, Vol. 6, 1962, pp. 152-154.

21. Ho, Y.C. and Lee, R.C.K., "Identification of Linear Dynamic Systems," *Journal of Information and Control*, Vol. 8, 1965, pp. 93-110.

22. Sakrison, D.J., "Stochastic Approximation: A Recursive Methoa for Solving Regression Problems," *Advances in Communication Systems*, Vol. 2, 1966, pp. 51-106.

23. Sage, A.P. and Melsa, J.L., *System Identification*, Academic Press, New York, 1967.

24. Albert, A.E. and Gardner, L.A., Jr., *Stochastic Approximation and Nonlinear Regression*, The M.I.T. Press, Cambridge, Mass., 1967.

25. Sinha, N.K. and Griscik, M.P., "A Stochastic Approximation Method," *IEEE Trans. on Systems, Man, and Cybernetics*, Vol. SMC-1, No. 4, October 1971, pp. 338-343.

26. Kiefer, E. and Wolfowitz, J., "Stochastic Estimation of the Maximum of a Regression Function," *Ann. Math. Statistics*, Vol. 23, No. 3, 1952.

27. Ho, Y.C. and Newbold, P.M., "A Descent Algorithm for Constrained Stochastic Extrema," *Journal of Optimization Theory and Applications*, Vol. 1, No. 3, 1967, pp. 215-231.

28. Saridis, G.N., Nikolic, Z.J., and Fu, K.S., "Stochastic Approximation Algorithms for System Identification, Estimation and Decomposition of Mixtures," *IEEE Trans. on Systems Science and Cybernetics*, Vol. SSC-5, No. 1, January 1969, pp. 8-15.

29. Tsypkin, Y.S., "Adaptation, Training and Self-Organizing in Automatic Systems," *Automatika i Telemekhanika*, Vol. 27, No. 1, January 1966, pp. 23-61.

30. Mendel, J.M. and Fu, K.S., *Adaptive, Learning and Pattern Recognition Systems*, Academic Press, New York, 1970.

31. Knopp, K., *Theory and Application of Infinite Series*, Hafner, New York, 1947.

32. Deyst, J.J., Jr., and Price, C.F., "Conditions for Asymptotic Stability of the Discrete, Minimum Variance, Linear Estimator," *IEEE Trans. on Automatic Control*, Vol. AC-13, No. 6, December 1968, pp. 702-705.

33. Price, C.F., "An Analysis of the Divergence Problem in the Kalman Filter," *IEEE Trans. on Automatic Control*, Vol. AC-13, No. 6, December 1968, pp. 699-702.

34. Carney, T.M., "On Joint Estimation of the State and Parameters for an Autonomous Linear System Based on Measurements Containing Noise," Ph.D. Thesis, Rice University, 1967.

35. Mehra, R.K., "Identification of Stochastic Linear Dynamic Systems," 1969 IEEE Symposium on Adaptive Processes, November 1969, Philadelphia, Pennsylvania.

36. Sage, A.P., *Optimum Systems Control*, Prentice-Hall, Inc., Englewood Cliffs, New Jersey, 1968.

37. Lion, P.M., "Rapid Identification of Linear and Nonlinear Systems," *AIAA Journal*, Vol. 5, No. 10, October 1967.

38. Kushner, H.J., "On the Convergence of Lion's Identification Method with Random Inputs," *IEEE Trans. on Automatic Control*, Vol. AC-15, No. 6, December 1970, pp. 652-654.

39. Astrom, K.J. and Eykoff, P., "System Identification – A Survey," *Automatica*, Vol. 7, No. 2, March 1971, pp. 123-162.

40. Price, C.F. and Koenigsberg, W.D., "Parameter Identification for Tactical Missiles," TR-170-3, The Analytic Sciences Corp., 1 December 1971.

41. Mehra, R.K., "Frequency-Domain Synthesis of Optimal Inputs for Linear System Parameter Identification," Harvard University, Technical Report No. 645, July 1973.

42. Kalman, R.E., "Fundamental Study of Adaptive Control Systems," Technical Report No. ASD-TR-61-27, Vol. I, Flight Control Laboratory, Wright-Patterson Air Force Base, Ohio, April 1962.

43. Athans, M. and Falb, P.L., *Optimal Control*, McGraw-Hill Book Co., Inc., New York, 1966.

44. Bryson, A.E., Jr. and Ho, Y. C., *Applied Optimal Control*, Blaisdell Publishing Co., Waltham, Mass., 1969.

45. Stear, E.B., Notes for a short course on Guidance and Control of Tactical Missiles, presented at the University of California at Los Angeles, July 21-25, 1969.

46. Potter, J.E., "A Guidance-Navigation Separation Theorem," Report RE-11, Experimental Astronomy Laboratory, M.I.T., August 1964.

47. Striebel, C.T., "Sufficient Statistics on the Optimal Control of Stochastic Systems," *Journal of Mathematical Analysis and Applications*, Vol. 12, pp. 576-592, December 1965.

48. Deyst, J.J., Jr., "Optimal Control in the Presence of Measurement Uncertainties," Sc.D. Thesis, January 1967, Department of Aeronautics and Astronautics, Massachusetts Institute of Technology, Cambridge, Massachusetts.

49. Deyst, J.J., Jr. and Price, C.F., "Optimal Stochastic Guidance Laws for Tactical Missiles," *AIAA Journal of Spacecraft and Rockets*, Vol. 10, No. 5, May 1973, pp. 301-308.

50. Fel'dbaum, A.A., *Optimal Control Systems*, Academic Press, New York, 1965.

51. Aoki, M., *Optimization of Stochastic Systems*, Academic Press, New York, 1967.

52. Yoshikawa, T. and Kobayashi, H., "Comments on 'Optimal Minimal Order Observer-Estimators for Discrete Linear Time-Varying Systems'," *IEEE Transactions on Automatic Control*, Vol. AC-17, No. 2, April 1972.

PROBLEMS

Problem 9-1

A system and measurement are described by the set of differential equations

$$\dot{x} = -\beta x + w, \qquad w \sim N(0,q)$$

$$z = x + v, \qquad v \sim N(0,r)$$

with measurement data processed to provide an estimate $\hat{x}$ according to

$$\dot{\hat{x}} = -\beta\hat{x} + k(z - \hat{x})$$

where k is based on erroneous *a priori* assumptions about q and r.

Describe a computational approach leading to the adaptive behavior of this estimator, in which the value of k can be improved as the system operates. Detail all the relevant equations and estimate, in terms of the given parameters, the length of time it would take to establish the correct value of k to within 5%.

Problem 9-2

The stochastic observer undergoes initial discontinuities in both the state estimates and estimation error covariance due to the exact information provided by the noise-free measurements at $t = 0^+$. Show that $\hat{\underline{x}}(0^+)$ and $P(0^+)$ satisfy Eqs. (9.2-39) and (9.2-40). [Hint: First show that

$$\hat{\underline{x}}(0^+) = \hat{\underline{x}}(0) + B_2(0)[\underline{z}_2(0) - H_2(0)\hat{\underline{x}}(0)]$$

$$P(0^+) = [I - B_2(0)H_2(0)]\ P(0)[I - B_2(0)\ H_2(0)]^T$$

and then determine the optimum gain, $B_2^{opt}(0)$, by minimizing the trace of $P(0^+)$ with respect to $B_2(0)$.]

Problem 9-3

Define the covariance of the observer error $\tilde{\underline{\xi}}$, to be

$$\Pi = E[\tilde{\underline{\xi}}\tilde{\underline{\xi}}^T]$$

Show that Π and P can be related by

$$\Pi = TPT^T$$

$$P = A\Pi A^T$$

Verify this result for the special case of the colored-noise stochastic observer, where A and T are specified according to Eq. (9.2-51).

Problem 9-4

The stochastic observer covariance propagation and optimal gains given by Eqs. (9.2-36) to (9.2-38) are written in terms of the $n \times n$ covariance matrix, P. As shown in Problem 9-3, the observer errors may be equivalently represented by the reduced-order $(n\text{-}m) \times (n\text{-}m)$ covariance matrix, Π. Starting with the differential equation for the observer error, $\tilde{\underline{\xi}}$, given in Eq. (9.2-30), derive the differential equation for the observer error covariance matrix, Π. By minimizing the trace of Π, derive expressions for the optimal gains of the stochastic observer in terms of the reduced-order covariance matrix, Π.

Problem 9-5

Prove that the modified Newton's algorithm

$$\hat{x}_{k+1} = \hat{x}_k - k_0 \frac{g(\hat{x}_k)}{g'(\hat{x}_k)}$$

will converge to the solution, x_0, of

$$g(x) = 0$$

from any initial guess, $\hat{x}_1$, where g(x) is a *known* function satisfying

$$0 \leqslant a \leqslant |g'(x)| \leqslant b < \infty$$

and k_0 is a constant which satisfies

$$0 < k_0 \leqslant \frac{a}{b}$$

(Hint: Find a recursion for the quantity, $x_i - x_0$.)

Problem 9-6

Demonstrate that conditions in Eqs. (9.3-7c) and (9.3-7d) are necessary in order that the sequence $\{\hat{x}_k\}$ generated by Eq. (9.3-5) converges to the solution of

$$g(x) = 0$$

(Hint: Evaluate $\hat{x}_i$ for the special case, $g(x) = x - a$, in terms of the gain sequence $\{k_k\}$.)

Problem 9-7

Verify that the gain sequences defined in Eq. (9.3-17) satisfy the conditions in Eqs. (9.3-7c) and (9.3-7d).

Problem 9-8

Derive Eqs. (9.3-37) through (9.3-40).

Problem 9-9

Suppose that a random sequence $\{x_k\}$ evolves according to

$$x_{k+1} = x_k + w_k; \qquad k = 1,2,\ldots$$

$$E[x_1] = 0$$

$$E[x_1^2] = \sigma_0^2$$

where $\{w_k\}$ is a sequence of independent zero mean random variables having uniform variance σ_1^2. Measurements of the form

$$z_k = x_k + v_k$$

are available, where v_k is a sequence of zero mean independent random variables having uniform variance σ_2^2. Using the algorithm

$$\hat{x}_{k+1} = \hat{x}_k + k_0(z_k - \hat{x}_k)$$

to estimate x_k, where k_0 is constant, derive conditions on k_0 such that the mean square estimation error is bounded. Compute the rms estimation error in terms of σ_0^2, σ_1^2, σ_2^2, and k_0.

Problem 9-10

Formulate and give the solution to the optimal control problem which is the dual of the estimation problem described in Example 4.6-1.

Problem 9-11

Prove that $\bar{J}$ in Eq. (9.5-17) can be expressed in the form given in Eq. (9.5-19) (Hint: Mimic the development beginning with Eq. (9.5-5), replacing $\underline{x}^T S \underline{x}$ by $E[\underline{x}^T S \underline{x}]$).

INDEX